Advanced Techniques for the Analysis of Protein and RNA

Advanced Techniques for the Analysis of Protein and RNA

Editor

Xiaolong Yang

MDPI

Basel • Beijing • Wuhan • Barcelona • Belgrade • Novi Sad • Cluj • Manchester

Editor
Xiaolong Yang
Queen's University
Kingston, ON
Canada

Editorial Office
MDPI AG
Grosspeteranlage 5
4052 Basel, Switzerland

This is a reprint of articles from the Special Issue published online in the open access journal *Chemosensors* (ISSN 2227-9040) (available at: https://www.mdpi.com/journal/chemosensors/special_issues/X9ULU18WU6).

For citation purposes, cite each article independently as indicated on the article page online and as indicated below:

Lastname, A.A.; Lastname, B.B. Article Title. *Journal Name* **Year**, *Volume Number*, Page Range.

ISBN 978-3-7258-1935-5 (Hbk)
ISBN 978-3-7258-1936-2 (PDF)
doi.org/10.3390/books978-3-7258-1936-2

Contents

chemosensors

Editorial

Advanced Techniques for the Analysis of Proteins and RNAs

Xiaolong Yang

Department of Pathology and Molecular Medicine, Queen's University, Kingston, ON K7L 3N6, Canada; yangx@queensu.ca

Proteins and RNAs, as fundamental components of cellular machinery, play pivotal roles in the intricate landscape of life [1,2]. They have a spectrum of functions, ranging from acting as crucial structural elements to orchestrating complex signaling pathways and cellular processes [3–10]. Throughout the dynamic stages of development and in biological responses to environmental factors, proteins and RNAs contribute significantly to the regulation of growth, differentiation, the immune response, and other essential biological processes [5,11–22]. Recognizing their multifaceted roles and enabling the rapid and accurate detection of proteins and RNAs, including mRNA and microRNA, are imperative to advancing our understanding of cellular dynamics and their implications in health and disease [11,23–28].

The advent of novel analytical techniques in recent years has led to an era of remarkable progress in this field. Technologies such as mass spectrometry [29–32], immunoassays [33–36], spectral analysis [37,38], chemosensors [39–41], and biosensors [41–47] have emerged as powerful tools that provide unprecedented insights into the functions and interactions of proteins and RNAs [48,49]. These and many other emerging technological advancements [50–52] have not only enhanced our ability to unravel the intricacies of cellular mechanisms, but also hold immense potential to reshape the landscape of healthcare and environmental science. The application of these cutting-edge techniques enables us to take a transformative approach to the diagnosis, monitoring, and treatment of a myriad of diseases, ultimately promising to elevate the quality of life of millions worldwide.

With the advancement of technologies, progress has recently been made in the analysis of proteins and RNAs. In this Special Issue of *Chemosensors* on "*Advanced Techniques for the Analysis of Proteins and RNAs*", I am pleased to present a comprehensive overview of the impactful research showcased in nine published papers. These studies represent cutting-edge methodologies and applications in the fields of biosensing, enzymatic assays, liquid chromatography, and colorimetric and fluorescent assays, providing valuable insights into the detailed analysis of proteins, RNA, and associated pollutants.

Two papers within this collection focus on the development of biosensors tailored for the detection of protein levels within living cells. The work by Wu et al. introduces a novel HiBiT biosensor meticulously designed to monitor the stability of YAP/TAZ proteins in cells [53]. Using this biosensor, the authors identified novel small molecules that regulate YAP/TAZ, offering a promising avenue to improving anti-triple-negative breast cancer (TNBC) therapy. This pioneering approach marks a significant stride in cancer drug discovery through the innovative application of biosensing technologies. In a parallel study, Sitkov et al. engineered an impedimetric biosensor featuring zinc oxide nanorods for the highly sensitive detection of biomarkers [54]. Their work underscores the biosensor's potential for utilization in point-of-care systems designed for the swift, multimodal detection of molecular markers across diverse diseases.

Several other papers employed peptides or enzymes with a high affinity for proteins to quantify proteins or protein-like reagents in vitro. Zimina et al. dedicated their research to designing a peptide ligand with a heightened affinity for proteins such as lactoferrin, a multifunctional protein derived from milk, employing in silico modeling and capillary

Citation: Yang, X. Advanced Techniques for the Analysis of Proteins and RNAs. *Chemosensors* **2024**, *12*, 12. https://doi.org/10.3390/chemosensors12010012

Received: 2 January 2024
Accepted: 8 January 2024
Published: 10 January 2024

electrophoresis techniques to study its binding specificity [55]. Shahzadi et al. explored the feasibility of leveraging ultra-thin 2D materials, renowned for their high affinity for proteins and DNA, for biosensing applications [56]. Their evaluation of the non-covalent functionalization of CVD-graphene using a pyrene-based supporter construct demonstrated its potential in biosensing applications, particularly in the detection of proteins like streptavidin. In their contribution, Esimbekova et al. introduced a trypsin-based chemoenzymatic assay tailored for the detection of pollutants and the safety assessment of food additives [57], showcasing the versatility of enzyme-based detection systems for the analysis of reagents akin to enzymatic proteins.

Additionally, Kuznetsov et al. presented a novel chromatographic sensor designed for fast protein and metabolite liquid chromatography, facilitating the evaluation of fish and meat freshness through the analysis of specific protein and DNA contents [58]. The proposed freshness index (H*) demonstrated a strong correlation with traditional freshness indicators. Moreover, Mamaeva et al. and Li et al. conducted insightful comparisons of various colorimetric and fluorometric chemosensors or assays for protein concentration determination, offering valuable insights into the comparative performance of different dyes and assays [59,60].

Finally, Lin et al. contributed a review article on microRNA biosensors tailored for the early detection of hepatocellular carcinoma, summarizing recent advances in biosensor technologies geared towards detecting these pivotal biomarkers [61].

The diverse range of contributions to this Special Issue collectively advance the field of analytical techniques for protein and RNA analysis, providing novel insights, methodologies, and applications that are poised to catalyze further research in these pivotal areas. I extend my sincere appreciation to all the abovementioned authors for their invaluable contributions to this Special Issue.

Funding: In addition, I would like to thank the Canadian Institute of Health Research (#186143, 119325), Cancer Research Society (#938324), and New Frontier Research Fund (#340043) for their financial support.

Acknowledgments: I thank all the authors who submitted their studies to the present Special Issue.

Conflicts of Interest: The authors declare no conflicts of interest.

References

1. Alberts, B. The cell as a collection of protein machines: Preparing the next generation of molecular biologists. *Cell* **1998**, *92*, 291–294. [CrossRef] [PubMed]
2. Nair, A.; Chauhan, P.; Saha, B.; Kubatzky, K.F. Conceptual Evolution of Cell Signaling. *Int. J. Mol. Sci.* **2019**, *20*, 3292. [CrossRef] [PubMed]
3. Ganser, L.R.; Kelly, M.L.; Herschlag, D.; Al-Hashimi, H.M. The roles of structural dynamics in the cellular functions of RNAs. *Nat. Rev. Mol. Cell Biol.* **2019**, *20*, 474–489. [CrossRef] [PubMed]
4. Westermarck, J.; Ivaska, J.; Corthals, G.L. Identification of protein interactions involved in cellular signaling. *Mol. Cell Proteom.* **2013**, *12*, 1752–1763. [CrossRef] [PubMed]
5. Taha, Z.; Janse van Rensburg, H.J.; Yang, X. The Hippo pathway: Immunity and cancer. *Cancers* **2018**, *10*, 94. [CrossRef]
6. Pfleger, C.M. The Hippo Pathway: A Master Regulatory Network Important in Development and Dysregulated in Disease. *Curr. Top. Dev. Biol.* **2017**, *123*, 181–228.
7. Ardestani, A.; Lupse, B.; Maedler, K. Hippo Signaling: Key Emerging Pathway in Cellular and Whole-Body Metabolism. *Trends Endocrinol. Metab.* **2018**, *29*, 492–509. [CrossRef]
8. Day, E.K.; Sosale, N.G.; Lazzara, M.J. Cell signaling regulation by protein phosphorylation: A multivariate, heterogeneous, and context-dependent process. *Curr. Opin. Biotechnol.* **2016**, *40*, 185–192. [CrossRef]
9. Li, X.; Tran, K.M.; Aziz, K.E.; Sorokin, A.V.; Chen, J.; Wang, W. Defining the Protein-Protein Interaction Network of the Human Protein Tyrosine Phosphatase Family. *Mol. Cell Proteom.* **2016**, *15*, 3030–3044. [CrossRef]
10. Xu, A.M.; Huang, P.H. Receptor tyrosine kinase coactivation networks in cancer. *Cancer Res.* **2010**, *70*, 3857–3860. [CrossRef]
11. Khan, M.G.M.; Wang, Y. Cell Cycle-Related Clinical Applications. *Methods Mol. Biol.* **2022**, *2579*, 35–46.
12. Goranov, A.I.; Cook, M.; Ricicova, M.; Ben-Ari, G.; Gonzalez, C.; Hansen, C.; Tyers, M.; Amon, A. The rate of cell growth is governed by cell cycle stage. *Genes Dev.* **2009**, *23*, 1408–1422. [CrossRef] [PubMed]
13. Tzur, A.; Kafri, R.; LeBleu, V.S.; Lahav, G.; Kirschner, M.W. Cell growth and size homeostasis in proliferating animal cells. *Science* **2009**, *325*, 167–171. [CrossRef]

14. Golubev, A. Transition probability in cell proliferation, stochasticity in cell differentiation, and the restriction point of the cell cycle in one package. *Prog. Biophys. Mol. Biol.* **2012**, *110*, 87–96. [CrossRef]
15. Miller, J.P.; Yeh, N.; Vidal, A.; Koff, A. Interweaving the cell cycle machinery with cell differentiation. *Cell Cycle* **2007**, *6*, 2932–2938. [CrossRef] [PubMed]
16. Hendriks, D.; Artegiani, B.; Kretzschmar, K. Editorial: Mechanisms of cellular differentiation, organ development, and novel model systems. *Front. Cell Dev. Biol.* **2022**, *10*, 970778. [CrossRef] [PubMed]
17. Newman, S.A. Cell differentiation: What have we learned in 50 years? *J. Theor. Biol.* **2020**, *485*, 110031. [CrossRef]
18. Yang, X.; Xu, T. Molecular mechanism of size control in development and human diseases. *Cell Res.* **2011**, *21*, 715–729. [CrossRef]
19. Nelson, C.M. Mechanical Control of Cell Differentiation: Insights from the Early Embryo. *Annu. Rev. Biomed. Eng.* **2022**, *24*, 307–322. [CrossRef]
20. Goddard, A.M.; Cho, M.G.; Lerner, L.M.; Gupta, G.P. Mechanisms of Immune Sensing of DNA Damage. *J. Mol. Biol.* **2023**, 168424. [CrossRef]
21. Chatterjee, N.; Walker, G.C. Mechanisms of DNA damage, repair, and mutagenesis. *Environ. Mol. Mutagen.* **2017**, *58*, 235–263. [CrossRef]
22. Yaribeygi, H.; Panahi, Y.; Sahraei, H.; Johnston, T.P.; Sahebkar, A. The impact of stress on body function: A review. *Excli. J.* **2017**, *16*, 1057–1072. [PubMed]
23. Ligasová, A.; Frydrych, I.; Koberna, K. Basic Methods of Cell Cycle Analysis. *Int. J. Mol. Sci.* **2023**, *24*, 3674. [CrossRef] [PubMed]
24. Roth, S.; Margulis, M.; Danielli, A. Recent Advances in Rapid and Highly Sensitive Detection of Proteins and Specific DNA Sequences Using a Magnetic Modulation Biosensing System. *Sensors* **2022**, *22*, 4497. [CrossRef] [PubMed]
25. Duffy, D.C. Short Keynote Paper: Single Molecule Detection of Protein Biomarkers to Define the Continuum from Health to Disease. *IEEE J. Biomed. Health Inf.* **2020**, *24*, 1864–1868. [CrossRef] [PubMed]
26. Meng, X.; Pang, X.; Yang, J.; Zhang, X.; Dong, H. Recent Advances in Electrochemiluminescence Biosensors for MicroRNA Detection. *Small* **2023**, e2307701. [CrossRef] [PubMed]
27. Shaterabadi, D.; Zamani Sani, M.; Rahdan, F.; Taghizadeh, M.; Rafiee, M.; Dorosti, N.; Dianatinasab, A.; Taheri-Anganeh, M.; Asadi, P.; Khatami, S.H.; et al. MicroRNA biosensors in lung cancer. *Clin. Chim. Acta* **2024**, *552*, 117676. [CrossRef] [PubMed]
28. Quazi, S. Application of biosensors in cancers, an overview. *Front. Bioeng. Biotechnol.* **2023**, *11*, 1193493. [CrossRef]
29. Boja, E.S.; Rodriguez, H. Mass spectrometry-based targeted quantitative proteomics: Achieving sensitive and reproducible detection of proteins. *Proteomics* **2012**, *12*, 1093–1110. [CrossRef]
30. Shuford, C.M.; Grant, R.P. Cheaper, faster, simpler trypsin digestion for high-throughput targeted protein quantification. *J. Mass Spectrom. Adv. Clin. Lab.* **2023**, *30*, 74–82. [CrossRef]
31. Shen, S.; Wang, X.; Zhu, X.; Rasam, S.; Ma, M.; Huo, S.; Qian, S.; Zhang, M.; Qu, M.; Hu, C.; et al. High-quality and robust protein quantification in large clinical/pharmaceutical cohorts with IonStar proteomics investigation. *Nat. Protoc.* **2023**, *18*, 700–731. [CrossRef] [PubMed]
32. Popp, R.; Li, H.; Borchers, C.H. Immuno-MALDI (iMALDI) mass spectrometry for the analysis of proteins in signaling pathways. *Expert Rev. Proteom.* **2018**, *15*, 701–708. [CrossRef] [PubMed]
33. Shin, H.; Oh, S.; Kang, D.; Choi, Y. Protein Quantification and Imaging by Surface-Enhanced Raman Spectroscopy and Similarity Analysis. *Adv. Sci. (Weinh.)* **2020**, *7*, 1903638. [CrossRef] [PubMed]
34. Dixit, C.K.; Kaushik, A. Nano-structured arrays for multiplex analyses and Lab-on-a-Chip applications. *Biochem. Biophys. Res. Commun.* **2012**, *419*, 316–320. [CrossRef] [PubMed]
35. Zhou, S.; Lu, X.; Chen, C.; Sun, D. An immunoassay method for quantitative detection of proteins using single antibodies. *Anal. Biochem.* **2010**, *400*, 213–218. [CrossRef]
36. Misiewicz-Krzeminska, I.; Corchete, L.A.; Rojas, E.A.; Martínez-López, J.; García-Sanz, R.; Oriol, A.; Bladé, J.; Lahuerta, J.J.; Miguel, J.S.; Mateos, M.V.; et al. A novel nano-immunoassay method for quantification of proteins from CD138-purified myeloma cells: Biological and clinical utility. *Haematologica* **2018**, *103*, 880–889. [CrossRef]
37. Gautier, A. Fluorescence-Activating and Absorption-Shifting Tags for Advanced Imaging and Biosensing. *Acc. Chem. Res.* **2022**, *55*, 3125–3135. [CrossRef]
38. Fukuyama, M.; Nakamura, A.; Nishiyama, K.; Imai, A.; Tokeshi, M.; Shigemura, K.; Hibara, A. Noncompetitive Fluorescence Polarization Immunoassay for Protein Determination. *Anal. Chem.* **2020**, *92*, 14393–14397. [CrossRef]
39. Crisp, S.J.; Dunn, M.J. Detection of proteins on protein blots using chemiluminescent systems. *Methods Mol. Biol.* **1994**, *32*, 233–237.
40. Khan, J. Synthesis and Applications of Fluorescent Chemosensors: A Review. *J. Fluoresc.* **2023**. [CrossRef]
41. Xia, N.; Chang, Y.; Zhou, Q.; Ding, S.; Gao, F. An Overview of the Design of Metal-Organic Frameworks-Based Fluorescent Chemosensors and Biosensors. *Biosensors* **2022**, *12*, 928. [CrossRef] [PubMed]
42. Kang, M.J.; Cho, Y.W.; Kim, T.H. Progress in Nano-Biosensors for Non-Invasive Monitoring of Stem Cell Differentiation. *Biosensors* **2023**, *13*, 501. [CrossRef]
43. Pipchuk, A.; Yang, X. Using Biosensors to Study Protein-Protein Interaction in the Hippo Pathway. *Front. Cell Dev. Biol.* **2021**, *9*, 660137. [CrossRef] [PubMed]
44. Azad, T.; Tashakor, A.; Hosseinkhani, S. Split-luciferase complementary assay: Applications, recent developments, and future perspectives. *Anal. Bioanal. Chem.* **2014**, *406*, 5541–5560. [CrossRef] [PubMed]

45. Mo, X.; Niu, Q.; Ivanov, A.A.; Tsang, Y.H.; Tang, C.; Shu, C.; Li, Q.; Qian, K.; Wahafu, A.; Doyle, S.P.; et al. Systematic discovery of mutation-directed neo-protein-protein interactions in cancer. *Cell* **2022**, *185*, 1974–1985.e12. [CrossRef] [PubMed]
46. Nouri, K.; Azad, T.; Lightbody, E.; Khanal, P.; Nicol, C.J.; Yang, X. A kinome-wide screen using a NanoLuc LATS luminescent biosensor identifies ALK as a novel regulator of the Hippo pathway in tumorigenesis and immune evasion. *FASEB J.* **2019**, *33*, 12487–12499. [CrossRef] [PubMed]
47. Azad, T.; Janse van Rensburg, H.J.; Lightbody, E.D.; Neveu, B.; Champagne, A.; Ghaffari, A.; Kay, V.R.; Hao, Y.; Shen, H.; Yeung, B.; et al. A LATS biosensor functional screen identifies VEGFR as a novel regulator of the Hippo pathway in angiogenesis. *Nat. Commun.* **2018**, *9*, 1061. [CrossRef]
48. Rao, V.S.; Srinivas, K.; Sujini, G.N.; Kumar, G.N. Protein-protein interaction detection: Methods and analysis. *Int. J. Proteom.* **2014**, *2014*, 147648. [CrossRef]
49. Pflieger, D.; Gonnet, F.; de la Fuente van Bentem, S.; Hirt, H.; de la Fuente, A. Linking the proteins--elucidation of proteome-scale networks using mass spectrometry. *Mass Spectrom. Rev.* **2011**, *30*, 268–297. [CrossRef]
50. Baysoy, A.; Bai, Z.; Satija, R.; Fan, R. The technological landscape and applications of single-cell multi-omics. *Nat. Rev. Mol. Cell Biol.* **2023**, *24*, 695–713. [CrossRef]
51. Hu, Y.; Cheng, K.; He, L.; Zhang, X.; Jiang, B.; Jiang, L.; Li, C.; Wang, G.; Yang, Y.; Liu, M. NMR-Based Methods for Protein Analysis. *Anal. Chem.* **2021**, *93*, 1866–1879. [CrossRef] [PubMed]
52. Brunner, A.D.; Thielert, M.; Vasilopoulou, C.; Ammar, C.; Coscia, F.; Mund, A.; Hoerning, O.B.; Bache, N.; Apalategui, A.; Lubeck, M.; et al. Ultra-high sensitivity mass spectrometry quantifies single-cell proteome changes upon perturbation. *Mol. Syst. Biol.* **2022**, *18*, e10798. [CrossRef] [PubMed]
53. Wu, L.; Ge, A.; Hao, Y.; Yang, X. Development of a New HiBiT Biosensor Monitoring Stability of YAP/TAZ Proteins in Cells. *Chemosensors* **2023**, *11*, 492. [CrossRef]
54. Sitkov, N.; Ryabko, A.; Kolobov, A.; Maximov, A.; Moshnikov, V.; Pshenichnyuk, S.; Komolov, A.; Aleshin, A.; Zimina, T. Impedimetric Biosensor Coated with Zinc Oxide Nanorods Synthesized by a Modification of the Hydrothermal Method for Antibody Detection. *Chemosensors* **2023**, *11*, 66. [CrossRef]
55. Zimina, T.; Sitkov, N.; Karasev, V.; Skorik, Y.; Kolobov, A.; Kolobov, A.; Bunenkov, N.; Luchinin, V. Design of Peptide Ligand for Lactoferrin and Study of Its Binding Specificity. *Chemosensors* **2023**, *11*, 162. [CrossRef]
56. Shahzadi, M.; Nisar, S.; Kim, D.-K.; Sarwar, N.; Rasheed, A.; Ahmad, W.; Afzal, A.M.; Imran, M.; Assiri, M.A.; Shahzad, Z.M.; et al. Highly Efficient, Non-Covalent Functionalization of CVD-Graphene via Novel Pyrene-Based Supporter Construct. *Chemosensors* **2023**, *11*, 83. [CrossRef]
57. Esimbekova, E.N.; Torgashina, I.G.; Nemtseva, E.V.; Antashkevich, A.A.; Sasova, P.Y.; Kratasyuk, V.A. Trypsin-Based Chemoenzymatic Assay for Detection of Pollutants and Safety Assessment of Food Additives. *Chemosensors* **2023**, *11*, 237. [CrossRef]
58. Kuznetsov, A.; Frorip, A.; Sünter, A.; Kasvand, N.; Korsakov, V.; Konoplev, G.; Stepanova, O.; Rusalepp, L.; Anton, D.; Püssa, T.; et al. Fast Protein and Metabolites (Nucleotides and Nucleosides) Liquid Chromatography Technique and Chemical Sensor for the Assessment of Fish and Meat Freshness. *Chemosensors* **2023**, *11*, 69. [CrossRef]
59. Mamaeva, A.A.; Martynov, V.I.; Deyev, S.M.; Pakhomov, A.A. Comparison of Colorimetric and Fluorometric Chemosensors for Protein Concentration Determination and Approaches for Estimation of Their Limits of Detection. *Chemosensors* **2022**, *10*, 542. [CrossRef]
60. Li, F.; Tan, J.; Yang, Q.; He, M.; Yu, R.; Liu, C.; Zhou, X. Multi-Endpoint Toxicity Tests and Effect-Targeting Risk Assessment of Surface Water and Pollution Sources in a Typical Rural Area in the Yellow River Basin, China. *Chemosensors* **2022**, *10*, 502. [CrossRef]
61. Lin, X.; Wang, K.; Luo, C.; Yang, M.; Wu, J. MicroRNA Biosensors for Early Detection of Hepatocellular Carcinoma. *Chemosensors* **2023**, *11*, 504. [CrossRef]

chemosensors

MDPI

Review

MicroRNA Biosensors for Early Detection of Hepatocellular Carcinoma

Xiaogang Lin [1,2,*], Ke Wang [1], Chunfeng Luo [1], Mengjie Yang [1] and Jayne Wu [3,*]

1 Key Laboratory of Optoelectronic Technology and Systems, Ministry of Education of China, Chongqing University, Chongqing 400044, China; kewang@stu.cqu.edu.cn (K.W.); 202108131150@stu.cqu.edu.cn (C.L.); 202208131095@stu.cqu.edu.cn (M.Y.)

2 Department of Clinical Laboratory, Chongqing University Jiangjin Hospital, School of Medicine, Chongqing University, Jiangjin, Chongqing 402260, China

3 Department of Electrical Engineering and Computer Science, The University of Tennessee, Knoxville, TN 37996, USA

* Correspondence: xglin@cqu.edu.cn (X.L.); jaynewu@utk.edu (J.W.)

Abstract: Hepatocellular carcinoma (HCC) is the main pathological type of liver cancer. Due to its insidious onset and the lack of specific early markers, HCC is often diagnosed at an advanced stage, and the survival rate of patients with partial liver resection is low. Non-coding RNAs (ncRNAs) have emerged as valuable biomarkers for HCC detection, with microRNAs (miRNAs) being a particularly relevant class of short ncRNAs. MiRNAs play a crucial role in gene expression regulation and can serve as biomarkers for early HCC detection. However, the detection of miRNAs poses a significant challenge due to their small molecular weight and low abundance. In recent years, biosensors utilizing electrochemical, optical, and electrochemiluminescent strategies have been developed to address the need for simple, rapid, highly specific, and sensitive miRNA detection. This paper reviews the recent advances in miRNA biosensors and discusses in detail the probe types, electrode materials, sensing strategies, linear ranges, and detection limits of the sensors. These studies are expected to enable early intervention and dynamic monitoring of tumor changes in HCC patients to improve their prognosis and survival status.

Keywords: hepatocellular carcinoma; microRNAs; electrochemistry; optics; electrochemiluminescence

Citation: Lin, X.; Wang, K.; Luo, C.; Yang, M.; Wu, J. MicroRNA Biosensors for Early Detection of Hepatocellular Carcinoma. *Chemosensors* **2023**, *11*, 504. https://doi.org/10.3390/chemosensors11090504

Academic Editors: Dario Compagnone and Alina Vasilescu

Received: 31 July 2023
Revised: 31 August 2023
Accepted: 11 September 2023
Published: 16 September 2023

1. Introduction

Liver cancer is a malignant tumor that poses a serious threat to human health and has a very high mortality rate. The pathological types of hepatocellular carcinoma include hepatocellular, biliary, and mixed cell types, of which hepatocellular carcinoma accounts for approximately 70% [1]. Hepatitis B virus (HBV), hepatitis C virus (HCV), alcohol, fatty liver, and genetic disorders may induce HCC [2]. Due to the insidious onset of HCC, most patients are usually diagnosed at a late stage, with a high frequency of metastasis and recurrence after surgery, resulting in poor prognosis and survival. Therefore, early diagnosis and detection of hepatocellular carcinoma is crucial to improve patient prognosis and survival. The main clinical methods for detecting hepatocellular carcinoma include computed tomography (CT) [3], ultrasonography (US) [4], magnetic resonance imaging (MRI) [5], and tissue biopsy [6]. These methods have limitations in detection sensitivity, expensive detection equipment, and are heavily operator-dependent and invasive to the patient's body [7]. Therefore, the development of analytical techniques that are simple to operate, inexpensive, highly sensitive, and selective is crucial for achieving early diagnosis of hepatocellular carcinoma. In recent years, the detection and analysis of tumor markers has become an effective tool in the diagnosis, prognosis, and treatment of cancer [8]. The detection of tumor markers in peripheral blood holds significant promise for clinical applications due to several key advantages, such as minimized invasiveness, continuous monitoring, and broad accessibility. Currently, several tumor markers have been evaluated in hepatocellular carcinoma, among which alpha-fetoprotein (AFP) is the most widely used in the diagnosis of primary liver cancer [9]. Despite the high specificity of AFP, its detection sensitivity and accuracy are not sufficient to detect patients with hepatocellular carcinoma in the early stages, and false negative results may occur [10].

With the extensive characterization of the protein-coding genome of liver tumors by researchers, ncRNAs were found to regulate hepatocarcinogenesis [11]. Classified based on length and shape characteristics, ncRNAs include miRNAs, long noncoding RNAs (lncRNAs), and circular RNAs (circRNAs) [12]. These transcripts have been shown to regulate the transcription, stability, or translation of protein-coding genes in the genome, although they are not transcribed into proproteins [13]. During hepatocellular carcinogenesis, investigators have found that the expression of miRNAs is disturbed in patients. Dysregulated miRNAs may affect HCC cell proliferation by directly interacting with key regulators of the cell cycle machinery [14]. Thus, miRNAs are potential HCC biomarkers and targets for intervention.

miRNAs are the smallest non-coding RNAs with an average size of 22 nucleotides [15]. The production of miRNAs in cancer cells, the process of obtaining miRNAs, and commonly used miRNA quantification methods are shown in Figure 1. At each step of HCC development, specific miRNAs are dysregulated. Accurate detection of miRNAs from complex biological samples is extremely challenging due to their short sequences, low levels in vivo, and easy degradation. Traditional detection methods include qRT-PCR, microarrays, and Northern blotting [16]. Among them, qRT-PCR offers high sensitivity and a wide dynamic range for miRNA expression profiles, but its application is limited due to susceptibility to contamination. Microarray technology has high throughput screening capability and can analyze a wide range of miRNAs, but requires complex probes and instruments, as well as specialized operators. Northern blotting is considered the "gold standard" for miRNAs, but has shortcomings in detection time and sample consumption, and has low sensitivity and throughput [17]. Biosensors offer significant advantages in terms of sample consumption, detection time, cost, portability, and complexity compared to the above-mentioned traditional techniques for miRNA detection [18]. Biosensor technology has become a cross-disciplinary field combining biology, chemistry, physics, medicine, electronics, and other disciplines [19]. In recent years, several biosensor-based techniques have emerged for the detection of miRNAs, such as electrochemistry [20], colorimetry [21], fluorescence [22], surface plasmon resonance (SPR) [23], surface-enhanced Raman scattering (SERS) [24], and electrochemiluminescence (ECL) [25]. These methods can also be used for in situ screening and mobile health monitoring.

In this review, the focus is on the impact of miRNAs as potential biomarkers for hepatocellular carcinoma. Recent advances in electrochemical and optical biosensors for the detection of hepatocellular carcinoma-associated miRNAs are reviewed. The target detector types, probe selection, electrode design, modification methods, sensing strategies, linear detection ranges, detection limits, and response times of these biosensors are discussed in detail. In addition, the characteristics and limitations of these biosensors are summarized. The organization is as follows: Section 2 provides an overview of the expression and roles of miRNAs associated with HCC. Sections 3 and 4 analyze the research progress of electrochemical and optical biosensors for the detection of HCC-associated miRNAs, respectively. Section 5 discusses the methods to improve the detection performance of miRNA biosensors and summarizes the challenges and prospects of miRNA biosensors for clinical applications in the future. Compared with some previous reviews on miRNA biosensors, this review carefully categorizes miRNA biosensors associated with HCC, and the biotechnology and sensing technologies covered in the review are more comprehensive.

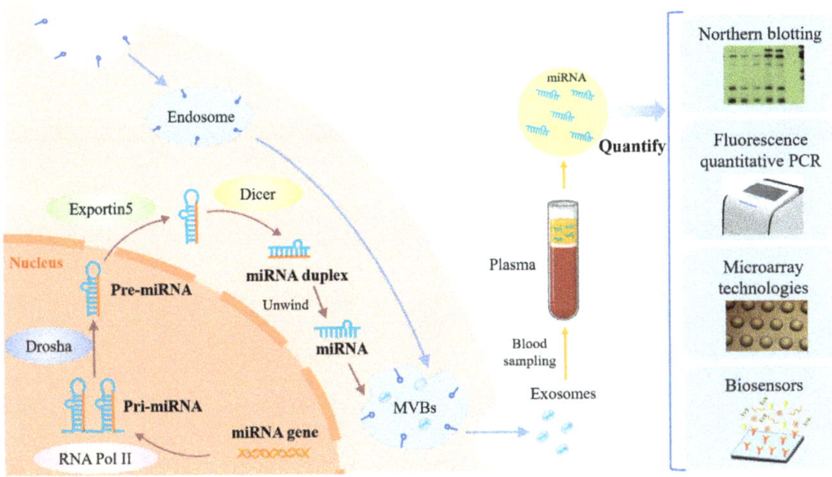

Figure 1. The principle of miRNA production in cancer cells, the process of obtaining miRNA, and commonly used miRNA quantification methods.

2. miRNAs Associated with HCC

Some miRNAs that are differentially expressed in HCC tumors compared to normal liver tissue include miRNA-21, miRNA-34a, miRNA-122, miRNA-125b, miRNA-141, miRNA-155, miRNA-223, miRNA-224, miRNA let-7a, and miRNA let-7b [26]. Among them, miRNA-21 is an important non-coding RNA affecting liver diseases, which can serve as a dual marker for early screening and prognosis of HCC [27]. miRNA-34a is an important tumor suppressor that can inhibit tumor progression and tumorigenesis [28]. However, miRNA-34a expression was significantly reduced in clinical HCC specimens, suggesting that miRNA-34a is a potential marker for HCC diagnosis and prognosis [29]. miRNA-122 is primarily expressed in liver tissue and plays a central role in various aspects of hepatocyte development and differentiation. It constitutes a substantial proportion of the total miRNA content within the liver, accounting for approximately 70% of the entire miRNA population in this organ [30]. Liang et al. [31] found that miRNA-125b inhibits the expression of the oncogene LIN28B and thus exerts tumor suppressor effects in HCC. However, downregulation of miRNA-125b was frequently observed in human hepatocellular carcinoma. miRNA-125b was under-expressed in most HCC cases and negatively correlated with the cell proliferation index in HCC. miRNA-141 plays an important role in cancer formation and progression. The downregulated miRNA-141 expression may be an important predictor of HCC [32]. miRNA-155 acts as a tumor suppressor in HCC and its expression level is significantly elevated in HCC tissues, with a marked increase of 1.5–6 times compared to normal liver tissues [33]. miRNA-223 was reported to be significantly lower in the serum of HCC patients compared to non-tumor livers by Elmougy et al. [34]. Eldeen et al. [35] demonstrated that miRNA-122 and miRNA-224 can be used as biomarkers for HCC diagnosis, and the detection of either of these miRNAs in combination with AFP will improve the accuracy of early HCC diagnosis. Qiu et al. [36] found a correlation between the expression level of miRNA let-7a and HBV replication. The downregulation of miRNA let-7a is associated with a decrease in HBV replication and may prevent the development of HCC. In addition, Wang et al. [37] showed that miRNA let-7b was able to inhibit the proliferation of HCC cells through Wnt/β-linker protein signaling in HCC cells, but miRNA let-7b was significantly downregulated in human HCC tissues. The expression of the above miRNA markers can respond to the development of HCC and has great potential in the early screening and diagnosis of HCC.

3. Electrochemical Biosensors

Electrochemical biosensors are devices that convert the biological signal generated by the specific binding of a recognition probe to a target to be measured into electrical signals such as voltage, current, and impedance [38]. Electrochemical biosensors are suitable for point-of-care (POC) detection due to the ease of miniaturization, automation, integration, and mass production. In recent years, nanotechnology has brought great opportunities for development in the field of electrochemical biosensors. The large surface volume ratio of nanomaterials helps to improve the detection sensitivity of biosensors [39]. Currently, the commonly used electrochemical detection methods mainly include voltammetry [40] and impedance methods [41]. Electrochemical biosensors for the detection of HCC-associated miRNAs will be discussed next according to the classification of electrochemical detection methods.

3.1. Voltammetry

Voltammetry is based on the relationship between the electrode potential and the current through the electrolytic cell to obtain analytical results. With the development of bioanalytical techniques, voltammetry is now mostly performed using a three-electrode system consisting of a working electrode, a counter electrode, and a reference electrode. The test methods mainly include cyclic voltammetry (CV) [42], square wave voltammetry (SWV) [43], differential pulse voltammetry (DPV) [44], and other methods.

Pathological studies have shown that miRNA-122 acts to repress oncogenes involved in different HCC features, and downregulation of miRNA-122 can cause tumor metastasis and hepatocellular carcinoma progression [45]. Therefore, simple and sensitive detection of miRNA-122 is highly relevant for the early diagnosis of HCC. Gao et al. [46] proposed an electrochemical biosensor based on the ion barrier effect for the detection of miRNA-122. A schematic diagram of the preparation and operation of the sensor is shown in Figure 2a. Prussian blue (PB) and gold nanoparticles (AuNPs) were first modified on the surface of a glassy carbon electrode (GCE) by a two-step electrodeposition method. The addition of Prussian blue was able to sensitize the GCE electrode to K^+, resulting in a significant change in the voltammetric signal. KNO_3 was chosen to provide K^+. The modification of AuNPs enabled the GCE electrode to immobilize thiolated DNA probes by the self-assembly of Au-S bonds. An ionic barrier effect was produced when the DNA probe hybridized specifically with the target miRNA-122, preventing the diffusion of K^+ from the solution to the electrode surface. In this way, the voltammetric signal at the electrode surface was suppressed, which achieved the quantitative detection of miRNA-122. The electrochemical response of the sensor was studied using DPV. Figures 2b and 2c, respectively, show the DPV response curve and calibration curve of the sensor. The sensor has a response time of 60 min and can analyze miRNA-122 in the concentration range of 0.1 fmol/L–1.0 nmol/L with a detection limit of 0.021 fmol/L. This biosensor based on the ion barrier effect has the advantages of simple operation, low cost, sensitive response, high specificity, and high stability. In addition, the method shows better detection in real human serum samples and can be used to analyze complex biological samples. However, the detection time of this method is long, which is not conducive to the realization of point-of-care detection.

Losada et al. [47] designed an electrochemical biosensing platform based on microfluidic sensing technology that can perform eight multiple measurements of miRNA-122. The platform consisted of a glass substrate containing gold microelectrodes and a polydimethylsiloxane (PDMS) layer containing microfluidic channels. The capture probe modified with thiols was incubated in the microfluidic channel, and the probe was able to form a self-assembled monolayer (SAM) by immobilizing it on the electrode surface with Au-S bonds. After rinsing the channel with 0.5 M NaCl, miRNA-122 was injected into the microfluidic channel to hybridize with the capture probe. In this study, CV was used for electrochemical measurements, and the detection time required 30 min. The electrochemical sensing platform has a linear working concentration of 10^{-18}–10^{-6} mol/L and a detection limit of 10^{-18} mol/L. This method has a wider linear detection range, lower detection limit, and shorter assay time for the detection of miRNA-122 compared to the method of Gao et al. [46]. In addition, microfluidic sensing technology makes the sensing platform miniaturized and more portable. It can also reduce costs and achieve high-throughput detection.

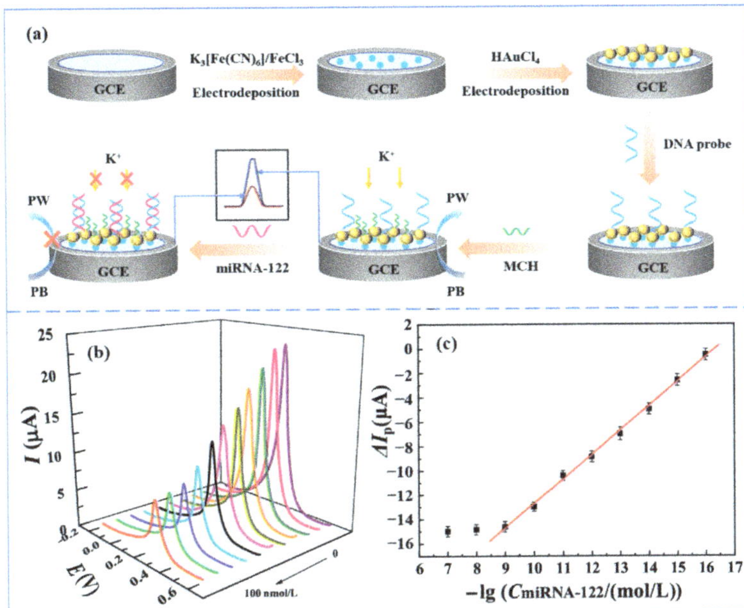

Figure 2. (**a**) Schematic diagram of the preparation and operation of the electrochemical biosensor based on the ion barrier effect for the detection of miRNA-122. (**b**) DPV response curves in the range of 0–100 nmol/L after hybridization of the biosensor with miRNA-122 (DPV responses curves at different concentrations are indicated by different colors. DPV response decreases with increasing concentration). (**c**) Correction curve for the amount of change in peak current versus the negative logarithm of miRNA-122 concentration. (Reproduced with permission from [46]).

In recent years, the signal amplification strategy based on 3D DNA walkers has shown great potential for the ultrasensitive detection of miRNAs. Yang et al. [48] designed an "on-super off" dual-mode photoelectrochemical (PEC) and ratiometric electrochemical (EC) biosensor based on this strategy for the detection of miRNA-224. A schematic diagram of the sensor is shown in Figure 3. The sensor applied methylene blue (MB) and ferrocene (Fc) to induce signal quenching and enhancement. CdS quantum dots (QDs) were used here as photoactive electrode materials due to their photoelectric conversion efficiency. The signal "on" state was achieved by immobilizing MB-labeled hairpin DNA (MB-DNA) through Cd-S bonding, which sensitized the CdS QDs and generated significant PEC signals. Hairpin MB-DNA was turned on after the introduction of DNA probes labeled with Ag nanocubes (Ag-DNA). Several ferrocene-labeled DNA (Fc-DNA) generated by amplification of the 3D DNA walker, Ag-DNA, and MB-DNA hybridized to form a "Y" shaped hairpin structure. This structure keeps the MB away from the CdS QDs and the Fc close to the CdS QDs, which results in a reduced PEC signal and achieves a signal "super-off" state. In addition, miRNA-224 detection was also accomplished on the ratiometric EC biosensor using SWV. As the concentration of miRNA-224 increased, the oxidized peak current of MB decreased, and the oxidized peak current of Fc increased. Quantification of miRNA-224 was achieved by evaluating the value of I_{MB}/I_{Fc}. The sensor has a linear detection range of 0.1–1000 fM with a detection limit of 0.019 fM in PEC and a detection range of 0.52–500 fM with a detection limit of 0.061 fM in ratiometric EC. CdS quantum dots exhibited excellent optoelectronic performance in the detection of miRNA-224 by this biosensor. The signal enhancing and quenching could be easily controlled by changing the structure of DNA, and the signal amplification strategy based on the 3D DNA walker significantly improved the sensitivity of detecting miRNAs.

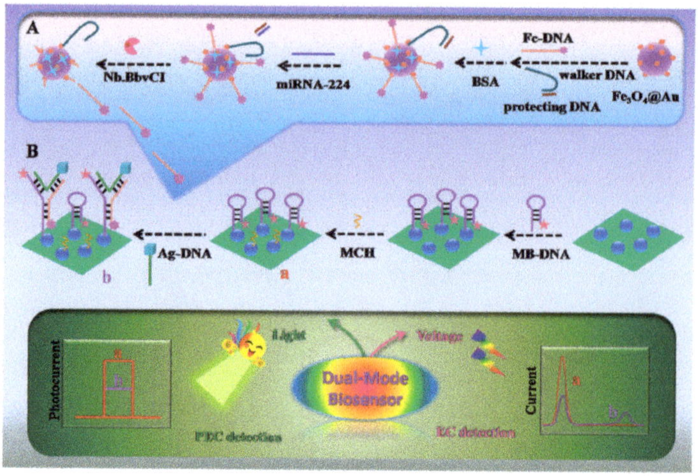

Figure 3. Schematic diagram of the dual model PEC and EC biosensor for miRNA-224 detection. (**A**). The process of 3D DNA walker-induced ring amplification. (**B**). Preparation process of the biosensor for the detection of miRNA-224. In the process from a to b, the hairpin structure probe transforms into a "Y" structure. (Reproduced with permission from [48]).

Homogeneous electrochemical biosensors are low-cost, simple to immobilize, and the detection process occurs in a homogeneous solution. Wu et al. [49] designed a homogeneous electrochemical biosensor based on MnO_2 nanosheets with dual enzyme activity for the detection of miRNA let-7a. In the absence of miRNA let-7a, the nucleic acid probe was tightly adsorbed on the surface of the 2D MnO_2 nanosheets, and the catalytic activity of the MnO_2 nanosheets was significantly inhibited. This led to the presence of a large amount of MB in the solution, which produced a very high DPV current peak. After the addition of miRNA let-7a, the phosphate group triggered the nucleobase pair shielding effect, and the probe was detached from the surface of MnO2 nanosheets after hybridization with miRNA let-7a. At this time, the surface-active sites of the MnO_2 nanosheets were significantly increased and were able to fully react with MB. As a result, a large amount of MB was eliminated, leading to a significant decrease in the DPV response. The linear detection range of this homogeneous electrochemical biosensor was 0.4–140 nM, and the detection limit was 0.25 nM. Although this homogeneous electrochemical biosensor was simple to prepare, the detection sensitivity was limited.

Azab et al. [50] prepared an miRNA let-7a biosensor with a sandwich structure based on nanomaterials. The schematic diagram of the biosensor is shown in Figure 4a. Chrysin and carbon nanotubes (CNTs) were, respectively, modified on the carbon paste electrode (CPE), which could improve the antioxidant property of the electrode and optimize the conductivity and biocompatibility of the electrode. Then AuNPs were employed to modify the electrode surface, enhancing both the active surface area of the electrode and the stability of the immobilized capture probe. The electrochemical response was monitored by DPV, and the optimal time for hybridization of this sensor is 30 min. Figure 4b shows the calibration curve of ΔI versus the logarithm of miRNA let-7a concentration. In this work, the current response increased with increasing miRNA let-7a concentration in the range of 1.0 zM to 11 nM, and the limit of detection was 1.0 zM. The introduction of nanomaterials such as CNTs and AuNPs has been instrumental in improving the sensitivity of the biosensor. In addition, the prepared biosensor has good applicability for miRNA let-7a detection in real serum samples. The biosensor has an excellent detection limit, which is highly favorable for early detection of clinical HCC.

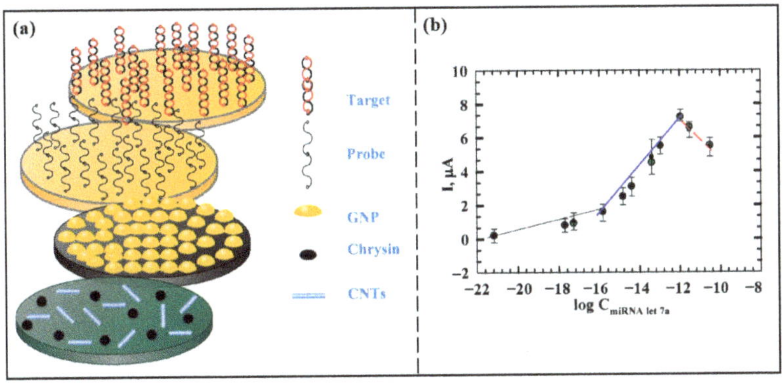

Figure 4. (**a**) Schematic diagram of the principle of miRNA let-7a detection by the sandwich-type biosensor. (**b**) The relationship between the ΔI peak and the logarithm of miRNA let-7a concentration. The ΔI response increased with miRNA let-7a concentration in two linear ranges with different slopes in the range of 1.0 zm to 11 nM. When the concentration was higher than 11 nM, the ΔI peak started to decrease due to the electrostatic repulsion between the probe and the target reducing the hybridization efficiency. (Reproduced with permission from [50]).

Cai et al. [51] reported an AuNP-modified graphene field effect transistor (FET) biosensor for the sensitive detection of miRNA let-7b. This FET biosensor was prepared by first dropwise addition of a reduced graphene oxide (R-GO) suspension on the FET surface, followed by modification of AuNPs on top of it. The PNA probes possess an electrically neutral backbone, which contributes to enhanced hybridization efficiency and reduced background noise in comparison to traditional DNA probes. This neutral nature of PNA probes mitigates the repulsive effects that can arise during the hybridization process [52]. In this study, after immobilizing the PNA probe on the surface of AuNPs through Au-S bonds, the excess active site was blocked using ethanolamine solution to minimize potential nonspecific binding. When the PNA probe hybridizes with miRNA let-7b, a distinct voltammetric response signal is generated due to the binding event. It was found that the developed FET biosensor could achieve detection limits as low as 10 fM in the linear range of 1 fM–100 pM. In addition, this highly sensitive and selective method was also successfully used for the detection of miRNA let-7b in serum samples. The PNA probe is highly promising for miRNA detection, and this PNA probe-based FET biosensor has the potential to be used as a point-of-care tool.

Erdem et al. [53] developed a method for electrochemical analysis of miRNA-34a based on the Zip nucleic acid (ZNA) probe. Figure 5a shows a schematic diagram of the electrochemical analysis based on the ZNA probe. ZNA probes were hybridized with target miRNA-34a in solution, and double-stranded products formed by hybridization were immobilized on the surface of magnetic beads (MBs) coated with streptavidin through biotin–streptavidin interaction. The double-stranded product was separated from the MBs by magnetic separation technique and then immobilized on the surface of pencil graphite electrodes (PGEs) for electrochemical measurements using DPV. Figure 5b–d show the DPV response curves and calibration plots of the ZNA probe hybridized with different concentrations of the target miRNA-34a. The detection limit of this method was 0.87 μg/mL in the linear range of 2–8 μg/mL. In addition, the electrochemical analysis method based on the ZNA probe is also suitable for the detection of miRNA-34a in real samples. The innovative use of ZNA probes in this sensor overcomes the electrostatic repulsion between probes and their complementary sequences, thus improving hybridization efficiency. However, the biosensor detects miRNA-34a for up to 60 min, which is not conducive to achieving point-of-care detection. In the future, with the advantage of ZNA nucleic acids combined with rapid enrichment methods such as the AC electrokinetics (ACEK) effect, the detection performance of the sensor will be improved, and the detection time will be shortened.

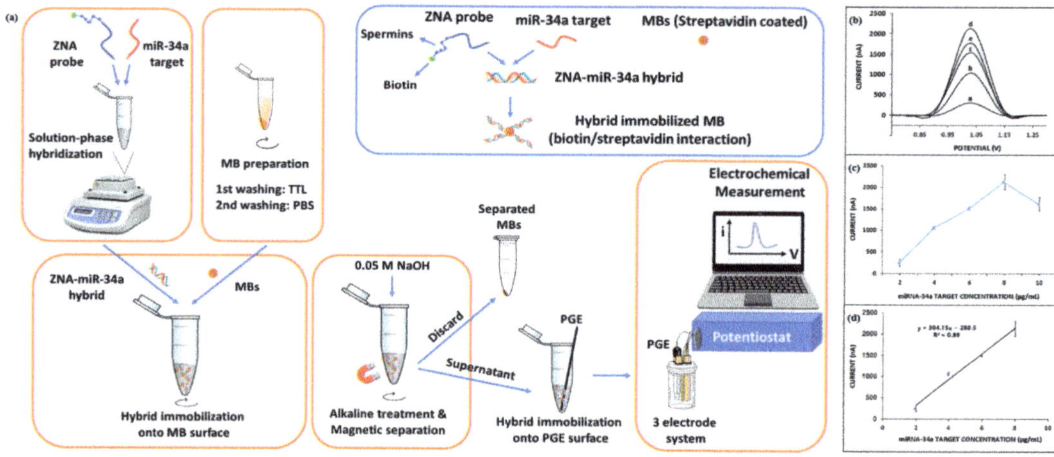

Figure 5. (**a**) Schematic of ZNA probe-based electrochemical analysis of miRNA-34a. (**b**) DPV response curves of different concentrations of miRNA-34a (a: 2 μg/mL, b: 4 μg/mL, c: 6 μg/mL, d: 8 μg/mL, and e: 10 μg/mL) detected by this biosensor. (**c**) Curve of average values of DPV peaks versus different concentrations of miRNA-34a. (**d**) Calibration curve of the average of DPV peaks versus 2–8 μg/mL miRNA-34a. (Reproduced with permission from [53]).

Zeng et al. [54] explored the photocurrent properties of yolk-in-shell Au@CdS and yolk-shell Au@CdS and established a sensitive and feasible PEC biosensor for the quantitative detection of miRNA-21 based on yolk-in-shell Au@CdS. The biosensor used HRP-labeled ssDNA combined with MBs to form MB-ssDNA-HRP as the signal probe, yolk-in-shell Au@CdS as the photoactive substrate, and benzo-4-chlorohexadienone (4-CD) precipitation as the signal quencher. In the presence of miRNA-21, miRNA-21 and two hairpin DNAs (H1, H2) could generate a large amount of H1-H2 double-stranded (dsDNA) by catalytic hairpin assembly (CHA) reaction. dsDNA binding to Cas12a-crRNA triggered the cleavage of MB-ssDNA-HRP by Cas12a, which led to the detachment of HRP from the MB surface. After magnetic separation, HRP was able to catalyze 4-chloro-1-naphthol (4-CN) to generate 4-CD precipitates that covered the yolk-in-shell Au@CdS surface, resulting in a significant decrease in its photocurrent response. The linear detection range of this PEC biosensor for miRNA-21 was 0.01 pM-10 nM, with a detection limit of 4.2 fM. In addition, stronger synergistic effects, SPR, and thermal electron transfer were found for yolk-in-shell Au@CdS by FDTD simulation combined with photocurrent/photothermal testing. Yolk-in-shell Au@CdS functional nanomaterials have great potential for early screening and diagnosis of various cancers.

Ouyang et al. [55] constructed an electrochemical biosensor for miRNA-21 detection by combining nanomaterials and hybridization chain reaction (HCR). First, Ti_3C_2 was obtained by etching Ti_3AlC_2 with HF, and then Ti_3C_2 was covered with Bi_2O_3 nanoparticles to form $Ti_3C_2@Bi_2O_3$ with an accordion-like structure. GCE was modified with $Ti_3C_2@B_{i2}O_3$ to enhance electrode conductivity and AuNPs to increase the active surface area. Then the thiol-modified capture probe (SH-CP) was immobilized on the electrode through Au-S bonds. The hairpin structure of the capture probe was opened when the target miRNA-21 was present, allowing miRNA-21 to hybridize specifically with the capture probe. The addition of primers H1 and H2 triggered the hybridization chain reaction, forming long double strands on the GCE surface. Many methylene blue (MB) molecules were embedded in the long double strand, which resulted in a significant DPV response of the biosensor at a potential of 0.19 V. This dual-signal amplification strategy based on nanomaterials and HCR can detect miRNA-21 in a wide linear range of 1 fM–100 pM with a detection limit as low as 0.16 fM. Compared with the qRT-PCR quantification technique, the biosensor detected miRNA-21 in the same linear range as qRT-PCR, but with a lower detection limit and higher sensitivity. In addition, the biosensor shows good applicability in human serum samples. However, the biosensor is dependent on primed DNA strands, which leads to higher costs and is not favorable for mass production.

3.2. Impedance Method

Electrochemical impedance spectroscopy (EIS) is a sensitive and versatile electrochemical sensing technique that finds extensive applications in the analysis of microscopic interfacial features associated with biomolecules [56]. By probing the impedance spectroscopic response of electrochemical systems, EIS provides valuable insights into various physical properties such as diffusion rates, reaction rates, and microstructural features [57]. Currently, EIS serves a dual role in electrochemical biosensors. It can be used to characterize the sensor construction process, and also to quantitatively detect biomolecules [58].

La et al. [59] proposed a signal amplification strategy induced by the insulating effect to enable sensitive impedance measurements of miRNA-21. The DNA probe encapsulated on the surface of the Au electrode was able to capture the target miRNA-21 and a biotin-modified miRNA (biotin–miRNA) with the same nucleic acid sequence. After self-assembly by biotin–F monomers, the streptavidin (SA)–biotin–FNP network was formed by binding SA. Biotin–miRNA adsorbed onto the SA–biotin–FNP network through biotin–SA interactions, forming an insulating layer on the electrode surface that hinders electron transport and consequently amplifies the impedance response. When the target miRNA-21 was present, miRNA-21 and biotin–miRNA competed for hybridization to the capture probe. With the reduction of biotin–miRNA captured on the electrode, the adsorbed SA–biotin–FNP network was also reduced, resulting in a significant decrease in the impedance signal. The linear range of this detection strategy for miRNA-21 detection is 0.1–250 fM, with detection limits as low as 0.1 fM. The SA–biotin–FNP network used in this method is relatively easy to prepare and the sensor has good applicability in real samples. However, the detection time of this sensor is greater than 2 h, which is not conducive to achieving rapid point-of-care detection.

Eksin et al. [60] developed a paper-based electrochemical impedance biosensor for quantitative detection of miRNA-155. This paper-based biosensor is shown in Figure 6a. The paper-based sensor consisted of a microfluidic channel and a working area where the working electrode, counter electrode, and reference electrode were placed. AuNPs-PE was formed by depositing AuNPs onto PE through the chronocurrent method. Figures 6b and 6c, respectively, show the EIS response curve and the calibration curve of this biosensor for the detection of miRNA-155. The thiol-modified DNA probe was immobilized on AuNPs-PE via Au-S bonds. When miRNA-155 was present, KCl solution containing $[Fe(CN)_6]^{3-/4-}$ was added dropwise, and miRNA-155 was quantified by measuring the change in charge transfer resistance (R_{ct}). The linear detection range of the sensor was 0–1.5 μg/mL in PBS with a detection limit of 33.8 nM, and 0–4 μg/mL in the fetal bovine serum (FBS) medium with a detection limit of 93.4 nM. The biosensor shows good selectivity for non-complementary and mismatched miRNA sequences. The paper-based electrochemical biosensor can selectively detect miRNAs even in complex media such as serum with a detection time of only 15 min and good stability, which makes it very suitable for POC detection applications.

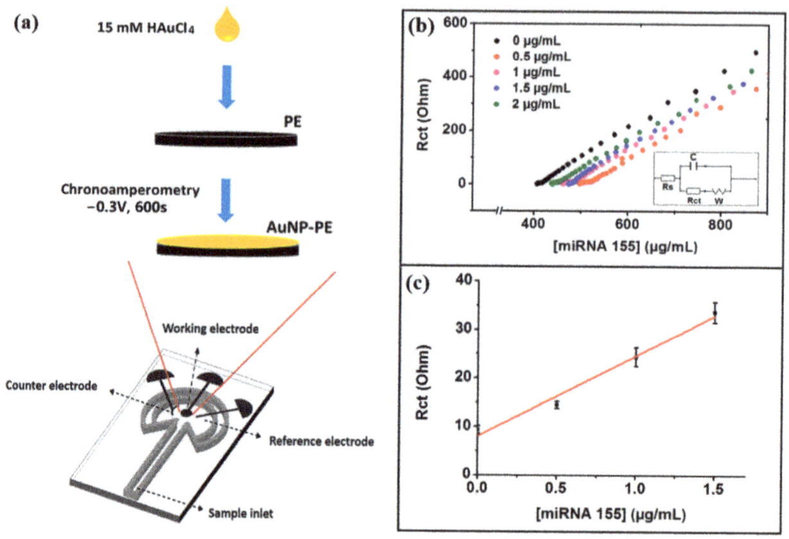

Figure 6. (**a**) Schematic diagram of a paper-based electrochemical biosensor. (**b**) Nyquist curves after hybridization of DNA probe and different concentrations of miRNA-155. (**c**) Calibration curve of R_{ct} versus 0–1.5 μg/mL miRNA-155. (Reproduced with permission from [60]).

Yarali et al. [61] developed an electrochemical biosensor for the detection of miRNA-155 and miRNA-21 associated with early cancer diagnosis using molybdenum disulfide (MoS_2) modified paper-based electrodes for the first time. Block crystals and sheets of MoS_2 were, respectively, fabricated and modified on the surface of paper-based electrodes to explore their performance in miRNA detection. The capture probe was modified on the MoS_2-modified electrode, and different concentrations of target miRNA solutions were added dropwise for hybridization. The electrochemical response of the sensor was measured by EIS technology, and the entire miRNA detection process was completed within 30 min. The linear detection range of this biosensing platform is 1–200 ng/mL. In the PBS buffer, the LOD for miRNA-155 was calculated to be 17.0 ng/mL and the LOD for miRNA-21 was 9.2 ng/mL through linear fitting. In the FBS medium, the LOD of miRNA-155 was 1.0 ng/mL and the LOD of miRNA-21 was 17.0 ng/mL. The electrically active surface area of bulk MoS_2 was larger compared to that of nanosheets, and thus the detection limit of the paper-based electrode modified by bulk MoS_2 was lower. In addition, they are effective in distinguishing non-target sequences with single base mismatches. The biosensor has a low manufacturing cost and can perform highly sensitive and selective quantitative analysis of miRNAs at low sample volumes, offering great potential for the detection of miRNA biomarkers in human serum.

3.3. Other Methods

In addition to the studies mentioned above, several other near-commercial miRNA biosensors have been developed. Jin et al. [62] combined magnetic nanobeads with metal–organic frameworks loaded with glucose oxidase (MOFs@GOX) and constructed a novel self-powered electrochemical sensor based on a photocatalytic zinc–air battery (ZAB-SPES) for the detection of miRNA let-7a. ZAB-SPES has a high-power density of 22.8 μW/cm^2, which is 2–3 times higher than that of commonly used photofuel cells. Gao et al. [63] reported a flexible graphene field effect transistor (Gr-FET) biosensor. The biosensor was able to achieve an miRNA detection limit as low as 10 fM within 20 min. Gr-FET-based biosensors will have prospective applications in wearable electronic devices for health monitoring and disease diagnosis. Xu et al. [64] integrated EBFCs on a flexible paper tape carrier to establish an ingenious sensor technology for the detection of tumor markers in complex samples. Multivariate detection was realized by receiving real-time instantaneous current values via a smartphone. This smartphone-based paper tape sensor platform provides an opportunity for early cancer diagnosis and lays the foundation for the construction of flexible wearable platforms.

Table 1 summarizes the characteristics of these electrochemical biosensors for the detection of HCC-associated miRNAs. The main characteristics include receptor type, electrode material, electrochemical method, linear detection range, detection limit, sensitivity, and response time. It was found that the above electrochemical biosensors had wide detection limits and high sensitivity. However, most of the sensors have poor immunity to interference and still have a long response time, which is not conducive to rapid bedside detection.

Table 1. Electrochemical biosensors for the detection of hepatocellular carcinoma-associated miRNAs.

Analyte	Receptor	Electrode	Electrochemical Method	Linearity Range	LOD	Sensitivity	Assay Time	Ref.
miRNA-122	DNA probe	GCE	DPV	0.1 fmol/L–1.0 nmol/L	0.021 fmol/L	—	60 min	[46]
miRNA-122	DNA probe	Au	CV	10^{-18}–10^{-6} mol/L	10^{-18} mol/L	—	30 min	[47]
miRNA-224	DNA probe	ITO	SWV	0.52–500 fM	0.061 fM	—	—	[48]
miRNA-let 7a	DNA probe	MnO_2	DPV	0.4–140 nM	0.25 nM	—	—	[49]
miRNA-let 7a	DNA probe	CPE	DPV	1.0 zM–11 nM	1.0 zM	—	30 min	[50]
miRNA-let 7b	PNA probe	AuNPs	Voltammetry	1 fM–100 pM	10 fM	—	30 min	[51]
miRNA-34a	ZNA probe	PGE	DPV	2–8 μg/mL	0.87 μg/mL	—	60 min	[53]
miRNA-21	DNA probe	Au@CdS	Photocurrent	0.01 pM–10 nM	4.2 fM	—	—	[54]
miRNA-21	DNA probe	GCE	DPV	1 fM–100 pM	0.16 fM	—	30 min	[55]
miRNA-21	—	Au	EIS	0.1–250 fM	0.1 fM	—	>2 h	[59]
miRNA-155	DNA probe	AuNPs-PE	EIS	0–1.5 μg/mL	33.8 nM	—	15 min	[60]
miRNA-21 miRNA-155	DNA probe	MoS_2-PE	EIS	0.025–0.75 μg/mL 0.05–0.15 μg/mL	9.2 ng/mL 17.0 ng/mL	1372.4 kOhm.mL.μg^{-1}.cm^{-2} 1361 kOhm.mL.μg^{-1}.cm^{-2}	30 min	[61]

4. Optical Biosensors

Optical signals are insensitive to noise interference, have good stability, and the spectral properties of different molecules to be tested are differentiated with high specificity. Therefore, optical biosensors can directly detect the molecules to be detected [65]. In addition, optical biosensors are easily miniaturized and have the potential to facilitate chip-level integration [66]. According to different working methods and principles, optical biosensors can be classified as colorimetric biosensors, fluorescent biosensors, SPR-based biosensors, and SERS-based biosensors. The optical biosensors for the detection of hepatocellular carcinoma-associated miRNAs are categorized and reviewed next.

4.1. Colorimetry

Colorimetry is a common method for the detection of biomolecules, which is evaluated by analyzing the change in absorbance or reflectance of the reagent [67]. Colorimetric biosensors have the advantages of naked eye determination, low cost, fast response, and ease of fabrication [68]. Colorimetric biosensors can change color in response to external physical or chemical factors, as well as through an enzyme-catalyzed chromogenic reaction inside the sensor, or with the help of metallic nanomaterials [65].

Shahsavar et al. [69] developed a novel colorimetric platform based on G-quadruplex spherical nucleic acid enzyme (SNAzyme) for the recognition of miRNA-155. Capture probe 1 and G-rich probes were attached to AuNPs via Au-S bonds to form G-quadruplexes. The G-tetramer was converted to SNAzyme in the presence of K^+ and hemin under buffered conditions of Tris-HCl 100 mM, KCl 150 mM pH 7. The target miRNA-155 was able to hybridize with the capture probe 1 to form a double strand, resulting in a significant decrease in the intensity of the colorimetric response. In the range of 1–100 nM, there is a linear relationship between the decrease in colorimetric response signal and the amount of miRNA-155. The detection limit for miRNA-155 using this method was 0.7 nM. In addition, the colorimetric sensing platform successfully realized the quantitative detection of miRNA-155 in human serum samples. SNAzyme has high thermal stability and resistance to nuclease degradation, which can improve the detection performance of the colorimetric sensing platform. However, this biosensor exhibited a linear detection range and detection limit at the nanomolar level, necessitating the incorporation of an amplification strategy to improve the sensitivity of the colorimetric biosensor.

Li et al. [70] designed a specific Y-type DNA probe for the colorimetric detection of miRNA-21 and miRNA-141. The detection scheme combining EXO III-assisted target cycling and HCR dual-signal amplification is shown in Figure 7. The Y-type DNA probe consists of the capture strand of target miRNAs and the HCR promoter strand. When two miRNAs were present, the capture probe specifically recognized target miRNAs, leading to the formation of a DNA double strand, resulting in two loops and the subsequent release of the HCR promoter strand Y3. EXO III can specifically cleave the DNA double strand and release target miRNAs again. Y3 can continuously cycle to initiate HCR amplification, thus realizing the amplification of colorimetric signals. The quantitative detection of miRNA-21 and miRNA-141 was based on UV–Vis absorption spectra. The detection limit of both miRNAs was 3 pM, and the linear range was 10 pM-0.4 nM. This colorimetric biosensor incorporating a dual-signal amplification strategy can simultaneously detect miRNA-21 and miRNA-141 with high sensitivity. The Y-shaped DNA probe provided a good application basis for the detection of miRNAs due to its simple preparation, high selectivity, and high stability. In addition, this colorimetric biosensor was successfully applied to the detection of miRNAs in human serum samples, and it is expected to be successfully applied to the diagnosis of clinical cancer in the future.

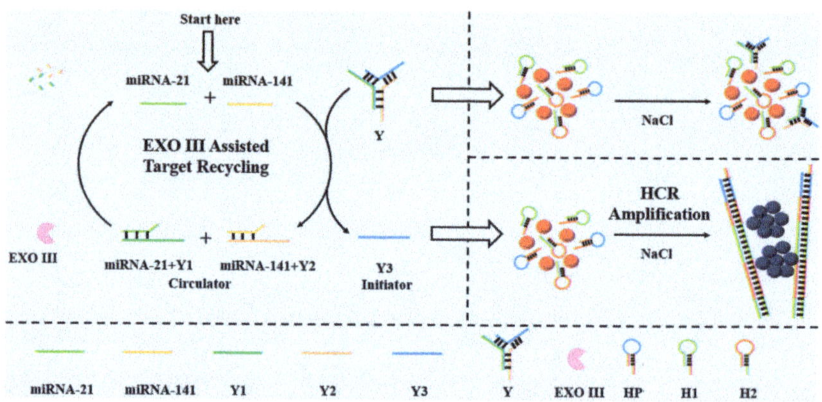

Figure 7. The Y-shaped probe combined with double amplification for colorimetric sensing of two miRNAs. (Reproduced with permission from [70]).

Yang et al. [71] proposed a colorimetric sensing strategy for miRNA-21 based on the combination of G-quadruplex (GQ) and CHA. The design principle of the sensor and the detection process are shown in Figure 8a,b. Multiple hairpin DNA H1 probes were attached to AuNPs. When miRNA-21 was present, hairpin DNA H1 was turned on to hybridize with miRNA-21 and hairpin DNA H2, and CHA would be triggered and continue to circulate. Upon addition of K^+ and hemin, the above precursors can self-assemble into a spherical DNAzyme. DNAzyme can catalyze redox reactions and color changes. As the concentration of miRNA-21 increased, the characteristic absorption signal of oxidized tetramethylbenzidine (TMBox) gradually increased and the color of the solution changed from burgundy to blue–violet. Figure 8c shows the UV–Vis absorption spectrum for the detection of miRNA-21, and Figure 8d shows the calibration curve. The developed method has a linear detection range of 100 fM to 20 nM and a detection limit of 90.3 fM. In addition, the sensing platform has been successfully used to detect miRNA-21 in human serum, providing a promising tool for the early diagnosis of hepatocellular carcinoma. Compared with the method of Li et al. [70] for detecting miRNA-21, this method has a wider linear range and lower detection limit. However, the detection throughput of this method is low, and only one miRNA can be detected at a time.

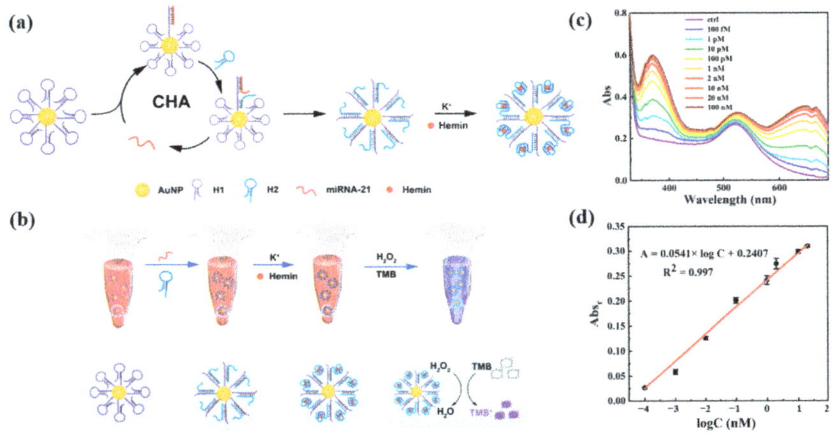

Figure 8. (**a**) Schematic diagram of CHA reaction and spherical DNAzyme self-assembly. (**b**) Schematic diagram of the colorimetric reaction principle of miRNA-21. (**c**) UV–Vis absorption spectra of the probe solutions after incubation with different concentrations of miRNA-21. (**d**) Calibration curve between absorbance change at 650 nm and the logarithm of the concentration (logC). (Reproduced with permission from [71]).

4.2. Fluorescence Method

Fluorescent biosensors are widely used in various fields such as biomedicine and environmental protection because of their high sensitivity, good selectivity, simple operation, and fast detection speed [72]. This method improves the sensitivity and selectivity of fluorescence detection by designing specific fluorescent probes or using fluorescent nanoparticles such as quantum dots and metal nanoparticles [73]. Fluorescent biosensors usually require fluorescence spectroscopy measurements with the help of specialized instruments such as fluorescence spectrophotometers.

In 2019, Wang et al. [74] designed a ratiometric fluorescent biosensor based on MnO_2 nanosheets for the detection and imaging of miRNA-21 in living cells. MnO_2 nanosheets were used as carriers for DNA probes H1 and H2. The recognition probe H1 of this biosensor was required to be labeled with the fluorescent donor FAM, and the amplification probe H2 was required to be labeled with the fluorescent acceptor TAMRA. As shown in Figure 9a, the target miRNA-21 in the cell hybridized with the recognition probe H1 and initiated CHA. The H1-H2 double-stranded body formed by hybridization prompted the fluorescence donor FAM to approach the fluorescence acceptor TAMRA, which induced a ratiometric fluorescence response. Figure 9b shows the fluorescence emission spectra of miRNA-21 at different concentrations, and Figure 9c demonstrates the linear fit curve about the ratio of $(F_A/F_D)_{positive}$ to $(F_A/F_D)_{negative}$ when different amounts of miRNA-21 were added. The ratio became larger with the increase in miRNA-21 concentration in the concentration range of 0.1 to 20 nM, with a detection limit of 73 pM. Li et al. [75] proposed a label-free fluorescence sensing strategy based on copper nanoclusters (CuNCs) to detect miRNA-21. The change in fluorescence of CuNCs can be used to quantify the concentration of the target miRNA-21. The DNA-CuNCs used in this biosensor were synthesized with high efficiency, which saved the preparation time of the sensor. The CuNCs have strong fluorescence properties and good biocompatibility, which improved the sensitivity of fluorescence sensing and provided a new idea for fluorescence detection of miRNAs. This strategy can quantify miRNA-21 in the range of 50–1000 pM with a detection limit of 18.7 pM. This strategy can quantitatively detect miRNA-21 in the range of 50–1000 pM with a detection limit of 18.7 pM, which is lower than the detection limit of the labeled fluorescence detection method proposed by Wang et al. [74] and has good specificity and selectivity. Compared with the labeled fluorescence detection method, the unlabeled method requires less material and is simpler to operate, which is expected to improve the detection sensitivity of miRNA fluorescence sensors.

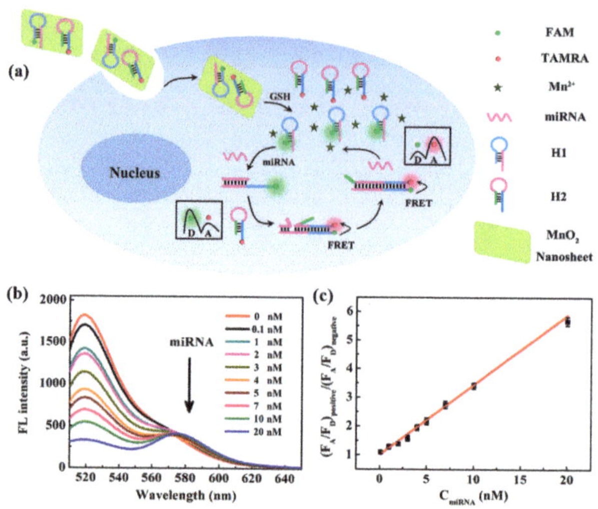

Figure 9. (**a**) Schematic diagram of the MnO_2 nanosheet-mediated ratiometric fluorescent biosensor for miRNA detection and imaging. (**b**) Fluorescence emission spectra of different concentrations of miRNA-21. (**c**) Variation of $(F_A/F_D)_{positive}/(F_A/F_D)_{negative}$ values with miRNA-21 concentration. $(F_A/F_D)_{positive}$ and $(F_A/F_D)_{negative}$, respectively, indicate the ratio of acceptor to donor fluorescence in the presence and absence of miRNA-21. (Reproduced with permission from [74]).

In 2021, Forte et al. [76] developed an advanced PNA microarray system for the detection of miRNA-122 using the PNA probe. The PNA microarray substrate consisted of a multilayer structure that included a silicon support, a mirror layer (Al-SiCu) produced by the CVD method, and a spin-coated agarose layer. The microarray coating was fully characterized by electron microscopy and SEM optical techniques. By combining the mirror effect of the aluminum membrane and the positive interference of the agarose membrane on the emission wavelength of the Cy5 fluorescent label, the system achieved direct optical signal enhancement. miRNA-122 was detected by the PNA microarrays with a sensitivity of approximately 1.75 μM^{-1} and a limit of detection of 0.043 nM. In addition, it was demonstrated by molecular dynamics simulations that the agarose substrate had a dsPNA-RNA interaction with positive contribution and was able to avoid potential non-specific binding. The Si/Al/agarose substrate is highly promising for the development of new microarray platforms for cancer diagnostic devices.

In 2022, He et al. [77] developed a fluorescent miRNA-21 sensing strategy based on carbon dots (CDs) and AuNPs. Positively charged CDs fluorophores (PEI-CDs) and DNA probe-modified AuNPs (AuNPs-cDNAs) were assembled by electrostatic interaction, resulting in fluorescence quenching. PEI-CDs were released when miRNA-21 was present, at which point fluorescence intensity was restored. The fluorescence intensity was linearly correlated with the logarithm of miRNA-21 concentration in the range of 1–1000 fM, and the detection limit was as low as 1 fM. The results of this method for detecting miRNA-21 in real serum samples were comparable to those of qRT-PCR. This fluorescent biosensor does not require complex labeling, effectively simplifying the process. Although it had a low detection limit for miRNA-21, the detection time still required 2 h, which was unfavorable for point-of-care detection. The detection time of this sensor can be effectively shortened if combined with a molecular enrichment strategy to accelerate the binding rate of miRNA-21 molecules to DNA probes.

In 2023, He et al. [78] proposed another novel miRNA-21 fluorescence sensing strategy based on CDs. The schematic diagram of this biosensor for the detection of miRNA-21 is shown in Figure 10a. A molecular beacon (MB) probe (CDs-MB-BHQ1) was constructed using CDs as fluorophores and BHQ1 labeled at the 5′ end of the DNA probe as a bursting agent. Not only can graphene oxide (GO) act as a co-bursting agent but it can also adsorb the MB probe. When miRNA-21 hybridized with the loop region of the MB probe, the hairpin MB probe opened, increasing the distance between the CDs and BHQ1. The MB probe was released from the GO, restoring the fluorescence intensity of the CDs. Figures 10b and 10c, respectively, show the fluorescence spectra and calibration curve of the sensor for miRNA-21 detection. The fluorescent biosensor can be used to determine miRNA-21 in the range of 0.5–800 pM with a detection limit of 500 fM. In addition, the MB probe can be used to detect miRNA-21 levels in real human serum. Although this sensing strategy is relatively novel, it requires labeling of DNA probes, which is complicated to operate. The detection limit of this strategy for miRNA-21 is not adequate for clinical applications, and efforts are needed to enhance the sensitivity of the fluorescent sensor.

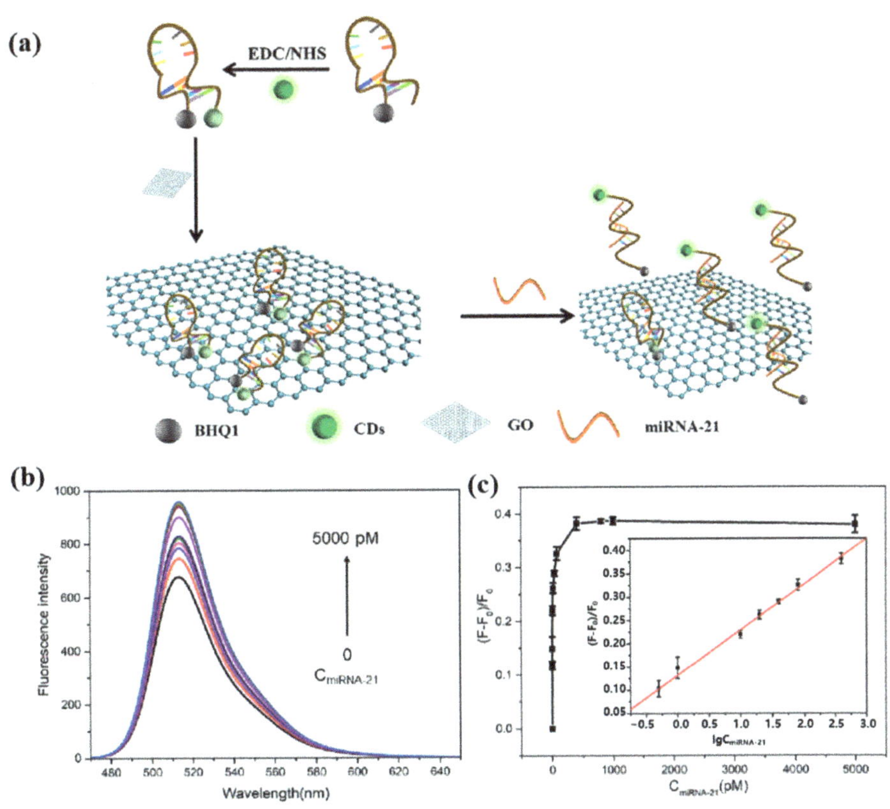

Figure 10. (**a**) Schematic illustration of CDs-MB-BHQ1/GO system for miRNA-21 detection. (**b**) Fluorescence spectra of CDs-MB-BHQ1/GO probe with the addition of various amounts of miRNA-21 (0, 0.5, 1, 10, 20, 40, 80, 400, 800, 1000, 5000 pM). (**c**) The relationship between $(F-F_0)/F_0$ and miRNA-21 concentration (Insert: calibration curve between $(F-F_0)/F_0$ and the logarithm of miRNA-21 concentration within the range of 0.5–800 pM). (Reproduced with permission from [78]).

4.3. *Surface Plasmon Resonance*

Surface plasmon resonance is a label-free optical detection technology. When visible or near-infrared light is incident on a metal surface, the refractive index (RI) near the metal surface can reflect the level of biomolecules [73]. Common methods used to excite surface plasmon include coupling of prisms, optical fibers, gratings, and nanoparticles [79]. The detection data of SPR biosensors can be collected and displayed in real time. It currently has a wide range of applications in biosensing, environmental monitoring, and clinical diagnosis.

Yu et al. [80] constructed a surface plasmon resonance biosensor for the combined detection of AFP and miRNA-125b markers associated with HCC. The SPR response was amplified using the double antibody sandwich method (DASM) and S9.6 antibody enhancement method to improve the sensitivity and specificity of the sensor. As shown in Figure 11a,b, the anti-AFP monoclonal antibody was modified on the surface of the CM5 chip by amide bonding, and the DNA probe paired with miRNA-125b was bound on the surface of the CM5 chip by biotin–streptavidin interaction. The linear range of AFP detection by DASM was 25–400 ng/mL, essentially covering the clinical AFP detection range (0–400 ng/mL). The S9.6 antibody enhancement method reached the detection limit of 123.044 pM for miRNA-125b in the linear range of 0–1000 pM. These results validate the feasibility of combined multi-marker detection in the early diagnosis of HCC. The detection technology of AFP as a clinical marker is relatively mature, and combining AFP with other miRNA markers can improve the accuracy and reliability of early cancer diagnosis. This combined detection strategy inspires the detection of various cancer markers.

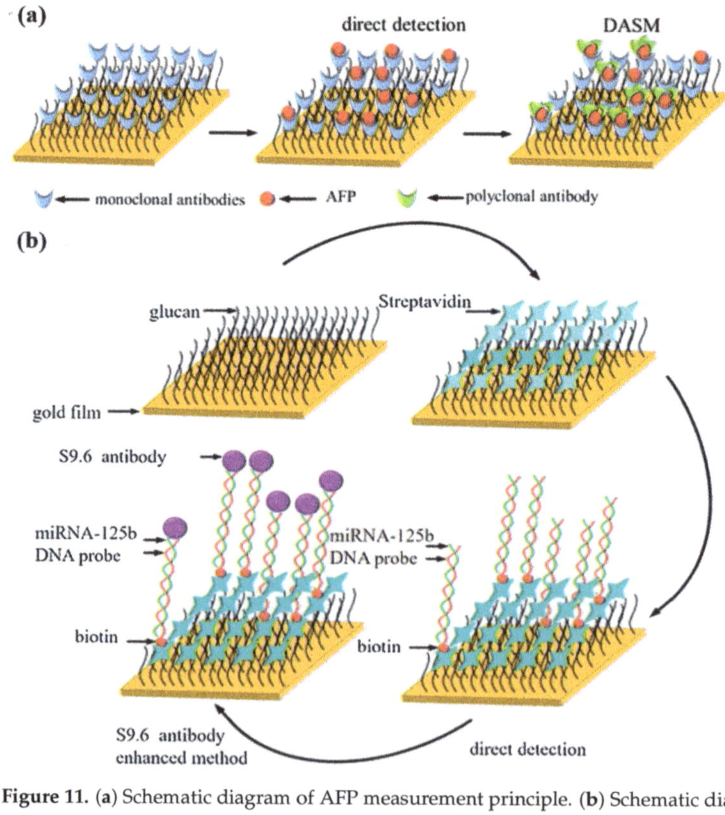

Figure 11. (**a**) Schematic diagram of AFP measurement principle. (**b**) Schematic diagram of miRNA-125b measurement principle. (Reproduced with permission from [80]).

In conventional surface plasmon resonance biosensors, the hybridization and enzymatic digestion reactions to detect biomarkers are performed on the chip surface. Huang et al. [81] proposed an innovative method to perform hybridization enzymatic digestion cyclic reactions in solution for miRNA-21 detection. The schematic diagram of this SPR biosensor is shown in Figure 12a. The target-free probe solution (bio-DNA-bio/DSN) was injected into the reference channel, the target-containing probe solution (bio-DNA-bio/miRNA-21/DSN) was injected into the detection channel, and SA-modified gold nanoparticles (SA-AuNPs) were injected into both channels. In the channel without miRNA-21, bio-DNA-bio adhered to the SA-modified chip via SA-bio binding and captured the SA-AuNPs, obtaining a stronger SPR signal. However, in the channel containing miRNA-21, the DNA hybridized with miRNA-21 was digested by DSN, and miRNA-21 was released into the next enzymatic cycle while many bio-DNA fragments were generated. The generated bio-DNA and undigested bio-DNA simultaneously bound to the SA on the surface of the chip, and the captured SA-AuNPs were reduced, leading to a decrease in the SPR signal. In Figure 12b, the SPR signal decreases with the increase in miRNA-21 concentration. Figure 12c demonstrates that the biosensor has a good linear correlation in the range of 1–15 fM. The SPR signal can be detected even when the concentration of miRNA-21 is as low as 1 fM. The sensing strategy of the SPR biosensor proposed in this study is relatively novel, and the SPR signal amplification can be realized by releasing miRNA-21 to participate in the cycle only through the DSN-digested DNA probes. Although the detection limit of this biosensor is ideal, its linear range is narrow compared to other studies of the same type, and the detection results are easily affected by errors.

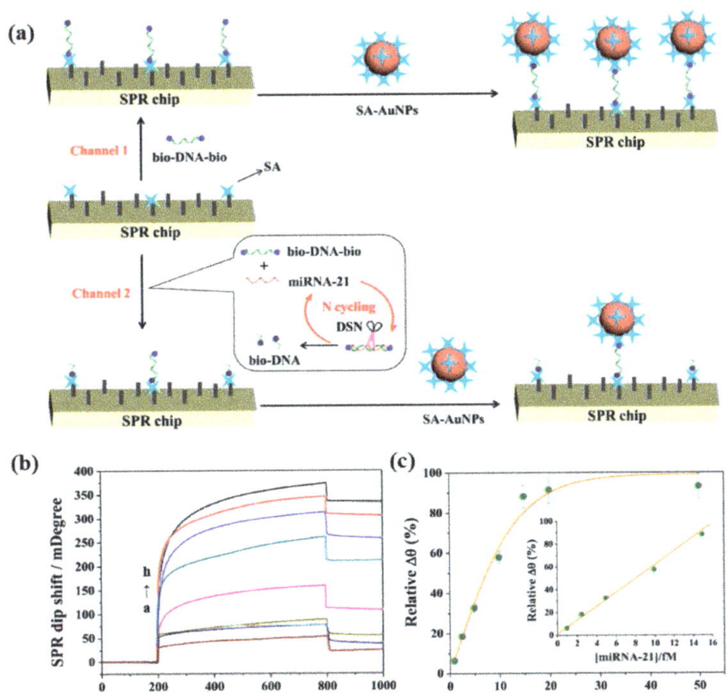

Figure 12. (**a**) Schematic representation of SPR biosensor for miRNA-21 detection based on dual-signal amplification of DSN and SA-AuNPs. (**b**) Representative SPR sensing plots for miRNA-21 assay at different concentrations (a: 0 fM, b: 1 fM, c: 2.5 fM, d: 5 fM, e: 10 fM, f: 15 fM, g: 20 fM, and h: 50 fM) with an enzymatic digestion time of 60 min. (**c**) Calibration curve of miRNA-21 concentrations versus relative Δθ. (Reproduced with permission from [81]).

Wang et al. [82] developed an in situ-prepared silver nanoparticles (AgNPs)-based SPR biosensor based on HCR for sensitive detection of miRNA let-7a. The dielectric constant property of AgNPs can significantly increase the angle of SPR. Three DNA probes (ON1, ON2, and ON3) were included in this strategy so that they could capture the miRNA let-7a, which subsequently triggered the HCR to produce a large amount of dsDNA on the SPR disk. ON2 and ON3 coexist in the reaction solution in the absence of the target miRNA, and only a small amount of dsDNA is attached to the SPR disk. Almost no AgNPs are produced on the dsDNA, resulting in smaller SPR angle changes. In the presence of target miRNA, many AgNPs were produced with the insertion of Ag^+ into the dsDNA strand by NaBH4 reduction, leading to a significant increase in the SPR angle. The SPR angle was proportional to the target miRNA concentration. The AgNPs-based SPR biosensor detected miRNA let-7a with a linear range of 0.001–0.1 pM and a detection limit of 0.35 fM, which is lower than that of other SPR biosensors using variable amplification tags. This SPR sensing strategy exhibited unmodified properties and excellent sensitivity, and it has great potential for health monitoring and early cancer diagnosis.

4.4. *Surface-Enhanced Raman Spectroscopy*

Surface-enhanced Raman spectroscopy has been widely used in different fields because of its high sensitivity, unique molecular vibrational fingerprinting, and ease of operation [83]. It is very suitable for the analysis and detection of trace biomolecules, including the sensitive detection of low-abundance miRNAs [84]. When the target biomolecules are adsorbed onto the surface of the SERS substrate, especially after binding to the receptors of the substrate, a distinct Raman signal can be observed [85]. In recent years, many researchers have combined SERS methods with biosensors for the detection of miRNAs.

Wang et al. [86] proposed a dual-mode microfluidic chip-based sensing strategy to detect miRNA-21 by combining fluorescence and SERS spectroscopy. miRNA-21 was detected by modifying AgNPs on glass slides to form a SERS-enhanced substrate. A customized molecular beacon (MB) is then modified. The 3′ end of the MB was decorated with a thiol group, which enables the MB to attach to the substrate surface. The 5′ end of the MB was labeled with 6-FAM, which serves as both a fluorophore and a Raman reporter. Target DNA was injected in parallel channels on a microfluidic chip prepared by PDMS, and fluorescence and SERS measurements were performed after placing the reaction for 1 h. MBs maintained their hairpin structure in the absence of target miRNA-21. Due to the proximity of 6-FAM labeling to AgNPs, the fluorescence of 6-FAM was quenched, and the Raman signal was enhanced. In the presence of the target miRNA-21, specific hybridization between the miRNA and MB will open the hairpin structure of MB. Due to the increased distance between 6-FAM and AgNPs, the fluorescence of 6-FAM was restored, and the SERS signal was weakened. This method can detect miRNA-21 in a linear range of 10^{-9}–10^{-7} M. Compared with the fluorescence or SERS sensing strategy alone, the combination of opposite variations of the two optical methods can improve the sensitivity and linearity of detection of miRNA-21. In addition, the introduction of microfluidic chips can shorten the reaction time and save the number of reagents while reducing the complexity of detection. However, the detection limit of this method is still unclear.

Si et al. [87] developed a novel SERS sensor array with nine sensing units based on a DNA hydrogel, which can simultaneously detect multiple cancer-related miRNAs in a single sample. After modifying SA on each sensing unit, a DNA hydrogel responsive to the target miRNA was added. Since the DNA hydrogel formed blocked the binding of the modified SA sensor unit to the SERS tag, no significant Raman signal could be observed. After the introduction of miRNA-21, the DNA probe in the DNA hydrogel of the SERS sensor unit hybridized with miRNA-21, and the DNA hydrogel was broken down accordingly. As a result, the SERS label was able to be captured onto the surface of the SA-modified sensing unit, resulting in a significant Raman signal. The detection limit of the sensor was calculated to be 0.11 nM in the range of 4–1200 nM. In this study, the smart combination of the SERS sensor array with barcodes highlights the applicability of the sensor array in multiple detection. This DNA hydrogel-based SERS sensor array is likely to be a promising candidate for early cancer screening and clinical diagnosis. However, the developed DNA hydrogel-based SERS sensor array has limited sensitivity for the detection of miRNA-21 and difficulty in detecting lower concentrations of miRNA.

Wu et al. [88] proposed a surface-enhanced Raman scattering biosensor based on magnetic nanoparticles, which consisted of a magnetic capture unit and a SERS tag for the ultrasensitive combined detection of miRNA-122, miRNA-223, and miRNA-21 biomarkers associated with HCC. Figure 13a,b show the synthesis process of the SERS tag and magnetic capture unit. The capture probe was modified on the silver shell on the surface of the magnetic bead to form a magnetic capture unit to enhance the SERS signal. After adding a mixture of target miRNA-122, miRNA-223, and miRNA-21 to hybridize with the capture probe, SERS labels, respectively, modified by rhodamine 6G (R6G), crystalline violet (CV), and 4-amino thiophenol (4-ATP) were attached to the magnetic capture unit to form a sandwich structure of capture unit/miRNAs/SERS labels. The detection process is shown in Figure 13c. The proposed strategy can simultaneously detect three miRNAs in a linear range from 1 fM to 10 nM. In human serum, the detection limits of miRNA-122, miRNA-223, and miRNA-21 were 349 aM, 374 aM, and 311 aM, respectively. The F-AuNPs constructed in this study have reusable SERS performance. This SERS biosensor allows simultaneous multiplexed detection of three miRNAs with ultra-high sensitivity with a detection limit as low as the aM level. In addition, the biosensor showed good utility for multiplex detection of three miRNAs in 92 clinical sera. This study provides a new approach for the early diagnosis of cancer, the staging of HCC patients, and the prognosis of cancer, which is highly valuable for clinical application.

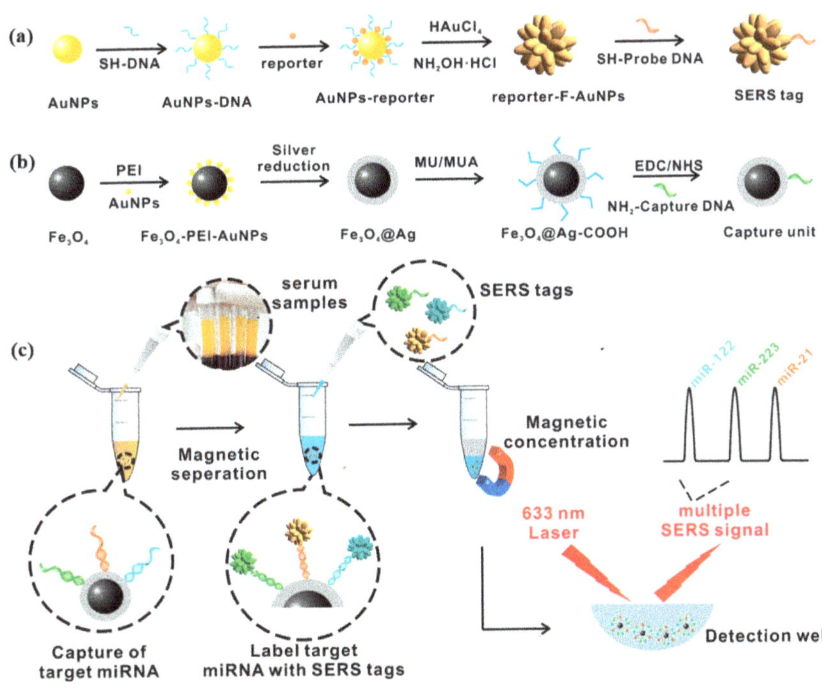

Figure 13. (**a**) Schematic diagram of the synthesis process of SERS tag. (**b**) Design and synthesis of capture substrate. (**c**) Multiple miRNA detection methods based on the capture substrate/miRNA/SERS tag sandwich structure. (Reproduced with permission from [88]).

Huang et al. [89] proposed a dual-mode biosensor based on SERS and fluorescence sensing strategy for the detection of miRNA-224 associated with HCC. The CHA strategy was constructed on gold nano-arrays (AuNAs) as shown in Figure 14a,b, where uniformly distributed hotspots on the AuNAs enhance the SERS signaling, and their wide surface area is very favorable for miRNA-224 adsorption. Cy3-labeled hairpin DNA H1 can capture miRNA-224. When Rox-functionalized hairpin DNA H2 was added, it was also able to hybridize with H1 to release miRNA-224, which initiated CHA cyclic amplification. The intensity of the Raman and fluorescence signals was altered by controlling the distance of Rox from the AuNAs. This dual-mode biosensor detected target miRNA-224 in the linear range of 1 fM to 1 nM, with a detection limit of 0.34 fM in SERS mode and 0.39 fM in FL mode. This dual-mode biosensor is also applicable and reliable in the analysis of human plasma samples. Figure 14c shows that the biosensor can distinguish between HCC patients and healthy individuals, monitor HCC patients before and after hepatectomy, and guide the different clinical liver cancer staging of BCLC. The AuNAs prepared in this study have a large surface area and good biocompatibility, which can be used as a generalized substrate for sensors that detect a wide range of biomarkers. In addition, the dual-mode detection results can provide double judgment and make the detection results more reliable.

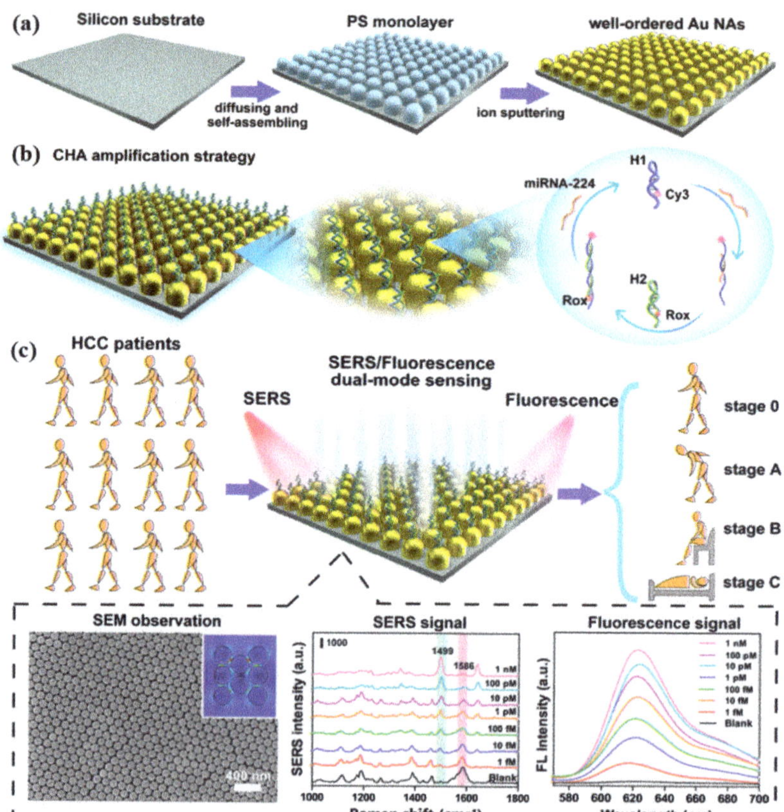

Figure 14. (**a**) Schematic diagram of the self-assembly process of AuNAs substrate. (**b**) Principle of AuNAs substrate for miR-224 detection combined with CHA amplification strategy. (**c**) SERS/FL dual-mode sensing procedure based on AuNAs substrate and its application in clinical staging. (Reproduced with permission from [89]).

4.5. Electrochemiluminescence

In the above study, electrochemical biosensors have excellent sensitivity, but the detection signal is prone to interference, which is not conducive to achieving accurate measurements. In contrast, optical biosensors have good stability and high signal-to-noise ratios, and each biomolecule has specific spectral properties, but their detection sensitivity for biomolecules is low. Therefore, electrochemiluminescence sensing technology has been investigated by many scholars. ECL is chemiluminescence triggered by an electrochemical reaction [90]. ECL technology combines the advantages of electrochemistry and chemiluminescence and is of great interest for early disease diagnosis and detection of hazardous substances [91]. So far, ECL has been widely used in various fields such as food safety, environmental monitoring, and medical diagnosis [92].

Due to the surface plasmon effect of non-metallic sulfides in the visible and near-infrared regions, Li et al. [93] designed a novel electrochemiluminescence sensing system based on SnS_2 nanomaterials incorporating the SPR effect for the detection of miRNA-21. SnS_2-PEI was used as the plasma source, and Figure 15a shows the preparation process of SnS_2-PEI. Figure 15b demonstrates the design principle of this ECL sensor, where the CHA cycle adsorbed more SA-PTCA by capturing more H2–biotin. There was a significant spectral overlap between the ECL emission spectrum of PTCA and the UV–Vis absorption spectrum of SnS_2-PEI, which can produce the SPR effect. Figure 15c shows the ECL response of this sensor, and Figure 15d shows the fitted relationship between the ECL response and miRNA-21 concentration. This ECL sensing system detects miRNA-21 in the linear range of 1 aM–1 nM with a detection limit as low as 0.6 aM. In addition, the ECL biosensor exhibited excellent selectivity, stability, reproducibility, and utility for detecting real samples. Compared to the electrochemical biosensors and optical biosensors described above, the ECL biosensor has a significantly improved sensitivity with a detection limit as low as the aM level for miRNA-21. In addition, the ECL biosensor has excellent selectivity, stability, and reproducibility, and can be used for the detection of real samples.

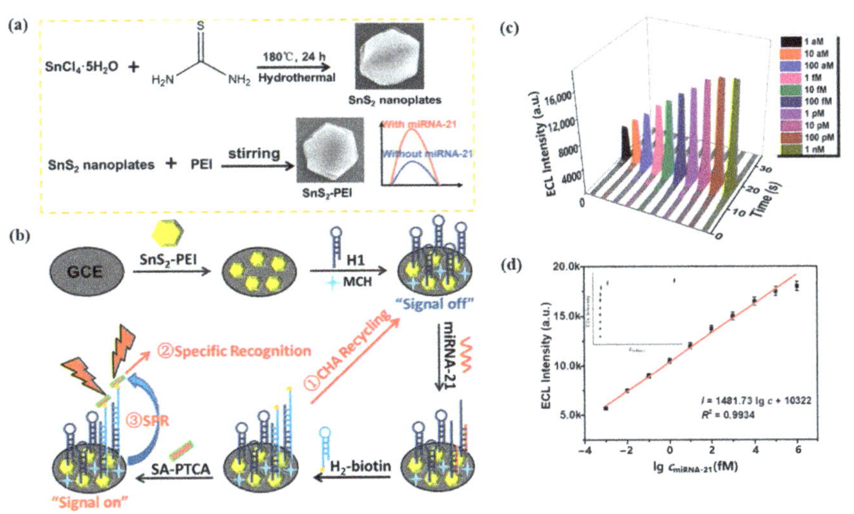

Figure 15. (**a**) Preparation of SnS_2-PEI. (**b**) Design of ECL biosensor for miRNA-21 assay. (**c**) Response of ECL to different concentrations of miRNA-21. (**d**) Quantitative calibration curve of miRNA-21 from 1 aM–1 nM. (Reproduced with permission from [93]).

In 2023, Wang et al. [94] developed an electrochemiluminescent biosensor based on three-dimensional (3D) DNA nanowalkers, which combined with DSN-mediated target cycling amplification to achieve sensitive detection of miRNA-21. Ferrocene-labeled DNA (Fc-DNA) was modified on Fe_3O_4 MBs to form 3D DNA tracks. When the target miRNA-21 appeared, DSN digested the hybridized double strand, releasing miRNA-21 and the bipedal DNA walker. miRNA-21 was again involved in the target cycle amplification, amplifying the response signal. As the DNA walker moved through the DNA track, the Fc-DNA hybridized to the DNA walker was cleaved by nucleic acid endonuclease and released Fc-DNA fragments. Fc-DNA fragments adsorbed to the surface of C-g-C_3N_4-modified electrodes, leading to quenching of the ECL signal from C-g-C_3N_4. The ECL biosensor has a wide linear range of 10 fM to 10 nM in miRNA-21 detection with a detection limit as low as 1.0 fM. This work provides an opportunity to construct miRNA biosensors based on DNA walkers. This bipedal walker significantly increases the walking efficiency, reduces the detection time, and improves the sensitivity of the assay as compared to the unipedal DNA walker. This work provides a new idea for constructing miRNA biosensors based on DNA walkers.

Graphene oxide quantum dots (GDYO QDs) are a derivative of graphene (GDY) with good electrical conductivity and luminescence properties. Lin et al. [95] constructed an ECL biosensor for the ultrasensitive determination of miRNA-21 based on GDYO QDs. The sensor utilized AuNPs/GDY as the electrode substrate material, which not only connected to the capture probe via Au-S bonds but also enhanced the conductivity of the electrode surface. Combined with DNA walker and HCR amplification technologies, the target miRNA-21 can drive DNA walker movement to generate many H1-H2 double strands. H3 and H4-GDYO QDs partially hybridized with H1-H2 via HCR, and the GDYO QDs enhanced ECL signals through electron transfer. This ECL biosensor can detect miRNA-21 in the linear range of 0.1 fM to 1 nM with a low detection limit of 0.023 fM. GDYO QDs prepared in this study have excellent conductivity, biocompatibility, and outstanding stability. Compared with other related studies, the ECL biosensor in this study detected miRNA-21 with a large linear range and a small detection limit.

Shen et al. [96] developed an ECL biosensor for the sensitive detection of miRNA-155 based on DNA nanoframe carrier luminophores (DNF-Dox-ABEI). Figure 16a demonstrates the preparation process of DNF-Dox-ABEI. Multiple luminescent molecules of Dox-ABEI were loaded on DNF-Dox-ABEI. As shown in Figure 16b, the double strand formed by the hybridization of DNA probes and cDNA could hybridize with miRNA-155. Then T7 exonuclease was able to digest the DNA probe, releasing miRNA-155 and cDNA at the same time. This process drove the targeting cycle, leading to signal amplification and the realization of the amplified ECL signal. Figure 16c shows the preparation process of TiO_2@Ag nanocomposites. The modified TiO_2@Ag on the GCE electrode exhibited excellent peroxide activity, and the generated reactive oxygen species further reacted with ABEI to generate an ECL signal. Figure 16d shows the schematic diagram of the ECL sensing strategy based on DNF-Dox-ABEI for the detection of miRNA-155. Figures 16e and 16f, respectively, show the standard ECL response and calibration curves for the detection of miRNA-155 by this sensor. The ECL biosensor can detect miRNA-155 in the linear range of 1.0 fM to 500.0 pM with a detection limit as low as 0.45 fM. Various signal enhancement strategies such as nanocomposites and target cycling amplification were applied in this study, which significantly improved the sensitivity of the sensor to detect miRNA-155. This work provides a potential cancer biomarker detection tool for early cancer diagnosis.

Table 2 summarizes the characteristics of these optical biosensors used to detect HCC-associated miRNAs. The main characteristics include receptor type, electrode material, optical method, spectral peak, linear detection range, detection limit, and response time. Most optical biosensors have poorer detection sensitivity and longer detection time than electrochemical biosensors. However, the SERS method has relatively superior sensitivity for miRNA detection.

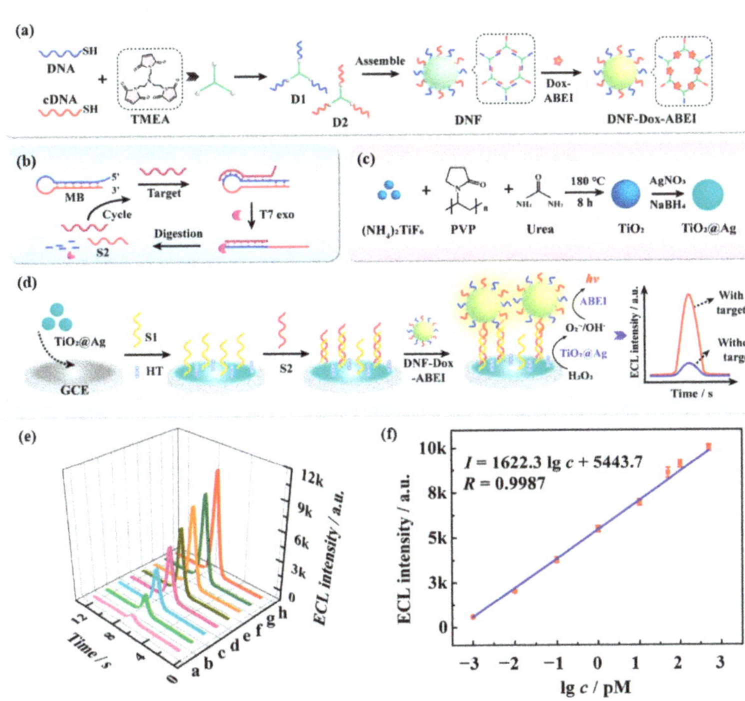

Figure 16. (**a**) Assembling of DNF-Dox-ABEI emitters. (**b**) Schematic description of T7 exo-TRA. (**c**) The synthesis procedure of TiO_2@Ag nanocomposites. (**d**) The construction process of the ECL biosensor. (**e**) Standard ECL response of the biosensor for different miRNA-155 concentrations (a: 1 fM, b: 10 fM, c: 100 fM, d: 1 pM, e: 10 pM, f: 50 pM, g: 100 pM, h: 500 pM). (**f**) Calibration plot to the ECL intensity and the logarithm of miRNA-155 concentrations. (Reproduced with permission from [96]).

Table 2. Optical biosensors for the detection of miRNAs associated with hepatocellular carcinoma.

Analyte	Receptor	Electrode	Optical Method	Spectral Peak	Linearity Range	LOD	Assay Time	Ref.
miRNA-155	DNA probe	—	Colorimetry	450 nm	1–100 nM	0.7 nM	—	[69]
miRNA-21 miRNA-141	DNA probe	—	Colorimetry	520 nm/650 nm	10 pM–0.4 nM	3 pM	—	[70]
miRNA-21	DNA probe	—	Colorimetry	650 nm	100 fM–20 nM	90.3 fM	20 min	[71]
miRNA-21	DNA probe	—	Fluorescent	520 nm/570 nm	0.1–20 nM	73 pM	40 min	[74]
miRNA-21	DNA probe	—	Fluorescent	605 nm	50–1000 pM	18.7 pM	—	[75]
miRNA-122	PNA probe	—	Fluorescent	650 nm	0.1–10 nM	0.043 nM	60 min	[76]
miRNA-21	DNA probe	—	Fluorescent	500 nm	1–1000 fM	1 fM	2 h	[77]
miRNA-21	DNA probe	—	Fluorescent	513 nm	0.5–800 pM	500 fM	—	[78]
miRNA-224	DNA probe	—	Fluorescent	618 nm	1 fM–1 nM	0.39 fM	1.5 h	[89]
miRNA-125b	DNA probe	—	SPR	—	0–1000 pM	123.044 pM	—	[80]
miRNA-21	DNA probe	—	SPR	—	1–15 fM	1 fM	60 min	[81]
miRNA let-7a	DNA probe	—	SPR	—	0.001–0.1 pM	0.35 fM	—	[82]
miRNA-21	DNA probe	—	Fluorescent +SERS	—	10^{-9}–10^{-7} M	—	1h	[86]
miRNA-21	DNA probe	—	SERS	—	4–1200 nM	0.11 nM	—	[87]
miRNA-122 miRNA-223 miRNA-21	DNA probe	—	SERS	615 cm^{-1} 918 cm^{-1} 1140 cm^{-1}	1 fM–10 nM	349 aM 374 aM 311 aM	—	[88]
miRNA-224	DNA probe	—	SERS	Cy3: 1586 cm^{-1} Rox: 1499 cm^{-1}	1 fM–1 nM	0.34 fM	1.5 h	[89]
miRNA-21	DNA probe	GCE	ECL	—	1 aM–1 nM	0.6 aM	—	[93]
miRNA-21	DNA probe	GCE	ECL	—	10 fM–10 nM	1.0 fM	—	[94]
miRNA-21	DNA probe	GCE	ECL	—	0.1 fM–1 nM	0.023 fM	—	[95]
miRNA-155	DNA probe	GCE	ECL	—	1.0 fM–500.0 pM	0.45 fM	—	[96]

5. Summary and Outlook

This paper reviews the biosensors used for the detection of miRNAs associated with hepatocellular carcinoma in recent years, mainly including electrochemical and optical biosensors. A detailed analysis was provided for the sensor probe types, electrode designs, sensing strategies, and detection effects. Among the electrochemical methods mentioned above, voltammetry is susceptible to noise interference despite its high detection sensitivity. Electrochemical impedance spectroscopy is a quasi-steady-state method, and the mathematical processing of the measurement results is relatively simple. Among the optical methods, colorimetry has the advantages of low cost and fast response, and can be determined by the naked eye, but is limited by the relative simplicity of the sample composition and the color development of the solution is not susceptible to interference. Fluorescence methods are characterized by high analytical sensitivity and selectivity but require the labeling of the probe and are prone to false positive or false negative results. SPR does not require the labeling of the substance to be measured and can monitor the dynamic process of molecular binding in real time and continuously, but it is sensitive to the composition of the sample and interferences such as temperature. SERS has high sensitivity and good reproducibility, with the disadvantage of low signal-to-noise ratio, which is difficult to realize and requires strict control of experimental parameters. In addition, electrochemiluminescent biosensors do not require a light source and have low background, high sensitivity, and good reproducibility.

With the continuous development of nanomaterials, it provides greater possibilities to realize the ultra-sensitive detection of microRNAs. Nanomaterials with multidimensional structures and composite nanomaterials usually have excellent conductive or luminescent properties and provide a larger active surface area for the attachment of capture probes, thus increasing the sensitivity of the assay. In addition, noble metal nanomaterials such as AuNPs and AgNPs have shown good biocompatibility as well as low toxicity in cancer marker detection. Microfluidic chips provide a pathway for the realization of portable and miniaturized detection of microRNAs. It utilizes a micrometer-scale structure to realize the detection process of the target within a microchannel or reaction chamber, with the advantages of controllable liquid flow, high throughput, and minimal consumption of samples and reagents. It has been recognized as an ideal technology for the development of diagnostic tests for point-of-care testing. In the above study, the combination of multiple signal amplification strategies was used to improve the sensitivity and reliability of the assay. Bimodal detection can provide dual judgment to the researcher and make the test results more reliable. By improving the probe structure, the capture efficiency of the probe was improved while reducing the background noise. In summary, the development of multi-modal biosensors, the application of multi-signal amplification strategies, the combined detection of multiple biomarkers, and the improvement of probe structures have greatly improved the sensitivity and specificity of miRNA biosensors.

However, the process of preparation and assay operation is also more complicated for many novel assays proposed so far. In order to design novel and efficient biosensors with simple preparation, further research is needed by relevant researchers. In addition, the long assay time of most biosensors is not conducive to point-of-care detection, so the introduction of fast enrichment detection strategies to shorten the binding time of the target and the probe needs to be considered. Whether the prepared biosensors can achieve detection in real samples is also one of the issues that researchers need to consider due to the complexity of the actual sample composition, which is also a necessary path for the developed biosensors to be applied in clinical testing. In the future, more new nanomaterials, microfluidic analysis techniques, and signal amplification strategies will emerge, and these materials and strategies will also lead to the development of more novel miRNA biosensors with high detection performance, which will provide more opportunities for realizing the clinical application of early cancer diagnosis.

Author Contributions: Methodology, formal analysis, investigation, data curation, writing—original draft, K.W., M.Y. and C.L.; conceptualization, validation, writing—review and editing, supervision, project administration, funding acquisition, X.L. and J.W.; resources, writing—original draft, X.L. and J.W. All authors have read and agreed to the published version of the manuscript.

Funding: This research was funded by the Natural Science Foundation of Chongqing, China (grant number CSTB2022NSCQ-MSX0560), the National foreign expert project (grant number G2022165024L), Graduate Research and Innovation Foundation of Chongqing, China (grant number CYS22108).

Institutional Review Board Statement: Not applicable.

Informed Consent Statement: Not applicable.

Data Availability Statement: Not applicable.

Conflicts of Interest: The authors declare no conflict of interest. The funders had no role in the design of the study; in the collection, analyses, or interpretation of data; in the writing of the manuscript; or in the decision to publish the results.

References

1. Massarweh, N.N.; El-Serag, H.B. Epidemiology of Hepatocellular Carcinoma and Intrahepatic Cholangiocarcinoma. *Cancer Control* **2017**, *24*, 1–11. [CrossRef] [PubMed]
2. Han, T.-S.; Hur, K.; Cho, H.-S.; Ban, H.S. Epigenetic Associations between LncRNA/CircRNA and MiRNA in Hepatocellular Carcinoma. *Cancers* **2020**, *12*, 2622. [CrossRef] [PubMed]
3. Al-Saeed, Y.; Gab-Allah, W.A.; Soliman, H.; Abulkhair, M.F.; Shalash, W.M.; Elmogy, M. Efficient Computer Aided Diagnosis System for Hepatic Tumors Using Computed Tomography Scans. *Comput. Mater. Contin.* **2022**, *71*, 4871–4894. [CrossRef]
4. Zhang, G.; Liu, D. Comparative the Clinical Value of Contrast-Enhanced Ultrasonography, Enhancement CT and MRI for Diagnosing of Liver Lesions. *Clin. Hemorheol. Microcirc.* **2022**, *80*, 241–251. [CrossRef] [PubMed]
5. Winder, M.; Grabowska, S.; Hitnarowicz, A.; Barczyk-Gutkowska, A.; Gruszczyńska, K.; Steinhof-Radwańska, K. The Application of Abbreviated MRI Protocols in Malignant Liver Lesions Surveillance. *Eur. J. Radiol.* **2023**, *164*, 110840. [CrossRef]
6. Huang, L.; Sun, H.; Sun, L.; Shi, K.; Chen, Y.; Ren, X.; Ge, Y.; Jiang, D.; Liu, X.; Knoll, W.; et al. Rapid, Label-Free Histopathological Diagnosis of Liver Cancer Based on Raman Spectroscopy and Deep Learning. *Nat. Commun.* **2023**, *14*, 48. [CrossRef]
7. Falahi, S.; Rafiee-Pour, H.-A.; Zarejousheghani, M.; Rahimi, P.; Joseph, Y. Non-Coding RNA-Based Biosensors for Early Detection of Liver Cancer. *Biomedicines* **2021**, *9*, 964. [CrossRef]
8. Liao, Z.; Zhou, Q.; Gao, B. AIEgens-Doped Photonic Crystals for High Sensitivity Fluorescence Detection of Tumor Markers. *Biosensors* **2023**, *13*, 276. [CrossRef]
9. Forner, A. Hepatocellular Carcinoma Surveillance with MiRNAs. *Lancet Oncol.* **2015**, *16*, 743–745. [CrossRef]
10. Wu, X.; Li, J.; Gassa, A.; Buchner, D.; Alakus, H.; Dong, Q.; Ren, N.; Liu, M.; Odenthal, M.; Stippel, D.; et al. Circulating Tumor DNA as an Emerging Liquid Biopsy Biomarker for Early Diagnosis and Therapeutic Monitoring in Hepatocellular Carcinoma. *Int. J. Biol. Sci.* **2020**, *16*, 1551–1562. [CrossRef]
11. Ghidini, M.; Braconi, C. Non-Coding RNAs in Primary Liver Cancer. *Front. Med.* **2015**, *2*, 36.
12. Rowe, M.M.; Kaestner, K.H. The Role of Non-Coding RNAs in Liver Disease, Injury, and Regeneration. *Cells* **2023**, *12*, 359. [CrossRef] [PubMed]
13. Ling, H.; Fabbri, M.; Calin, G.A. MicroRNAs and Other Non-Coding RNAs as Targets for Anticancer Drug Development. *Nat. Rev. Drug Discov.* **2013**, *12*, 847–865.
14. Kamel, R.R.; Amr, K.S.; Afify, M.; Elhosary, Y.A.; Hegazy, A.E.; Fahim, H.H.; Ezzat, W.M. Relation between MicroRNAs and Apoptosis in Hepatocellular Carcinoma. *Open Access Maced. J. Med. Sci.* **2016**, *4*, 31–37. [CrossRef] [PubMed]
15. O'Brien, J.; Hayder, H.; Zayed, Y.; Peng, C. Overview of MicroRNA Biogenesis, Mechanisms of Actions, and Circulation. *Front. Endocrinol.* **2018**, *9*, 402. [CrossRef]
16. Song, C.; Chen, W.; Kuang, J.; Yao, Y.; Tang, S.; Zhao, Z.; Guo, X.; Shen, W.; Lee, H.K. Recent Advances in the Detection of Multiple MicroRNAs. *TrAC Trends Anal. Chem.* **2021**, *139*, 116269. [CrossRef]
17. Chen, Y.-X.; Huang, K.-J.; Niu, K.-X. Recent Advances in Signal Amplification Strategy Based on Oligonucleotide and Nanomaterials for MicroRNA Detection-a Review. *Biosens. Bioelectron.* **2018**, *99*, 612–624. [PubMed]
18. Kilic, T.; Erdem, A.; Ozsoz, M.; Carrara, S. MicroRNA Biosensors: Opportunities and Challenges among Conventional and Commercially Available Techniques. *Biosens. Bioelectron.* **2018**, *99*, 525–546. [CrossRef]
19. Singh, R.P. Prospects of Nanobiomaterials for Biosensing. *Int. J. Electrochem.* **2011**, *2011*, 125487. [CrossRef]
20. Zhang, L.; Su, W.; Liu, S.; Huang, C.; Ghalandari, B.; Divsalar, A.; Ding, X. Recent Progresses in Electrochemical DNA Biosensors for MicroRNA Detection. *Phenomics* **2022**, *2*, 18–32. [PubMed]
21. Tian, R.; Zheng, X. Sensitive Colorimetric Detection of MicroRNA Based on Target Catalyzed Double-arm Hairpin DNA Assembling. *Anal. Sci.* **2016**, *32*, 751–755. [CrossRef] [PubMed]

22. Guk, K.; Hwang, S.G.; Lim, J.; Son, H.; Choi, Y.; Huh, Y.-M.; Kang, T.; Jung, J.; Lim, E.-K. Fluorescence Amplified Sensing Platforms Enabling MiRNA Detection by Self-Circulation of a Molecular Beacon Circuit. *Chem. Commun.* **2019**, *55*, 3457–3460. [CrossRef] [PubMed]
23. Nie, W.; Wang, Q.; Yang, X.; Zhang, H.; Li, Z.; Gao, L.; Zheng, Y.; Liu, X.; Wang, K. High Sensitivity Surface Plasmon Resonance Biosensor for Detection of MicroRNA Based on Gold Nanoparticles-Decorated Molybdenum Sulfide. *Anal. Chim. Acta* **2017**, *993*, 55–62. [CrossRef] [PubMed]
24. Wu, Y.; Li, Y.; Han, H.; Zhao, C.; Zhang, X. Dual Cycle Amplification and Dual Signal Enhancement Assisted Sensitive SERS Assay of MicroRNA. *Anal. Biochem.* **2019**, *564*, 16–20. [CrossRef]
25. Huang, Y.; Yao, Y.; Wang, Y.; Chen, L.; Zeng, Y.; Li, L.; Guo, L. Strategies for Enhancing the Sensitivity of Electrochemiluminescence Biosensors. *Biosensors* **2022**, *12*, 750. [CrossRef] [PubMed]
26. Toh, T.B.; Lim, J.J.; Chow, E.K. Epigenetics of Hepatocellular Carcinoma. *Clin. Transl. Med.* **2019**, *8*, 13. [CrossRef] [PubMed]
27. Zhang, J.; Li, D.; Zhang, R.; Gao, P.; Peng, R.; Li, J. The MiR-21 Potential of Serving as a Biomarker for Liver Diseases in Clinical Practice. *Biochem. Soc. Trans.* **2020**, *48*, 2295–2305. [CrossRef]
28. Kalfert, D.; Ludvikova, M.; Pesta, M.; Ludvik, J.; Dostalova, L.; Kholová, I. Multifunctional Roles of MiR-34a in Cancer: A Review with the Emphasis on Head and Neck Squamous Cell Carcinoma and Thyroid Cancer with Clinical Implications. *Diagnostics* **2020**, *10*, 563. [CrossRef]
29. Zhang, H.; Wang, Y.; Han, Y. MicroRNA-34a Inhibits Liver Cancer Cell Growth by Reprogramming Glucose Metabolism. *Mol. Med. Rep.* **2018**, *17*, 4483–4489. [CrossRef]
30. Coulouarn, C.; Factor, V.M.; Andersen, J.B.; Durkin, M.E.; Thorgeirsson, S.S. Loss of MiR-122 Expression in Liver Cancer Correlates with Suppression of the Hepatic Phenotype and Gain of Metastatic Properties. *Oncogene* **2009**, *28*, 3526–3536. [CrossRef] [PubMed]
31. Liang, L.; Wong, C.-M.; Ying, Q.; Fan, D.N.-Y.; Huang, S.; Ding, J.; Yao, J.; Yan, M.; Li, J.; Yao, M.; et al. MicroRNA-125b Suppressesed Human Liver Cancer Cell Proliferation and Metastasis by Directly Targeting Oncogene LIN28B2. *Hepatology* **2010**, *52*, 1731–1740. [CrossRef]
32. Shaker, O.G.; Khairy, A.M.; Ali, R.M.M.; Badr, A.M. MiRNA-141 and Its Target Long Non-Coding RNA HOTAIR as Diagnostic Marker in Hepatocellular Carcinoma on Top of Hepatitis C Virus. *Gene Rep.* **2020**, *21*, 100807. [CrossRef]
33. Ratnasari, N.; Lestari, P.; Renovaldi, D.; Raditya Ningsih, J.; Qoriansas, N.; Wardana, T.; Hakim, S.; Signa Aini Gumilas, N.; Indrarti, F.; Triwikatmani, C.; et al. Potential Plasma Biomarkers: MiRNA-29c, MiRNA-21, and MiRNA-155 in Clinical Progression of Hepatocellular Carcinoma Patients. *PLoS ONE* **2022**, *17*, e0263298. [CrossRef] [PubMed]
34. Elmougy, F.A.F.; Mohamed, R.A.; Hassan, M.M.; Elsheikh, S.M.; Marzban, R.N.; Ahmed, F.M.; Elaraby, R.E. Study of Serum MicroRNA19a and MicroRNA223 as Potential Biomarkers for Early Diagnosis of Hepatitis C Virus-Related Hepatocellular Carcinoma. *Gene Rep.* **2019**, *15*, 100398. [CrossRef]
35. Shehab-Eldeen, S.; Nada, A.; Abou-Elela, D.; El-Naidany, S.; Arafat, E.; Omar, T. Diagnostic Performance of MicroRNA-122 and MicroRNA-224 in Hepatitis C Virus-Induced Hepatocellular Carcinoma (HCC). *Asian Pac. J. Cancer Prev.* **2019**, *20*, 2515–2522. [CrossRef]
36. Qiu, D.; Chen, J.; Liu, J.; Luo, Z.; Jiang, W.; Huang, J.; Qiu, Z.; Yue, W.; Wu, L. Expression of MicroRNA Let-7a Positively Correlates with Hepatitis B Virus Replication in Hepatocellular Carcinoma Tissues. *Exp. Biol. Med.* **2017**, *242*, 939–944. [CrossRef]
37. Wang, Y.; Mo, Y.; Wang, L.; Su, P.; Xie, Y. Let-7b Contributes to Hepatocellular Cancer Progression through Wnt/β-Catenin Signaling. *Saudi J. Biol. Sci.* **2018**, *25*, 953–958. [CrossRef]
38. Wang, K.; Peng, Z.; Lin, X.; Nian, W.; Zheng, X.; Wu, J. Electrochemical Biosensors for Circulating Tumor DNA Detection. *Biosensors* **2022**, *12*, 649. [CrossRef]
39. Singh, A.; Sharma, A.; Ahmed, A.; Sundramoorthy, A.K.; Furukawa, H.; Arya, S.; Khosla, A. Recent Advances in Electrochemical Biosensors: Applications, Challenges, and Future Scope. *Biosensors* **2021**, *11*, 336. [CrossRef]
40. Gulaboski, R. Future of Voltammetry. *Maced. J. Chem. Chem. Eng.* **2022**, *41*, 151–162. [CrossRef]
41. Kanoun, O. Impedance Spectroscopy: From Laboratory Instrumentation to Field Sensors. *IEEE Instrum. Meas. Mag.* **2020**, *23*, 4–7. [CrossRef]
42. Feng, X.; Gan, N.; Zhang, H.; Li, T.; Cao, Y.; Hu, F.; Jiang, Q. Ratiometric Biosensor Array for Multiplexed Detection of MicroRNAs Based on Electrochemiluminescence Coupled with Cyclic Voltammetry. *Biosens. Bioelectron.* **2016**, *75*, 308–314. [CrossRef] [PubMed]
43. Aamri, M.E.; Mohammadi, H.; Amine, A. Novel Label-Free Colorimetric and Electrochemical Detection for MiRNA-21 Based on the Complexation of Molybdate with Phosphate. *Microchem. J.* **2022**, *182*, 107851. [CrossRef]
44. Torul, H.; Yarali, E.; Eksin, E.; Ganguly, A.; Benson, J.; Tamer, U.; Papakonstantinou, P.; Erdem, A. Paper-Based Electrochemical Biosensors for Voltammetric Detection of MiRNA Biomarkers Using Reduced Graphene Oxide or MoS_2 Nanosheets Decorated with Gold Nanoparticle Electrodes. *Biosensors* **2021**, *11*, 236. [CrossRef]
45. Amr, K.S.; Elmawgoud Atia, H.A.; Elazeem Elbnhawy, R.A.; Ezzat, W.M. Early Diagnostic Evaluation of MiR-122 and MiR-224 as Biomarkers for Hepatocellular Carcinoma. *Genes Dis.* **2017**, *4*, 215–221. [CrossRef]
46. Gao, F.; Chu, Y.; Ai, Y.; Yang, W.; Lin, Z.; Wang, Q. Hybridization Induced Ion-Barrier Effect for the Label-Free and Sensitive Electrochemical Sensing of Hepatocellular Carcinoma Biomarker of MiRNA-122. *Chin. Chem. Lett.* **2021**, *32*, 2192–2196. [CrossRef]
47. Gonzalez-Losada, P.; Freisa, M.; Poujouly, C.; Gamby, J. An Integrated Multiple Electrochemical MiRNA Sensing System Embedded into a Microfluidic Chip. *Biosensors* **2022**, *12*, 145. [CrossRef]

48. Yang, R.; Jiang, G.; Liu, H.; He, L.; Yu, F.; Liu, L.; Qu, L.; Wu, Y. A Dual-Model "on-Super off" Photoelectrochemical/Ratiometric Electrochemical Biosensor for Ultrasensitive and Accurate Detection of MicroRNA-224. *Biosens. Bioelectron.* **2021**, *188*, 113337. [CrossRef]
49. Wu, J.; Lv, W.; Yang, Q.; Li, H.; Li, F. Label-Free Homogeneous Electrochemical Detection of MicroRNA Based on Target-Induced Anti-Shielding against the Catalytic Activity of Two-Dimension Nanozyme. *Biosens. Bioelectron.* **2021**, *171*, 112707. [CrossRef]
50. Azab, S.M.; Elhakim, H.K.A.; Fekry, A.M. The Strategy of Nanoparticles and the Flavone Chrysin to Quantify MiRNA-Let 7a in Zepto-Molar Level: Its Application as Tumor Marker. *J. Mol. Struct.* **2019**, *1196*, 647–652. [CrossRef]
51. Cai, B.; Huang, L.; Zhang, H.; Sun, Z.; Zhang, Z.; Zhang, G.-J. Gold Nanoparticles-Decorated Graphene Field-Effect Transistor Biosensor for Femtomolar MicroRNA Detection. *Biosens. Bioelectron.* **2015**, *74*, 329–334. [CrossRef] [PubMed]
52. Cadoni, E.; Manicardi, A.; Madder, A. PNA-Based MicroRNA Detection Methodologies. *Molecules* **2020**, *25*, 1296. [CrossRef] [PubMed]
53. Erdem, A.; Eksin, E. Zip Nucleic Acid-Based Genomagnetic Assay for Electrochemical Detection of MicroRNA-34a. *Biosensors* **2023**, *13*, 144. [CrossRef]
54. Zeng, R.; Xu, J.; Lu, L.; Lin, Q.; Huang, X.; Huang, L.; Li, M.; Tang, D. Photoelectrochemical Bioanalysis of MicroRNA on Yolk-in-Shell Au@CdS Based on the Catalytic Hairpin Assembly-Mediated CRISPR-Cas12a System. *Chem. Commun.* **2022**, *58*, 7562–7565. [CrossRef] [PubMed]
55. Ouyang, R.; Jiang, L.; Xie, X.; Yuan, P.; Zhao, Y.; Li, Y.; Tamayo, A.I.B.; Liu, B.; Miao, Y. Ti_3C_2@Bi_2O_3 Nanoaccordion for Electrochemical Determination of MiRNA-21. *Microchim. Acta* **2023**, *190*, 52. [CrossRef]
56. Bahadır, E.B.; Sezgintürk, M.K. A Review on Impedimetric Biosensors. *Artif. Cell. Nanomed. Biotechnol.* **2016**, *44*, 248–262. [CrossRef]
57. Ciucci, F. Modeling Electrochemical Impedance Spectroscopy. *Curr. Opin. Electrochem.* **2019**, *13*, 132–139. [CrossRef]
58. Brett, C.M.A. Electrochemical Impedance Spectroscopy in the Characterisation and Application of Modified Electrodes for Electrochemical Sensors and Biosensors. *Molecules* **2022**, *27*, 1497. [CrossRef]
59. La, M.; Zhang, Y.; Gao, Y.; Li, M.; Liu, L.; Chang, Y. Impedimetric Detection of MicroRNAs by the Signal Amplification of Streptavidin Induced In Situ Formation of Biotin Phenylalanine Nanoparticle Networks. *J. Electrochem. Soc.* **2020**, *167*, 117505. [CrossRef]
60. Eksin, E.; Torul, H.; Yarali, E.; Tamer, U.; Papakonstantinou, P.; Erdem, A. Paper-Based Electrode Assemble for Impedimetric Detection of MiRNA. *Talanta* **2021**, *225*, 122043. [CrossRef]
61. Yarali, E.; Eksin, E.; Torul, H.; Ganguly, A.; Tamer, U.; Papakonstantinou, P.; Erdem, A. Impedimetric Detection of MiRNA Biomarkers Using Paper-Based Electrodes Modified with Bulk Crystals or Nanosheets of Molybdenum Disulfide. *Talanta* **2022**, *241*, 123233. [CrossRef] [PubMed]
62. Jin, Y.; Wu, Z.; Li, L.; Yan, R.; Zhu, J.; Wen, W.; Zhang, X.; Wang, S. Zinc-Air Battery-Based Self-Powered Sensor with High Output Power for Ultrasensitive MicroRNA Let-7a Detection in Cancer Cells. *Anal. Chem.* **2022**, *94*, 14368–14376. [CrossRef] [PubMed]
63. Han, L. Ultrasensitive Label-Free MiRNA Sensing Based on a Flexible Graphene Field-Effect Transistor without Functionalization. *ACS Appl. Electron. Mater.* **2020**, *2*, 1090–1098.
64. Xu, J.; Liu, Y.; Li, Y.; Liu, Y.; Huang, K.-J. Smartphone-Assisted Flexible Electrochemical Sensor Platform by a Homology DNA Nanomanager Tailored for Multiple Cancer Markers Field Inspection. *Anal. Chem.* **2023**, *95*, 13305–13312. [CrossRef]
65. Yan, T.; Guo, C.; Wang, C.; Zhu, K. Optical Biosensing Systems for a Biological Living Body. *View* **2023**, *4*, 20220059. [CrossRef]
66. Chen, C.; Wang, J. Optical Biosensors: An Exhaustive and Comprehensive Review. *Analyst* **2020**, *145*, 1605–1628. [CrossRef]
67. Yan, T.; Zhang, G.; Chai, H.; Qu, L.; Zhang, X. Flexible Biosensors Based on Colorimetry, Fluorescence, and Electrochemistry for Point-of-Care Testing. *Front. Bioeng. Biotechnol.* **2021**, *9*, 753692. [CrossRef]
68. Zhu, D.; Liu, B.; Wei, G. Two-Dimensional Material-Based Colorimetric Biosensors: A Review. *Biosensors* **2021**, *11*, 259. [CrossRef]
69. Shahsavar, K.; Shokri, E.; Hosseini, M. Sensitive Colorimetric Detection of MiRNA-155 via G-Quadruplex DNAzyme Decorated Spherical Nucleic Acid. *Microchim. Acta* **2022**, *189*, 357. [CrossRef]
70. Li, Z.-H.; Yang, M.; Zhao, C.-X.; Shu, Y. Bifunctional Y-Shaped Probe Combined with Dual Amplification for Colorimetric Sensing and Molecular Logic Operation of Two MiRNAs. *Talanta* **2023**, *259*, 124480. [CrossRef]
71. Yang, X. Target-Catalyzed Self-Assembled Spherical G-Quadruplex/Hemin DNAzymes for Highly Sensitive Colorimetric Detection of MicroRNA in Serum. *Anal. Chim. Acta* **2023**, *1247*, 340879. [CrossRef]
72. Son, M.H.; Park, S.W.; Sagong, H.Y.; Jung, Y.K. Recent Advances in Electrochemical and Optical Biosensors for Cancer Biomarker Detection. *BioChip J.* **2023**, *17*, 44–67. [CrossRef]
73. Camarca, A.; Varriale, A.; Capo, A.; Pennacchio, A.; Calabrese, A.; Giannattasio, C.; Murillo Almuzara, C.; D'Auria, S.; Staiano, M. Emergent Biosensing Technologies Based on Fluorescence Spectroscopy and Surface Plasmon Resonance. *Sensors* **2021**, *21*, 906. [CrossRef]
74. Wang, S.; Wang, L.; Xu, X.; Li, X.; Jiang, W. MnO2 Nanosheet-Mediated Ratiometric Fluorescence Biosensor for MicroRNA Detection and Imaging in Living Cells. *Anal. Chim. Acta* **2019**, *1063*, 152–158. [CrossRef]
75. Li, Y.; Tang, D.; Zhu, L.; Cai, J.; Chu, C.; Wang, J.; Xia, M.; Cao, Z.; Zhu, H. Label-Free Detection of MiRNA Cancer Markers Based on Terminal Deoxynucleotidyl Transferase-Induced Copper Nanoclusters. *Anal. Biochem.* **2019**, *585*, 113346. [CrossRef]
76. Forte, G.; Ventimiglia, G.; Pesaturo, M.; Petralia, S. A Highly Sensitive PNA-microarray System for MiRNA122 Recognition. *Biotechnol. J.* **2022**, *17*, 2100587. [CrossRef]

77. He, M.; Shang, N.; Zheng, B.; Yue, G.; Han, X.; Hu, X. Ultrasensitive Fluorescence Detection of MicroRNA through DNA-Induced Assembly of Carbon Dots on Gold Nanoparticles with No Signal Amplification Strategy. *Microchim. Acta* **2022**, *189*, 217. [CrossRef]
78. He, M.; Zheng, B.; Shang, N.; Xiao, Y.; Wei, Y.; Hu, X. Synergistic Effect Enhancing the Energy Transfer Efficiency of Carbon Dots-Based Molecular Beacon Probe for Ultrasensitive Detection of MicroRNA. *Microchem. J.* **2023**, *190*, 108593. [CrossRef]
79. Nurrohman, D.T.; Chiu, N.-F. A Review of Graphene-Based Surface Plasmon Resonance and Surface-Enhanced Raman Scattering Biosensors: Current Status and Future Prospects. *Nanomaterials* **2021**, *11*, 216. [CrossRef]
80. Yu, H.; Han, R.; Su, J.; Chen, H.; Li, D. Multi-Marker Diagnosis Method for Early Hepatocellular Carcinoma Based on Surface Plasmon Resonance. *Clin. Chim. Acta* **2020**, *502*, 9–14. [CrossRef] [PubMed]
81. Huang, Y.; Sun, T.; Liu, L.; Xia, N.; Zhao, Y.; Yi, X. Surface Plasmon Resonance Biosensor for the Detection of MiRNAs by Combining the Advantages of Homogeneous Reaction and Heterogeneous Detection. *Talanta* **2021**, *234*, 122622. [CrossRef]
82. Wang, X.; Hou, T.; Lin, H.; Lv, W.; Li, H.; Li, F. In Situ Template Generation of Silver Nanoparticles as Amplification Tags for Ultrasensitive Surface Plasmon Resonance Biosensing of MicroRNA. *Biosens. Bioelectron.* **2019**, *137*, 82–87. [CrossRef]
83. Muhammad, M.; Huang, Q. A Review of Aptamer-Based SERS Biosensors: Design Strategies and Applications. *Talanta* **2021**, *227*, 122188. [CrossRef] [PubMed]
84. Li, C.; Li, S.; Qu, A.; Xu, C.; Xu, L.; Kuang, H. Dimensional Surface-Enhanced Raman Scattering Nanostructures for MicroRNA Profiling. *Small Struct.* **2021**, *2*, 2000150. [CrossRef]
85. Wang, H.-X.; Zhao, Y.-W.; Li, Z.; Liu, B.-S.; Zhang, D. Development and Application of Aptamer-Based Surface-Enhanced Raman Spectroscopy Sensors in Quantitative Analysis and Biotherapy. *Sensors* **2019**, *19*, 3806. [CrossRef] [PubMed]
86. Wang, Z.; Zong, S.; Wang, Z.; Wu, L.; Chen, P.; Yun, B.; Cui, Y. Microfluidic Chip Based Micro RNA Detection through the Combination of Fluorescence and Surface Enhanced Raman Scattering Techniques. *Nanotechnology* **2017**, *28*, 105501. [CrossRef]
87. Si, Y.; Xu, L.; Wang, N.; Zheng, J.; Yang, R.; Li, J. Target MicroRNA-Responsive DNA Hydrogel-Based Surface-Enhanced Raman Scattering Sensor Arrays for MicroRNA-Marked Cancer Screening. *Anal. Chem.* **2020**, *92*, 2649–2655. [CrossRef] [PubMed]
88. Wu, J.; Zhou, X.; Li, P.; Lin, X.; Wang, J.; Hu, Z.; Zhang, P.; Chen, D.; Cai, H.; Niessner, R.; et al. Ultrasensitive and Simultaneous SERS Detection of Multiplex MicroRNA Using Fractal Gold Nanotags for Early Diagnosis and Prognosis of Hepatocellular Carcinoma. *Anal. Chem.* **2021**, *93*, 8799–8809. [CrossRef] [PubMed]
89. Huang, X.; Tian, H.; Huang, L.; Chen, Q.; Yang, Y.; Zeng, R.; Xu, J.; Chen, S.; Zhou, X.; Liu, G.; et al. Well-Ordered Au Nanoarray for Sensitive and Reproducible Detection of Hepatocellular Carcinoma-Associated MiRNA via CHA-Assisted SERS/Fluorescence Dual-Mode Sensing. *Anal. Chem.* **2023**, *95*, 5955–5966. [CrossRef]
90. Fiorani, A.; Valenti, G.; Iurlo, M.; Marcaccio, M.; Paolucci, F. Electrogenerated Chemiluminescence: A Molecular Electrochemistry Point of View. *Curr. Opin. Electrochem.* **2018**, *8*, 31–38. [CrossRef]
91. Hou, L. Electrochemiluminescent Biosensors for the Detection of MicroRNAs: A Review. *Int. J. Electrochem. Sci.* **2019**, *14*, 2489–2508. [CrossRef]
92. Fiorani, A.; Merino, J.P.; Zanut, A.; Criado, A.; Valenti, G.; Prato, M.; Paolucci, F. Advanced Carbon Nanomaterials for Electrochemiluminescent Biosensor Applications. *Curr. Opin. Electrochem.* **2019**, *16*, 66–74. [CrossRef]
93. Li, J.; Cai, R.; Tan, W. A Novel ECL Sensing System for Ultrahigh Sensitivity MiRNA-21 Detection Based on Catalytic Hairpin Assembly Cascade Nonmetallic SPR Effect. *Anal. Chem.* **2022**, *94*, 12280–12285. [CrossRef]
94. Wang, M.-Y.; Jing, W.-J.; Wang, L.-J.; Jia, L.-P.; Ma, R.-N.; Zhang, W.; Shang, L.; Li, X.-J.; Xue, Q.-W.; Wang, H.-S. Electrochemiluminescence Detection of MiRNA-21 Based on Dual Signal Amplification Strategies: Duplex-Specific Nuclease -Mediated Target Recycle and Nicking Endonuclease-Driven 3D DNA Nanomachine. *Biosens. Bioelectron.* **2023**, *226*, 115116. [CrossRef] [PubMed]
95. Lin, Y.; Wu, J.; Wu, Y.; Ma, R.; Zhou, Y.; Shi, J.; Li, M.; Tan, X.; Huang, K. An All-Graphdiyne Electrochemiluminescence Biosensor for the Ultrasensitive Detection of MicroRNA-21 Based on Target Recycling with DNA Cascade Reaction for Signal Amplification. *Analyst* **2023**, *148*, 1330–1336. [CrossRef]
96. Shen, B.; Wu, Q.; Fan, Y.; Wu, H.; Li, X.; Zhao, X.; Wang, Y.; Ding, S.; Zhang, J. TiO_2@Ag Nanozyme Enhanced Electrochemiluminescent Biosensor Coupled with DNA Nanoframework-Carried Emitters and Enzyme-Assisted Target Recycling Amplification for Ultrasensitive Detection of MicroRNA. *Chem. Eng. J.* **2022**, *445*, 136820. [CrossRef]

chemosensors

MDPI

Review

Paper-Based Loop Mediated Isothermal Amplification (LAMP) Platforms: Integrating the Versatility of Paper Microfluidics with Accuracy of Nucleic Acid Amplification Tests

Debayan Das [1], Manaswini Masetty [2] and Aashish Priye [2,*]

1 Department of Chemical Engineering, National Institute of Technology (NIT) Durgapur, Mahatma Gandhi Rd., A-Zone, Durgapur 713209, India
2 Department of Chemical and Environmental Engineering, University of Cincinnati, Cincinnati, OH 45221, USA
* Correspondence: priyeah@uc.edu; Tel.:+1-513-556-3261

Abstract: Paper-based diagnostics offer a promising alternative to traditional diagnostic methods for point-of-care use due to their low cost, ease of use, portability, rapid results, versatility, and low environmental impact. While paper-based serology tests in the form of lateral flow assays can provide rapid test results for past pathogen exposure, they currently lack the accuracy and sensitivity offered by molecular diagnostic tests such as the polymerase chain reaction (PCR). Loop-mediated isothermal amplification (LAMP)—an isothermal nucleic acid amplification test (NAAT)—provides PCR-like performance while simultaneously reducing the instrumentation and assay complexity associated with PCR. In this review, we discuss a newly emerging class of paper-based LAMP platforms that integrates the versatility of paper microfluidics with the accuracy of NAATs. Since its first adoption in 2015, we have discussed all paper-based LAMP platforms in terms of the paper substrates, reagent incorporation techniques, paper platform design, heating hardware, detection methods, and sensitivity and specificity of paper-based LAMP assays. We conclude by identifying the current challenges and future prospects of paper-based NAATs.

Keywords: LAMP; RT-LAMP; paper-based diagnostics; point-of-care diagnostics

Citation: Das, D.; Masetty, M.; Priye, A. Paper-Based Loop Mediated Isothermal Amplification (LAMP) Platforms: Integrating the Versatility of Paper Microfluidics with Accuracy of Nucleic Acid Amplification Tests. *Chemosensors* **2023**, *11*, 163. https://doi.org/10.3390/chemosensors11030163

Academic Editors: Vahid Hamedpour and Andreas Richter

Received: 24 January 2023
Revised: 20 February 2023
Accepted: 23 February 2023
Published: 28 February 2023

1. Introduction

Early stage diagnosis is a critical component of infectious disease management systems to mitigate and arrest the propagation of pathogen outbreaks [1]. This has largely been applicable in developing countries where difficult living conditions and limited health care cause diseases to propagate more swiftly [2–4]. In general, the capacity to perform early detection of pathogenic infections has been primarily hindered by the difficulty in identifying and isolating infected people, owing to a lack of rapid, portable, and accurate diagnostic tests [5,6]. Serological tests play an essential role in rapidly testing infectious diseases [7]. However, designing antibodies can be challenging, leading to slower development of antigen-based assays. Furthermore, immunoassays often lack the sensitivity for direct detection of pathogens. Detection of host immune response (i.e., pathogen-specific IgG and IgM) requires seroconversion, which may lag behind the relevant window for both therapies and patient isolation by several days. Antibody tests can also lead to false positives as they may be detecting the antibodies for the pathogen that infected the body in the past [7,8]. In contrast, molecular diagnostics test such as the polymerase chain reaction (PCR) are considered the gold standard in pathogen detection as they offer high sensitivity and specificity resulting highly reliable diagnostics even in asymptomatic patients [9–11].

In this regard, detection methodologies based on nucleic acid amplification tests (NAATs) are expected to play a major in inhibiting the spread of viral and bacterial infections. With resource-limited settings, especially in developing countries, implementing molecular detection techniques such as NAAT at any point-of-care (POC) location is extremely challenging [12]. Conventional NAATs such as the PCR relies on thermal cycling, with a temperature regulation within a few degrees around reaction temperature (~55 °C–~95 °C), along with fluorescence measurements for assay readout [13,14]. These constraints require the use of sophisticated and bulky laboratory equipment. Furthermore, the thermal cycling requirement for PCR makes the test slow, expensive, and power intensive, making them sub-optimal for point of care (POC) diagnosis. To address these limitations, isothermal nucleic acid amplification techniques have become extremely popular recently [10,15,16]. The isothermal operation enables a greatly simplified and low-powered thermal system while simultaneously reducing the testing time. They rely on using multiple specialized primers to initiate strand displacement activity without requiring high denaturation temperatures in PCR [10]. Recently, several promising isothermal amplification strategies have been proposed which can primarily be classified based on their working temperature ranges, enzymes used, necessity for pre-heating, and the overall efficiency. Isothermal amplification techniques such as NASBA (nucleic acid sequence-based amplification) [17,18], 3SR (self-sustained sequence replication) [19,20], TMA (transcription-mediated amplification) [21], SDA (strand displacement amplification) [22,23], and SMART (a simple method for amplifying RNA targets) [24,25] requires the pre-heating procedure (65 °C to 95 °C) followed by the application of isothermal reaction temperature ranging from 37 °C to 41 °C. Although this step reduces the total number of thermal cycles compared to PCR, implementing even two temperature cycles at any POC setting is still challenging [13]. Alternatively, researchers have also developed amplification techniques that can be executed at a single temperature without any pre-heating step. These include processes such as HDA (helicase-dependent amplification) [26,27], RPA (recombinase polymerase amplification) [28,29], RCA (rolling circle amplification) [30,31], RAM (ramification amplification) [32], MDA (multiple displacement amplification) [33,34], and LAMP (loop-mediated isothermal amplification) [35,36]. While promising, isothermal processes such as NASBA, 3SR, TMA, and SDA require two to three enzymes, increasing the preservation and application cost for their use in the POC setting [37]. TMA and SDA have demonstrated an overall decrease in amplification efficiency compared to NASBA and 3SR. In addition, complications toward nuclease selection and higher inefficiency for long target sequences are the other drawbacks of the SDA process [37]. The SMART amplification process suffers from its dependence on multiple probes and enzymes for hybridization steps, thus adding to the complexity of the amplification process [38]. The HDA process utilizes expensive enzymes and thus incurs higher costs to the overall amplification process. Although RPA provides several advantages, such as simple reaction scheme, low reaction temperature, high efficiency, and high specificity, the greater length of the primers and probe limits amplification of longer templates. On the other hand, the RCA process requires circular nucleic acid templates restricting their use for particular target nucleic acid templates. RAM is an improved version of RCA, which results in an exponential amplification of target nucleic acid sequences as compared to RCA. However, RAM utilizes padlock probe whose complex secondary structure not only interferes with the assay detection but also impairs the assay detection efficiency. One of the significant drawbacks of the MDA amplification technique is the high rate of allele dropout (ADO) and extensive preferential amplification. Among all these techniques, LAMP—operating isothermally around 65 °C—has emerged as the most promising and widely used nucleic acid isothermal amplification techniques (Figure 1A,B). Although the primer design of LAMP is complex [39], the usage of 4 to 6 primers spanning 6–8 distinct sequences enhance the specificity as compared to the other amplification procedures. Moreover, the LAMP amplification process utilizes only one enzyme (*BST Polymerase*) unlike the other isothermal amplification processes. Finally, the amplification efficiency of LAMP is significantly higher as the entire process yields nearly 10^9-fold amplification in less than 30 min [35,36].

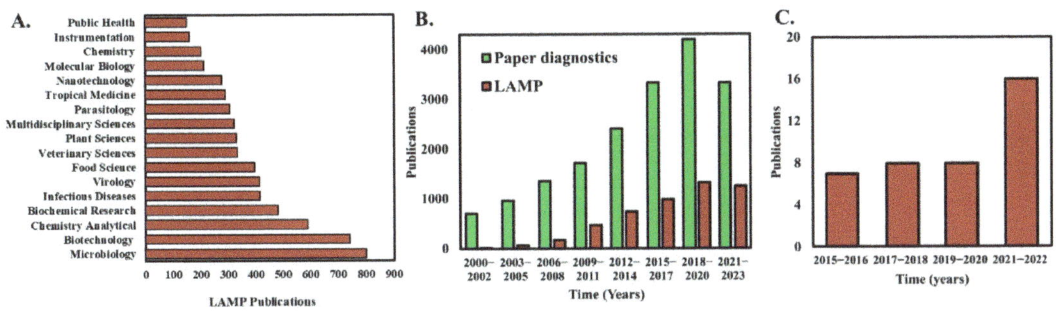

Figure 1. (**A**) Analysis of the number of publications that apply LAMP categorized by the topic area. (**B**) Analysis of annual publications in peer reviewed journals with keywords "paper diagnostics" and "Loop-mediated isothermal amplification" in the title or abstract of the publication. (**C**) The number of publications that demonstrate and use paper-based LAMP assays. Publication data were compiled on 13 January 2023 from Web of Science database using the keywords "Loop-mediated isothermal amplification" in (**A**), "Loop-mediated isothermal amplification" and "Paper diagnostics" in (**B**).

Even with the advantages and simplicity of isothermal LAMP operation, the assay chip brings additional complexity to an otherwise straightforward NAAT [40]. For example, most implementation of LAMP still requires peripheral components such as fluid pumping systems and valves for fluid manipulation, increasing the cost and operational complexity of the system [41]. Alternatively, paper-microfluidics offer several advantages that traditional sample-in-tube assays do not offer such as their low cost, accessibility, biocompatibility, and portability [42]. Furthermore, the porous paper substrates enable capillary-driven passive fluid transport to reliably move liquid samples and reagents without the need for external fluid pumps [43]. Consequently, the number of publications that harness paper for diagnostics have steadily been increasing over the last decade (Figure 1B). Paper-based loop-mediated isothermal amplification (LAMP) is a method that combines the simplicity and low cost of paper-based diagnostic tests with the specificity and sensitivity of LAMP to detect nucleic acid sequences in samples. In this method, the LAMP reaction is performed on a small piece of paper, such as filter paper or chromatography paper, instead of in a test tube. The paper is then incubated at the appropriate temperature, typically ~65 °C, for the LAMP reaction to occur. Recently, researchers have explored the potential of simpler paper-based LAMP devices, well-known for their robustness, cost-effectiveness, and user-friendliness [44–46]. A critical aspect of paper-based LAMP detection is the paper's material selection [40,47]. The nonspecific binding of DNA molecules to the paper's fibers as well as paper self-fluorescence, negatively influence the noise level and thus the limit of detection (LOD). The most used paper membranes are nitrocellulose-based FTA (Flinders Technology Associates), cellulose, or glass-fiber-based [47]. Another factor in determining the efficiency of paper-based LAMP is the pre-drying of LAMP reagents with the paper matrix. Paper-based LAMP systems that utilize pre-dried reagents eliminate the need for sample preparation steps [46,48]. Dry reagents are less sensitive to storage conditions, thus preventing problems with temperature-dependent transport or storage. Taken together, paper-based LAMP devices offer significant potential to be commercially used as accurate and cost-effective NAATs, especially in resource-limited settings.

Few review articles related to paper-based devices and nucleic acid amplification have been published in the past few years. Choi et al. [49] reviewed the advances and challenges of fully integrated paper-based point-of-care NAATs with a discussion on the implementation of NAATs on low-cost paper substrates. Tian et al. [50] illustrated the latest developments of integrated μPADs (paper-based analytical devices) and highlighted the accomplishments and challenges of each component, including sample collection/pre-treatment, signal transduction, and amplification followed by detection. This review paper provides a comprehensive yet compact discussion on the implementation of paper-based LAMP for the first time. Notably, the first work on paper-based LAMP was reported by the Whitesides group [45] in their work entitled "Paper Machines for molecular diagnostics" where they detected the *malb* gene in *E. coli*. Their work was directed toward developing an integrated paper-based LAMP device that involved the functional steps of viral nucleic acid capture, purification and isothermal amplification, and thereafter, leading to real-time fluorescence detection. Although this was the only work on paper-based LAMP reported in 2015, a steady increase in publication records can be found since then, showcasing the significance of this technology. The last couple of years, especially since the COVID-19 pandemic era, have seen a significant increase in paper-based LAMP research and publications [51–54] (Figure 1C).

First, this review focuses on the development of paper-based LAMP technology in the past decade. In the next section, details about the fabrication of the paper-based assay are discussed, including—(1) LAMP-assay preparation which reviews the different methods for preparing the Clinical/DNA samples; (2) paper membranes which illustrate the different paper membranes used for carrying out LAMP; (3) incorporation of LAMP reagents on paper matrix which discusses the imbibition techniques to introduce LAMP reagents on paper substrate; (4) design of the device which reviews the design and fabrication of paper-based LAMP platforms; (5) heating techniques which discuss the various heating instrument that have been used for paper-based LAMP assays; (6) detection techniques which discuss the various assay detection methodologies utilized for paper-based LAMP, and (7) sensitivity and specificity where the robustness and limit of detection (LOD) of paper-based LAMP assays is reviewed. Finally, the conclusive remarks, challenges, and future scope of paper-based LAMP technology is presented.

2. Fabrication and Design of Paper-Based LAMP Assays

Most NAAT systems function in conjunction with existing laboratory facilities and equipment. Conventional LAMP assays (i.e., LAMP performed in polypropylene PCR tubes) rely on the use of several manual fluid handling steps, primarily with pipettes [40]. Paper-based LAMP systems eliminate these steps resulting in a more straightforward and cost-effective NAAT assay system [9,50]. Here we discuss several aspects of paper-based LAMP assay systems.

2.1. Design of the Paper-Based LAMP Platform

Researchers have proposed various paper-based LAMP devices that integrate the processes of extraction, purification, and amplification. Although few researchers have focused on first extracting the nucleic acid (RNA/DNA) and then amplifying it in the presence of primers and LAMP buffer mixture on paper substrates, the majority of them have used the hybrid technique of detection in lateral flow assays or dipstick after carrying out the conventional tube-based LAMP. Even though FTA card had been used extensively for nucleic acid extraction (Figure 2A) [55,56], it was the implementation by Connelly et al. [45] that first demonstrated the use of the FTA card as an effective paper-based LAMP platform. Their platform—aptly called "paper machines"—comprised of a movable reaction paper disc in a magnetic slide strip-based arrangement (Figure 2B). The reaction paper disk is first overlapped on the paper-based sample port where the clinical sample is added. The excess sample is absorbed into the paper-based sample port, and the sample imbibed reaction paper disc is next hauled to the paper-based wash port. Here, washing is carried out, and the retentate on the reaction paper disc is mainly the nucleic acid. Finally, the reaction paper disc is slid past, and the LAMP master mix is added to execute the LAMP reaction. Rodriguez et al. [57] presented a paper-based LAMP device involving amplification at the sample inlet zone followed by Lateral flow assay-based detection at the right half. The sample inlet port has an absorbent pad beneath the amplification pad which absorbs the excess liquid. It was washed with ethanol, and the wash liquid wicked through to the absorbent pad, removing impurities and leaving behind the purified precipitated DNA.

The LAMP reaction mix is placed directly onto the sample port where the purified DNA remains, and the chip's bottom tab is folded over the designated perforation to act as a cover film for the sample port and prevent evaporation during the heating step. The eluted products wick through the LFD strip toward the right (Figure 2C). Choi et al. [58] came up with an integrated device consisting of an amplification region and a lateral flow assay (LFA)-based detection part (Figure 2D). The amplification region exhibits a glass-fiber membrane protected by a PVC adhesive tape and adhesive PVC backing pad. One end of this backing pad was attached to the lateral flow assay. Later, Choi et al. [59] added the step of cell lysis and extraction of nucleic acid on the FTA paper membrane mounted on a glass-fiber membrane and multiple-layered PVC backing pad. Amplification reagents were added to the glass-fiber membrane, and this was then moved into the covered heating compartment of the handheld heating device for amplification. After amplification, this was directly pasted on the lateral flow assay sample pad, where external fluid was added for carrying over the sample toward the detection section. They also modified the LFA design, where they added PDMS drops and shunt paper layers to delay the flow and enhance the sensitivity.

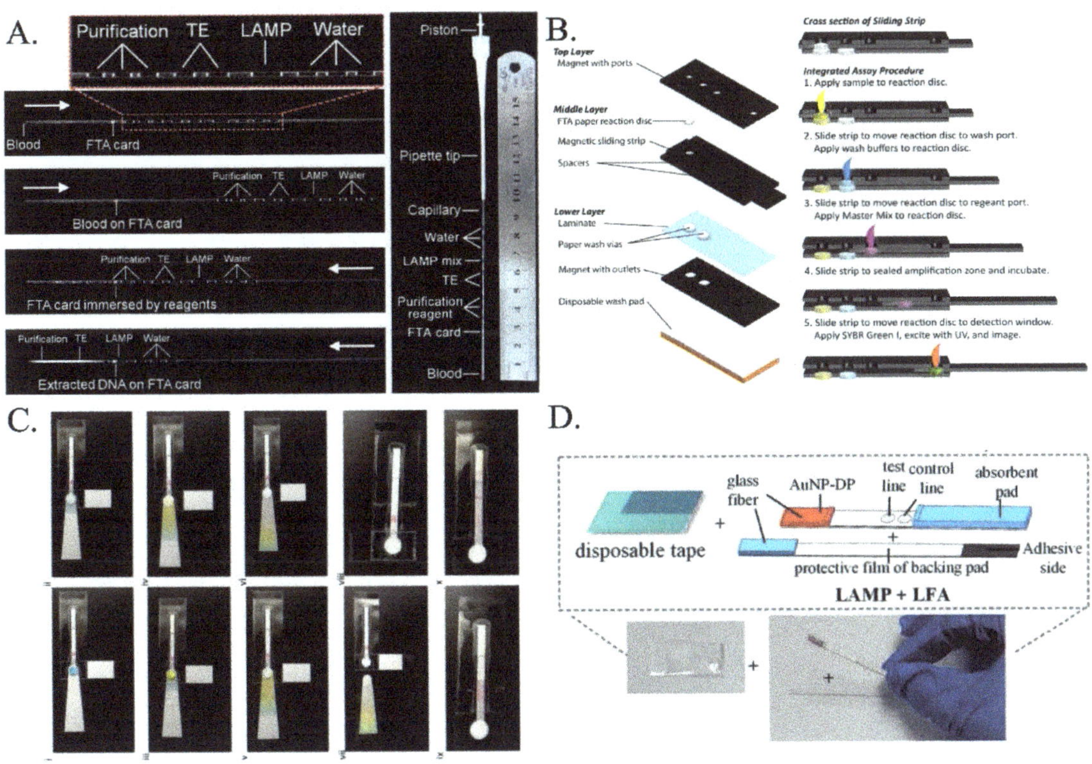

Figure 2. (**A**) The microcapillary with preloaded reagents connected to a pipet tip forms all in one LAMP LFA. The white arrow indicates the direction of the fluids [from [56], copyright 2014 American Chemical Society]. (**B**) Schematic of sliding-strip device shows the three major layers of the sliding-strip architecture and their components [from [45], copyright 2015 American Chemical Society]. (**C**) Fluidic demonstration of chip operation demonstrates paper-based LAMP assay including sample lysis, capillary fluid transport, ethanol wash, buffer wash, extraction of purified DNA followed by LAMP reaction [from [57], copyright 2016 Royal Society of Chemistry]. (**D**) An integrated paper-based device incorporating LAMP and LFA [from [58], copyright 2016 Royal Society of Chemistry].

Champlauk et al. [60] established a bendable chip consisting of a paper-based reaction and detection pads connected by a fishing line. Both the amplification and detection pads are circular discs (Figure 3A). The detection part also includes a square paper pad which is coated with phosphate-buffered saline (PBS). The entire design is a fishing wire-based network coupled with paper pads for amplification and detection. Ru Choi et al. [61] proposed an integrated paper-based biosensor consisting of four layers (1) top PVC layer—lateral flow assay detection; (2) second glass fiber—for carrying out LAMP; (3) third nitrocellulose membrane layer—for sample addition and nucleic acid extraction; (4) fourth absorbent pad—for sample purification and washing. The LAMP paper device given by Seok et al. [46] and Batule et al. [48] had two layers of paper membranes sandwiched together. The bottom layer was essential in distributing the inserted sample and helping it vertically imbibe the four paper-based LAMP zones. Jiang et al. [62] developed a new concept named VLEAD which consisted of four reservoirs performing the function of lysis, wash, and purification, followed by buffering, mixing, and detection in the vertical direction in one of the reservoirs (Figure 3B). Li et al. [63] came up with a unique concept of fabricating 12 paper-based reaction pads laminated between two magnetic plates to minimize evaporation rates. An origami-paper-based DNA purification (using chitosan matrix)—LAMP amplification-detection device was fabricated by Tung Trieu and Lee [64] (Figure 3C). Wang et al. [54] proposed a new methodology in which a protruded stick can be pressed on a sponge pad-based amplification system that transfers the amplicons into the interconnected detection zones. Garneret et al. [51] proposed a movable system named COVIDISC consisting of a rectangular-shaped extraction membrane that was overlapped on two paper-reaction pads (one for test and the other for control) (Figure 3D). Recently, Rofman et al. [65] developed a design involving the sample transfer from the pumping pad to the target pad via a superhydrophobic membrane. Overall, it is seen that designs ranging from combined and lateral flow-based paper pads to disc-shaped paper arrangements have all been successful in amplifying and detecting nucleic acid targets via LAMP.

2.2. Paper-Based LAMP Assay Preparation

The sample preparation step involves processing LAMP reaction either from extracted nucleic acid or directly from crude clinical samples [35,36]. Optimal paper-based LAMP should (i) enable integrated sample preparation steps such as extracting DNA/RNA from crude samples, (ii) enable on-paper reagent storage, and (iii) enable automated fluid transport and assay readout. In general, LAMP reaction requires four to six target-specific primers. In addition, for the execution of LAMP, a master reaction consisting of tris-hydrochloric acid (pH 8.8), potassium chloride, ammonium sulfate, magnesium sulphate, Tween 20, betaine, Bst DNA polymerase, and reverse transcriptase (only for the detection of viral pathogens), and dNTPs is required [36]. In addition to this, detection reagents such as intercalating dyes or fluorescent probes are also added.

Efficient extraction and purification of nucleic acids from complex biological samples are challenging in paper-based NAATs. First, paper-based methods often require smaller sample volumes, potentially limiting the amount of genetic material that can be extracted and processed. The porous structure of the paper can lead to incomplete recovery of genetic material and contamination from impurities. Additionally, achieving consistent and reproducible nucleic acid extractions can be difficult due to variations in the physical and chemical properties of the paper substrate, which can impact the efficiency of sample preparation steps. Incomplete sample recovery may reduce the sensitivity of the assay, and the quality of the genetic material extracted may be more variable compared to the traditional methods, affecting the accuracy and reproducibility of the assay. Most proof-of-concept work uses nucleic acid extraction kits or laboratory-based sample processing to isolate nucleic acids from crude biological samples before performing paper-based LAMP. Alternatively, researchers have also opted for spiking DNA templates in clinical samples before performing paper-based LAMP reactions. On the other hand, a few researchers used paper membranes to demonstrate a sample-to-answer test that performs nucleic acid extraction followed by LAMP reaction and detection on paper. Connelly et al. [45] introduced the sliding strip-based paper-LAMP assay, which can perform all the steps of sample preparation involving *E. coli* cell lysis, DNA isolation, and purification, as well as LAMP amplification and detection. Rodriguez et al. [57] developed single-step chaotropic cell lysis and DNA extraction methodology based on the alcohol precipitation method for extracting DNA from the clinical cervical sample. The entire procedure, along with ethanol washing, was executed on the paper membrane, which left DNA precipitate on the surface of the paper substrate. LAMP mixture was added next to initiate the reaction, followed by the detection of the target amplicons. Kaarj et al. [44] demonstrated ZIKA virus lysis and subsequent filtration followed by detection on a LAMP-paper assay. They demonstrated that the input sample could even be human urine, human blood plasma, or solution spiked with the ZIKA virus. Jiang et al. [62] introduced a paper-based unit for sample lysis, RNA enrichment, and purification, followed by RT-LAMP. Overall, it is found that these multi-functional single-step processes are relatively cost-effective and can be easily operated. Even though such sample-in answer-out type paper-based LAMP has been demonstrated, systems that can seamlessly integrate single-step extraction, purification, and detection process on paper are rare and considerable future scope exists toward engineering these systems.

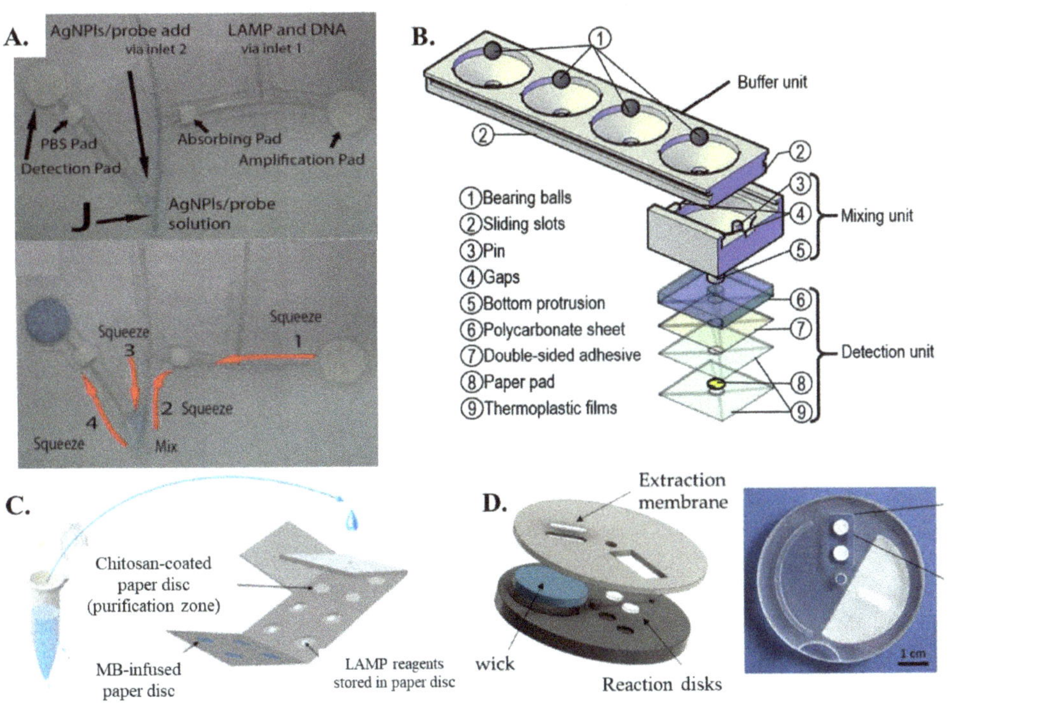

Figure 3. (**A**) The bendable chip has two parts, one for DNA amplification and one for detection that are connected with channels made of fishing line [from [60], copyright 2016 Wiley-VCH GmbH]. (**B**) Exploded view of VLEAD consisting of three components, buffer unit at the top, mixing unit in the middle, and detection unit at the bottom [from [62], copyright 2018 Wiley-VCH GmbH]. (**C**) Schematic representation of the concept of sample preparation for the performance of the PMA–LAMP assay on the origami paper microdevice [from [64], copyright 2018 American Chemistry Society]. (**D**) A-COVIDISC workflow decomposed in three steps: (1)—injection, washing (fluids flow through the capture membrane and get absorbed by capillarity in the absorbent wick (in blue)), drying. (2)—Disk rotation and elution. (3)—Disk counter-rotation, coverage of the reaction zone by a PCR sealing film, heating, amplification, and readout [from [51], CC BY License].

2.3. Paper Membranes

Since LAMP reaction and detection, along with nucleic acid extraction/purification, is being carried out on paper membranes, the composition of the paper matrix plays a very important role in determining the efficiency of the amplification reaction. Researchers have used wide variety of paper pads ranging from polyethersulfone membrane (PES) and FTA to Whatman filter papers and chromatography papers [66]. However, the broad range of classification involves paper membranes composed of cellulose and nitrocellulose [47]. Cellulose-based paper matrices are primarily composed of hydroxyl groups and a relatively lesser amount of carboxyl group, which renders them negatively charged. These are essentially hydrophilic in nature, so the retention time of fluids on these paper membranes is comparatively less. The primarily used cellulose paper membranes are chromatography and filter papers. The other widely used cellulose-based paper membrane is FTA (fast technology analysis) cards [9,45]. FTA cards are good retainers of dried lysis agents and stabilizers, which are desirable for a single-step nucleic acid extraction prior to performing the LAMP reaction [45].

When the nitrate groups replace the hydroxy group in the cellulose, nitrocellulose is created, which is usually cast into membrane sheets after dissolving into solvents. The porosity of the paper membranes can be controlled by changing the rate of evaporation of the added solvent. In addition, the permeability of the nitrocellulose membranes can be adjusted by varying the concentration ratio of nitrocellulose and the organic solvent. Nitrocellulose membranes are primarily hydrophobic in nature, and the extent of binding of proteins on these membranes is dependent on the mechanism of electrostatic or hydrophobic interaction. It has been observed that dsDNA generally does not bind, whereas denatured single stranded DNA easily binds on the nitrocellulose membranes. Recently, there has been significant development in enhancing the hydrophilicity of the nitrocellulose membrane by adding several surfactants at different compositions [67]. Other than cellulose and nitrocellulose membranes, PES (polyether sulphone), PC (polycarbonate), and glass fiber membranes have been used for executing LAMP [66]. Linnes et al. [66] compared the extent of LAMP amplification on cellulose chromatography (CHR), PES, NC, PC, and unbound glass fiber paper membranes. Among all the membranes, PES demonstrated positive LAMP amplification every time, while PC, on rare occasions, exhibited false negative LAMP amplification. Cellulose sometimes demonstrated successful, positive LAMP amplification, whereas NC and unbound glass fiber membranes failed to show any positive amplification. Seok et al. [46] used PES as transfer pads while glass fiber paper membranes were implemented for the detection of multiple pathogens using the LAMP. Glass fibers have larger pores and are biocompatible, which makes them highly effective. Moreover, LAMP reagents are provided with sufficient reaction space on the glass fibers and these paper substances are devoid of any water-soluble material which may hamper LAMP. It was inferred that the addition of PVA (polyvinyl alcohol) on glass fiber membranes was deemed to be more effective in amplification and visual detection. They also mentioned that other membranes that contain hydrophilic fibers, such as polyether sulfone and cellulose acetate fibers, inhibited the LAMP reaction.

Several paper-based LAMP devices consist of a combination of different types of paper membranes. These are more common for devices involving lateral flow assay-based detection just after the LAMP reaction. Choi et al. [58] developed the integrated lateral flow assay for DNA amplification to detection. The developed device consisted of a glass fiber pad for LAMP, a glass fiber pad for LFA, a nitrocellulose membrane, an absorbent pad, and two PVC backing pads. Rodriguez et al. [57] combined PES and cellulose blotting paper as the sample pad for isolating and purifying the DNA from the sample. After initiating the LAMP amplification process in the combined membrane setup, the amplified products are swept away into the commercially brought lateral flow assay strip consisting of glass fiber-based conjugate pad followed by the nitrocellulose-based detection pad and, finally, the absorbent pad. Choi et al. [59] developed an integrated paper-based biosensor consisting of four layers. The top PVC layer is the lateral flow layer supported by a PVC backing pad, which consists of a glass fiber, a nitrocellulose membrane, and an absorbent pad. The second layer is composed of glass fiber for a highly specific and sensitive nucleic acid amplification technique (i.e., LAMP). The third layer consists of a piece of FTA card with a diameter of 0.25 cm for sample addition and nucleic acid extraction. The bottom layer is composed of an absorbent pad for sample purification and washing. Batule et al. [48] developed a handheld paper device by integrating multiple pads for loading, transferring, and binding the sample. Specifically, an asymmetric polyethersulfone membrane was used as the transfer pad, and a GF/C grade glass pad, glass fiber, SS DNA-modified glass fiber, and an absorbent pad were used to fabricate loading, sample, binding pads, and the absorbent pad, respectively. The paper-strip-based extracted viral RNAs were directly added to the RT-LAMP reaction buffer, and then viral RNA containing RT-LAMP buffer was pipetted into the sample hole of the ready-to-use paper chip, consisting of a PES transfer pad and glass fiber LAMP reaction pad. Other than the lateral flow assay, Champlauk et al. [60] used a bendable plastic laminated chip consisting of separate paper-based LAMP amplification and detection zones connected by a fishing line. The amplification zone was made up of amplification and absorption pads, while the detection zone comprised a PBS (phosphate buffered saline) pad and detection pad. All the paper pads were made up of Whatman filter paper no. 1 grade membrane except the PBS pad, which was prepared from a square polystyrene pad. Thus, it is well understood that combining paper pads for performing separate functions in a paper-based LAMP amplification device renders them more effective.

2.4. Incorporation of LAMP Reagents on Paper Matrix

There are several ways through which LAMP reagents can be integrated and stored onto paper matrices. Connelly et al. [45] applied purification buffer and nuclease-free water on the FTA-based LAMP detection pad. The disc was then dried completely by being placed in an oven at 65 °C for 5 min and 10 μL of LAMP Master Mix was applied. Champlauk et al. [60] fed the Whatman filter paper-based LAMP amplification pad with the LAMP reagents and DNA sample through one of the inlet tubes. In both of their works on paper-based LAMP assay, Choi et al. [58,59] pipetted out the mixture of the samples and LAMP reagents onto the glass fiber pad for LAMP protected by a disposable tape. Linnes et al. [66] first added the nucleic acid sample and then followed it up with the LAMP reagents onto the circular discs made from different paper membranes. Zhang et al. [56] inserted FTA card in the icLAMP microcapillary system followed by three segments of purification reagent, two segments of TE buffer, one segment of LAMP reaction mix, and three segments of water droplets. Hence, no reagent is directly added to the FTA card. Rodriguez et al. [57] added the liquid LAMP reaction mix directly onto the sample port of the PES membrane (LAMP amplification pad). Seok et al. [46] and Batule et al. [48] used dried LAMP reagent pre-imbibed on the glass pad for LAMP amplification. The solution-treated glass pad was dried for 20 min at 37 °C in a drying oven. Then, the reaction buffer without primer, polymerase, and HNB was pipetted onto a glass pad and heated for 60 min at 63 °C. Jiang et al. [62] used a fluid-control ball valve to trigger the release of reagent from the buffer to the mixing unit in the coffee mug-paper-based LAMP amplification unit. In order to detect ZIKA virus on paper assay, Kaarj et al. [44] inserted 15 μL of RT-LAMP reaction mixture directly on the excised paper and covered it with glass slide to sandwich the paper, followed by sealing with Parafilm M to prevent evaporation. Li et al. [63] performed a multiplex LAMP assay in the paper-based chip where isothermal amplification buffer and the bacteria genome specimen were added, and then 5 μL of mineral oil was applied to each hole to prevent liquid evaporation during the amplification. Naik et al. [68] added LAMP reaction mixture containing the bacterial culture to a 5 mm diameter paper disc, heat-sealed into a plastic pouch, and incubated it at 60 °C for varying durations. Trieu and Lee [64] developed an origami all-in-one paper structure which consisted of a chitosan-based DNA purification pad and a separate LAMP reaction pad. Due to the foldable structure, the purification pad was overlapped on the reaction pad thereby initiating the LAMP reaction. Wang et al. [54] developed an on-chip RT-LAMP assay where amplification mixtures were loaded onto the sponge-like PVA pad. Then the sample-loaded RT-LAMP chip was sealed using transparent pressure-sensitive adhesive tape and placed on the thermal plate at 50 °C for 35 min and then at 65 °C for 40 min. A slightly different process was adopted by Suea-Ngam et al. [69] for their developed paper-based LAMP assay where LAMP primer solution was first added to the paper disc without the forward inner primer. Thereafter, the polymerase, target DNA and LAMP buffer was added to the paper substrate. Choopara et al. [70] inserted LAMP mixture onto the reaction pad, dried in a sterile air flow and stored at low temperatures. To use the paper-based LAMP device, the reaction pad was placed on the reaction layer of the sandwich-like bottom base, 1 µL of DNA sample and 14 µL of sterile water were pipetted onto the reaction pad, covered by the clear top seal to prevent evaporation during the LAMP incubation. Trinh et al. [71] mixed agarose gel and LAMP reagents for an effective storage at low gelling temperature. When this mixture was deposited on the surface, it solidified at room temperature. Thereafter, the DNA solution was added. Paper-based LAMP reactions can also operate effectively at different ambient humidity conditions [72]. The various methods used to incorporate LAMP reagents on paper have been summarized in Table 1. Overall, it is found that adding the LAMP reagents directly onto the paper substrate followed by drying is an effective procedure for achieving LAMP-based detection.

Table 1. Methods to integrate LAMP reagents on paper matrices.

Reference	Paper Matrix	LAMP Reagent Integration Method
Connelly et al. [45]	FTA	Purification buffer and nuclease-free water applied on detection pad, followed by drying and application of LAMP Master Mix
Chaumpluk et al. [60]	Whatman filter paper	LAMP reagents and DNA sample added through inlet tube to amplification pad
Choi et al. [58,59]	Glass fiber pad	Mixture of samples and LAMP reagents pipetted onto pad, protected by disposable tape
Linnes et al. [66]	Circular paper discs	Nucleic acid sample added followed by LAMP reagents
Zhang et al. [56]	FTA card in icLAMP microcapillary system	Segments of purification reagent, TE buffer, LAMP reaction mix, and water droplets added in sequence, with no reagent directly added to FTA card
Rodriguez et al. [57]	PES membrane	Liquid LAMP reaction mix added directly onto sample port of amplification pad
Seok et al. [46], Batule et al. [48]	Glass pad	Dried LAMP reagent pre-imbibed on pad, followed by pipetting of reaction buffer without primer, polymerase, and HNB, and heating
Jiang et al. [62]	Coffee mug-paper-based unit	Fluid-control ball valve used to trigger release of reagent from buffer to mixing unit
Kaarj et al. [44]	Excised paper	15 µL of RT-LAMP reaction mixture inserted directly onto paper, covered with glass slide, sealed with parafilm M
Li et al. [63]	Paper-based chip	Isothermal amplification buffer and bacteria genome specimen added, followed by application of mineral oil to each hole to prevent evaporation
Naik et al. [68]	Paper disc	LAMP reaction mixture containing bacterial culture added, heat-sealed into plastic pouch, and incubated at 60 °C for varying durations
Trieu and Lee [64]	Origami all-in-one paper structure	Chitosan-based DNA purification pad overlapped on LAMP reaction pad to initiate reaction
Wang et al. [54]	Sponge-like PVA pad	Amplification mixtures loaded onto pad, sealed with transparent pressure-sensitive adhesive tape, and placed on thermal plate
Suea-Ngam et al. [69]	Paper disc	LAMP primer solution added first, followed by addition of polymerase, target DNA, and LAMP buffer
Choopara et al. [70]	Reaction pad	LAMP mixture inserted onto pad, dried, and stored at low temperatures, then DNA sample and sterile water pipetted onto pad and covered by clear top seal for LAMP incubation
Trinh et al. [71]	Agarose gel	Agarose gel and LAMP reagents mixed and deposited on surface, solidified at room temperature, followed by addition of DNA solution

3. Heating Technique

The primary advantage of LAMP over PCR is the simplicity of the heating hardware. PCR thermal cyclers comprise a metal heating block whose temperature is controlled by power-intensive Peltier heaters. Even for heating small fluidic volumes, most of the thermal energy is wasted in heating the metallic blocks with a high heat capacity and do not actively participate in the reaction, resulting in a typical power consumption of anywhere from ~150 W–470 W for PCR protocol. Additionally, cooling fans must be switched on at each thermal cycling step to reduce the reaction temperature to primer annealing temperatures. The repeated heating and cooling steps severely inhibit the PCR-based NAAT devices to be operated with low powered and simple hardware. The isothermal operation of LAMP significantly reduces the thermal management and power consumption of LAMP-based NAATs, making them amenable for portable and off-the-grid use. LAMP reactions require isothermal conditions between 63 °C and 67 °C for optimal amplification. Typically, a temperature of 65 °C is used in most cases as LAMP. In the context of paper-based LAMP platforms, the initial implementation of heating utilized ovens, incubators, and hot plates [68] to maintain the paper-based assay at ~65 °C (Figure 4A). For example, Connelly et al. [45] used a laboratory incubator for heating the paper-based LAMP platform. Rodriguez et al. [57] placed their paper-based chip face-down on a 63 °C heat block or hot plate for 30 min. Seok et al. [46] used a heat block powered by Peltier elements attached to the bottom of the multiplex paper NAAT device. Kaarj et al. [44] used a hot plate for heating the segmented and cut RT-LAMP assays. Apart from this heating apparatus, researchers also used a biological incubator for heating the paper-based LAMP device to the required isothermal temperature. Davidson et al. [52] developed a paper assay for detecting SARS-CoV-2 using RT-LAMP, and this assay was heated to 65 °C in a standard 75 L biological incubator for 60 min. Later, more portable and integrated heating units were integrated with the paper-based LAMP platforms. For example, Choi et al. [58] developed a handheld system consisting of a closed heating compartment for amplification, a non-heating testing compartment for target analyte detection, an integrated battery, an integrated temperature controller, and a charger (Figure 4B). In this setup, the heating compartment was comprised of an aluminum enclosure with external insulation, whereas the testing compartment was comprised of chemical-resistant polyformaldehyde. A battery was integrated into the system with a programmable temperature controller (5 °C to 100 °C), with a resolution of ±0.1 °C. A temperature sensor was installed in the aluminum enclosure, with the temperature displayed externally. Li et al. [63] used a homemade heating device containing a DC power supply, a temperature controller, and a copper heating block in their study for detecting antibiotic resistance genes on paper membranes. Wang et al. [73] designed an integrated thermal and image box to conduct the RT-LAMP reaction to detect prostate cancer biomarkers. Their box included a thermal plate (area: 9 cm^2 and thickness: 0.15 cm) powered by a 12 V lithium-ion portable battery capable of generating a heating power of 0.6 w/cm^2. Since the operating temperature of LAMP is not very high, researchers have also incorporated inexpensive off-the-shelf kitchen gadgets to carry out the isothermal reaction. For example, Jiang et al. [62] used a battery-powered coffee mug warmer to maintain the water bath temperature at the LAMP operating temperature and detected Zika virus using a paper-based LAMP assay (Figure 4C) whereas Rofman et al. incorporated an extremely low powered thin film heater integrated beneath the paper substrate [65] (Figure 4D). The various methods of heating paper-based LAMP reactions have been summarized in Table 2. While these efforts are bringing us closer to portable operation, there still exists a gap in engineering more efficient and low-powered isothermal heaters to perform LAMP reactions.

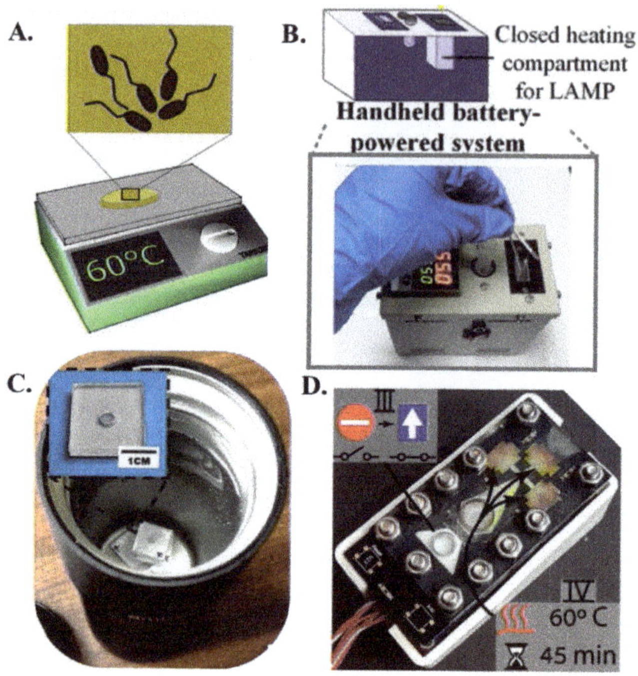

Figure 4. Heating methods (**A**). A hot plate incubator [from [68], copyright 2019 Elsevier]. (**B**). A hand-held battery-powered system for paper-based amplification and detection of nucleic acid [from [58], copyright 2016 American Chemical Society]. (**C**). ZIKV detection using VLEAD where RT-LAMP is carried out in a water bath inside an Ember coffee mug [from [62], copyright 2018 Wiley-VCH GmbH]. (**D**). Paper-based lab-on-a-chip device allowing implementation of multistep assays. The lower stacks serve for power and logic and the top is the functional PCB containing heating elements and driving electrodes [from [65], copyright 2022 Royal Society of Chemistry].

Table 2. Heating methods for paper-based LAMP platforms.

Reference	Heating Method
Connelly et al. [45]	Laboratory incubator
Rodriguez et al. [57]	Heat block or hot plate
Seok et al. [46]	Heat block powered by Peltier elements
Kaarj et al. [44]	Hot plate
Davidson et al. [52]	Biological incubator
Choi et al. [58]	Handheld system with closed heating compartment
Li et al. [63]	Homemade heating device
Wang et al. [73]	Integrated thermal and image box
Jiang et al. [62]	Battery-powered coffee mug warmer
Rofman et al. [65]	Thin film heater

4. Detection Methodologies

The detection in paper-based LAMP assays/devices has been mostly carried out using optical readouts (colorimetric or fluorescence-based) [9,37]. Visual detection methods of LAMP are simple to use, easy to read, and do not require specialized equipment, making it a cost-effective detection method. One common method is the use of a color-changing indicator, such as a pH indicator, that changes color in the presence of amplified DNA, suggesting a positive LAMP reaction. The user can quickly interpret a positive or negative assay readout with the naked eye or use an optical detector for more quantitative analysis. Many researchers have used color-changing dyes on paper-LAMP assays. Seok et al. [46],

Hongwarittorrn et al. [74], and Batule et al. [48] utilized the color change by hydroxynapthol blue (HNB) dye for identifying positive LAMP reaction, which generally depends on the concentration of magnesium ions and the resulting pyrophosphate generation. Kaarj et al. used phenol red colorimetric pH indicator along with their RT-LAMP master mix to detect positive ZIKV samples [44] (Figure 5A). Tung Trieu and Lee [64] used methylene blue (MB) dye for qualitatively detecting the LAMP amplification in a paper-based assay containing sodium sulphite. In the absence of an amplified target, colored MB completely reacts with excessive sodium sulphite to produce colorless leuco-methylene blue (LMB). Phenol red was used as a colorimetric dye for the detection of DNA amplicons by Davidson et al. [52]. Trinh et al. [75] used a unique colorimetric dye named Chemosensor L, ((Z)-4-((2,3 dihydroxybenzylidene)amino)benzenesulfonamide) for amplicon detection. Overall, colorimetric assays can be used with paper-based LAMP assays to provide simple visual detection but fail to yield high signal to noise in positive and negative samples for robust discrimination.

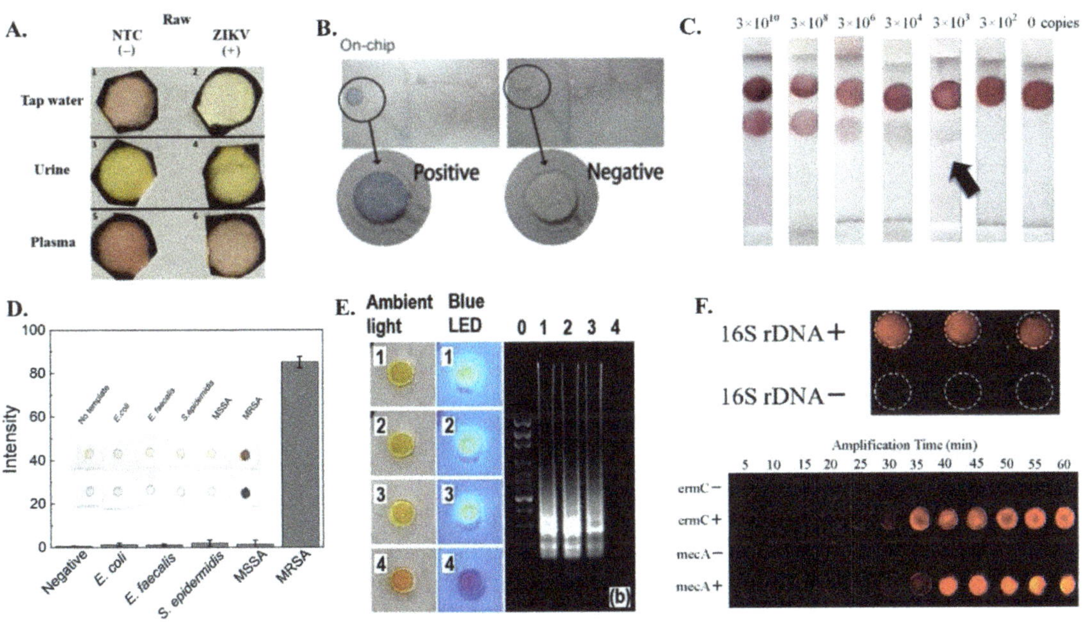

Figure 5. Detection methods (**A**) Colorimetric detection of ZIKV RT-LAMP assay on samples with tap water, urine and plasma [from [44], copyright 2018 Springer Nature]. (**B**) Colorimetric signal detection on On-chip [from [60] copyright 2016 Wiley-VCH GmbH]. (**C**) The integrated paper-based LFA device can achieve the colorimetric detection limit of as low as 3 × 103 copies at the optimum incubation time of 60 min [from [58], copyright 2016 Royal Society of Chemistry]. (**D**) The application of gold nanoparticles (AgNPIs) with paper pad enable quantitative assay readout [from [69], copyright 2021 Wiley-VCH GmbH]. (**E**) Smartphone photos of the detection units under the ambient light and blue LED flashlight to detect 1, 0.5, and 0.1 PFU of ZIKV spiked in human urine samples while device 4 is a negative control. [from [62], copyright 2018 Wiley-VCH GmbH]. (**F**) (top) Fluorescence image of the paper-based chip for 16S rDNA detection; (bottom) Real-time amplification of multiple LAMP for mecA and ermC. Fluorescence intensity of amplification changes with time. [from [63], copyright 2018 American Chemical Society].

Due to their operational simplicity, the paper-based LAMP platforms have often combined colorimetric detection with lateral flow assay (LFA). These systems frequently use nanoparticles which can be used for colorimetric detection by exploiting the unique optical properties of their small size. By functionalizing the nanoparticles with specific biomolecules, they can be used to detect the presence of specific analytes through color changes. The on-chip signal detection in the paper-LAMP device designed by Champlauk et al. [60] used a paper-based LAMP assay in conjunction with silver nanoparticles (AgNP) to facilitate a colorimetric readout of amplified LAMP products (Figure 5B). Choi et al. [58,59] used a coupled detector probe (DP)-gold nanoparticle (AuNP) for capturing the nucleic acid target (Figure 5C). The streptavidin lining at the test line effectively captures the biotinylated amplicons, while the remaining DP-AuNP gets captured at the control line. A slightly different approach was adopted by Rodriquez et al. [57], where they used streptavidin-conjugated gold nanoparticles bound to the FITC (fluorescein isothiocyanate) probe-amplicons, biotin-based control line, and anti-FITC test line. A positive detection will result in the binding of the FITC bound amplicons on the anti-FITC line while excess streptavidin-conjugated gold nanoparticles will be captured at the biotin control line. Ru Choi et al. [61] also used gold nanoparticle conjugates for detection on lateral flow assay. Suea-Ngam et al. [69] used silver nanoparticles for detection in their paper-LAMP device (Figure 5D). It was found that free primers produced orange-yellow color, whereas silver nanoparticles were less etched and red in color. Similar to Rodriquez et al. [57], FITC-labeled probe was also used for detection purposes by Zhu et al. [76]. Jawla et al. [77] demonstrated a lateral flow assay where the amplified product was tagged with biotin and FITC, which combined streptavidin-gold nanoparticles in the conjugate pad downstream. Both Linnes et al. [66] and Saxena et al. [78] have used FAM signals to detect LAMP products on paper.

Fluorescence dyes can be used to detect LAMP products by binding to the double-stranded DNA produced during the amplification process. This binding results in a change in the fluorescence intensity of the dye, which can be detected using a fluorometer or even a smartphone [79,80]. The use of fluorescence dyes (such as SYBR Green, EvaGreen, Rox, etc.,) allows for real-time monitoring of the LAMP reaction and can provide a rapid and sensitive method for detecting specific nucleic acid sequences. This allows for early detection of LAMP products and can provide quantitative information about the amount of amplified DNA present. In the context of paper-based LAMP, SYBR Green fluorescent intercalating dye has been used by Connelly et al. [45] for visualizing the amplicons. Hiltunen et al. [81] used EvaGreen fluorophore for detection on an aluminum-coated paper substrate and PMMA. EvaGreen has been found to be fluorescent at ambient temperature, and it was effective in the detection of positive LAMP reactions on paper substrates. Jiang et al. [62] and Liu et al. [82] incorporated SYBR Green 1 dye in the paper-LAMP chambers for visualizing the products generated from LAMP (Figure 5E). Li et al. [63] introduced a new fluorescent detection agent in the form of a transition metal complex $[Ru(phen)_2dppz](PF_6)_2$ in their paper-based LAMP assay. Their fluorescent detection method enabled them to achieve high signal to noise ratio enabling them to not only robustly discriminate between positive and negative samples but also enable real time and quantification amplification on paper (Figure 5F). DNA binding Picogreen dye was used to generate fluorescence on paper by Naik et al. [68] to detect LAMP amplicons. Calcein has been used by Wang et al. [73] and Zhou et al. [83] since the presence of manganese ions initially inhibits the fluorescence of calcein. However, LAMP results in the formation of pyrophosphate, which further reacts with manganese, and eventually, it separates out from calcein generating bright fluorescence. Garneret et al. [51] and Rofman et al. [65] used SYTO82 in their paper-based assay for amplicon detection and visualization. While several researchers have incorporated fluorescent dyes in their paper-based LAMP assays, they are known to inhibit the LAMP reaction to some extent, and their use is not optimal for resource-limited settings.

5. Sensitivity and Specificity of Paper-Based LAMP Systems

The sensitivity and specificity of LAMP reactions depend on several factors, including the initial concentration of target nucleic acid, the specificity of the primers, and the reaction conditions. Generally, LAMP is considered to have similar sensitivity as traditional PCR and can detect as low as a few copies of target DNA. In some studies, LAMP has been shown to detect as low as ten copies of target DNA, and can even detect single copies of a viral genome in clinical samples. Ideally, a molecular diagnostic assay should have high sensitivity and high specificity [84]; however, paper-based assays generally exhibit lower sensitivity than their liquid assay counterparts. Sensitivity is often expressed in terms of limit of detection (LOD) with lower LOD assays corresponding to higher sensitivity [84].

The first paper-based LAMP assay developed by Connelly et al. [45] demonstrated a LOD corresponding to one copy of dsDNA and 5 *E. coli* cells. Champlauk et al. [60] targeted six areas of the *aflR* gene in the *Aspergillus* strain for amplification and demonstrated a LOD of 100 *aflR* copies with 94.47% specificity. At the same time, 43.75% of contamination was identified in 14 of 32 herbal samples tested. In the integrated LAMP-LFA paper-based platform, Choi et al. [58] demonstrated a LOD of 3×10^3 copies of dengue viral RNA. Later, in their PDMS drop-infused modified design of LFA, Choi et al. [59] demonstrated a LOD of 20 pM, with a 2.5-fold signal enhancement for the case involving 5 PDMS drops. Also, these assays also demonstrated good specificity in detecting the Hepatitis B virus. Rodriguez et al. [57] demonstrated a LOD of 10^4 HPV (human papilloma virus) copies in their combined paper-based LAMP-LFA device. Ru Choi et al. [61] achieved a LOD of 10 CFU/mL *E. coli* copies in milk and 10^3 CFU/mL *E. coli* copies in spinach from the paper-LAMP biosensor device resulting in a higher sensitivity in comparison to the existing paper-based assays. For specificity analysis, *S. pneumonia* was targeted, and this was only detected when the other *HBV* and *E. coli* samples showed negative results. Seok et al. [46] demonstrated that the sensitivity of their paper-based LAMP assay for detecting *S. pneumonia* was in the range 0.7 pg–700 pg of DNA mass with an LOD of 0.7 pg. Jiang et al. [62] demonstrated that the detection limit of ZIKA virus on their paper-device is 0.5 PFU of 140 mL of ZIKA spiked urine sample. In the same year, Kaarj et al. [44] found the detection limit as 1 genome ZIKA sample per ml for their paper-based LAMP assay. Li et al. [63] showed an LOD of 100 copies for *mecA* gene and 285 copies for *ermC* gene from MRSA (methicillin resistant *Staphylococcus aureus*). Liu et al. [82] demonstrated a LOD of 10^{-4} dilution for their proposed LAMP-LFA device. A detection sensitivity of almost 100 CFU/mL was found in the paper substrate used for LAMP by Naik et al. [68]. An *E. coli* bacterial concentration as low as 1000 CFU/mL was found as the detection limit on paper substrates by Tung Trieu and Lee [64]. Zhu et al. [76] displayed the LOD of *Prymnesium parvum* g DNA as 0.03 ng/µL in their LAMP-LFA device. Batule et al. [48] showed that the detection limit of ZIKA virus was 1 and 10 copies from 1×PBS and 100% human serum sample, respectively, utilizing the paper-based setup used by Seok et al. [46]. Wang et al. [73] found the detection sensitivity of Prostate cancer antigen 3 (PCA 3) to be around 0.34 fg/µL in the PVA-paper-LAMP-based setup. Suea-Ngam et al. [69] concluded that when FIP primers were immobilized on the paper surface, two fold sensitivity increased as compared to the solution-based LAMP. Davidson et al. [52] found 100% sensitivity with a detection limit of 200 copies of SARS-CoV-2 virus particles per µL. In addition, the specificity of SARS-CoV-2 detection was found to be around 76%. Jawla et al. [77] found that pig DNA template concentration of 10 fg produced an observable amplification signal in their paper LAMP-LFA device. Nearly a detection limit of 100 CFU/mL of vancomycin-resistant *Enterococcus* in milk was achieved by Trinh et al. [71]. Wang et al. [54] achieved a sensitivity of 3 SARS-CoV-2 RNA copies in their paper-LAMP-ELISA assay whereas 1 SARS-CoV-2 copy per ml was found to be the LOD in the paper-based device demonstrated by Garneret et al. [51]. Saxena et al. [78] found the limit of detection to be around 10 agµL^{-1} for the N-gene (corresponds to 1.61×10^2 copies of customized DNA fragment), *ORF1ab*-gene (2.13×10^2 copies of customized DNA fragment) and E-gene (8.69×10^2 copies of customized DNA fragment), and 100 agµL^{-1} (7.44×10^3 copies of customized DNA fragment) for the S-gene. Zhou et al. [83] displayed a pathogen (*E. coli*, *Salmonella* sp., and *Staphylococcus aureus*) detection sensitivity of 2.8×10^{-5} ng per µL and 10 CFU/mL in spiked milk samples. The detection methods employed along with the limit of detection values for various paper-based LAMP platforms have been summarized in Table 3. Overall, almost all the paper-based LAMP assay systems demonstrated high enough sensitivity and specificity to be applicable for clinical diagnostics.

Table 3. Paper-based LAMP platforms, targets, detection methods and limit of detection.

Reference	LAMP Platform	Detection Method	Targets	Limit of Detection
Connelly et al. [45]	A paper microfluidic device that enables a central patterned paper strip to slide in and out of fluidic path and allows sample preparation, isothermal amplification and detection.	Fluorescence detection using SYBR Green I	Whole, live *E. coli* cells (malB gene) in human plasma	5 cells
Linnes et al. [66]	Lateral flow detection strip with polyethersulphone membrane	Colorimetric-Biotin Streptavidin chemistry	Four separate DNA and RNA targets (Bordetella pertussis, Chlamydia trachomatis, Neisseria, gonorrhoeae, and Influenza A H1N1)	NA
Choi et al. [59]	Lateral flow assay strip which incorporates a piece of nitro cellulose paper-based shunt and a polydimethylsiloxane barrier to the strip	Gold Nano Particle Streptavidin chemistry	Hepatitis B Virus	10^2 IU/mL (International units per milliliter)
Choi et al. [58]	An integrated paper-based biosensor incorporating nucleic acid extraction, amplification and visual detection	Gold Nano Particle Streptavidin chemistry	Escherichia coli and Streptococcus pneumonia	10–1000 CFU/mL
Roy et al. [85]	Paper microchip fabricated in a cellulose paper and a small wax chamber	Colorimetric- Leuco crystal violet	Sus scrofa (porcine) and Bacillus subtilis (bacteria) DNA	1 picogram/µL and 10 picogram/µL respectively
Kaarj et al. [44]	Paper microfluidic chip	Colorimetric-Phenol red	ZIKV RNA	1 copy/uL
Kim et al. [86]	Polyethersulfone (PES) paper embedded with a polymethyl methacrylate (PMMA) platform for simultaneous DNA amplification and colorimetric detection	Colorimetric-eriochrome black T (EBT)	DNA of Staphylococcus aureus	1 femtogram/mL
Lin et al. [87]	Digital LAMP directly on polycarbonate membrane	Primer-probe-primer-quencher fluorescence chemistry	*E. coli*, *E. faecalis*, and Salmonella Typhi DNA; also MS2 virus in wastewater	11 to 1.1×10^5 copies/µL
Naik et al. [68]	Paper-based LAMP	Fluorescence detection using PicoGreen	Escherichia coli (MG1655) and Mycobacterium smegmatis (mc2155) cells	NA
Varsha et al. [88]	Penta- cloverleaf modelled paper-based (Whatman filter paper) LAMP device	Colorimetric- Leuco crystal violet	Multiple gene targets in Leptospira	50 attogram/µL
Jawla et al. [77]	Paper-based LAMP lateral flow assay strip with nitro cellulose membrane	Colorimetric-Biotin Streptavidin chemistry	Cattle DNA	0.1 picogram
Choopara et al. [70]	Paper-based (cellulose membrane) LAMP device consisting of a sandwich-like bottom base, a reaction pad in the center, and a clear top seal	Fluorescence detection using SYBR Green I	MRSA (gene mecA)	10 attogram—equivalent to 1 copy

6. Conclusions and Future Perspectives

While a direct comparison of paper-based LAMP and microfluidic LAMP platforms for point-of-care diagnostics cannot be made without defining precise comparison criteria, generally, microfluidic LAMP platforms offer improved assay performance including higher sensitivity, specificity, and reproducibility. This may be due to several factors, such as the ability to control and optimize reaction conditions in microfluidic platforms, the use of specialized materials and fabrication methods to reduce variability and increase sensitivity, and the integration of additional diagnostic features such as on-chip sample preparation and detection. However, microfluidic LAMP platforms can be more complex and costly to develop and manufacture compared to paper-based platforms. Previously, paper membranes have been used in microfluidic-based systems only for extracting nucleic acids [55,56]. Additionally, paper membranes, specifically lateral flow assays, have been widely used for detection after carrying out tube/solution-based LAMP [89]. The present review article discusses emerging trends in developing various paper-based LAMP assays/devices in recent years. While still in its infancy, the future prospects of paper-based LAMP are promising as it offers cost-effective, rapid, and reliable detection capabilities for a wide range of applications. It can be used as point-of-care diagnostics, allowing for rapid and accurate diagnosis at the patient's bedside or in remote or field settings. The ability to multiplex the assay helps detecting multiple targets at the same time and increases its diagnostic utility. In addition, paper-based LAMP can be used for environmental monitoring, detecting microorganisms in water, soil, and other environmental samples. It can also be used for food safety, providing a rapid and specific detection of food-borne pathogens in food products. Additionally, it can be used in research applications, such as monitoring the presence of specific microorganisms in a sample or for genotyping. Due to its relatively low cost, it is an attractive option for large-scale testing in low- and middle-income countries, where the cost of traditional diagnostic tests can be prohibitive. Overall, paper-based LAMP has a wide range of applications and great potential to improve global health, food safety, and environmental monitoring.

However, several bottlenecks should be addressed to enable wider adoption of this technology. The heating setup to actuate paper-based LAMP assays should be optimally engineered so that the heat source is miniaturized, low-powered, and ideally—electricity free. The detection technique should be revised to incorporate inexpensive dyes (possibly carbon dots) as opposed to costly fluorophores. Smartphone-based readout coupled with improved image analysis should be developed to enable quantitative analysis with paper-based LAMP assays in contrast to the traditional yes/no result. Paper-based assays should be adapted and optimized to incorporate multiplexed LAMP to enable differential diagnostics. This will be essential in detecting multiple pathogens in a clinical sample. Paper-based sample preparation protocols should be further analyzed to incorporate a wider range of clinical samples. Storage of LAMP reagents in their dried form on the paper substrate should be explored in more detail with a range of drying and lyophilizing protocols. This should greatly make paper-based assay platforms more robust and increase their shelf life. Even with these challenges, researchers are pouring in efforts from several directions to make the next generation of paper-based NAAT devices more robust and scalable so that they can address the global diagnostic demands in a cost-effective manner.

Author Contributions: Conceptualization, D.D. and A.P.; formal analysis, D.D., M.M. and A.P.; data curation, D.D., M.M. and A.P.; writing—original draft preparation, D.D. and A.P.; writing—review and editing, D.D. and A.P.; supervision, A.P.; funding acquisition, A.P. All authors have read and agreed to the published version of the manuscript.

Funding: This work was funded partly by a grant from Department of Defense (CDMRP grant number: W81XWH2210071; PI: Aashish Priye) and University of Cincinnati (Faculty scholar research award, grant number: 1018268; PI: Aashish Priye). This was an invited publication with a waiver on the article processing charges.

Institutional Review Board Statement: Not applicable.

Informed Consent Statement: Not applicable.

Data Availability Statement: No new data was generated.

Conflicts of Interest: The authors declare no conflict of interest.

References

1. Sun, H.; Xiong, L.; Huang, Y.; Chen, X.; Yu, Y.; Ye, S.; Dong, H.; Jia, Y.; Zhang, W. AI-aided on-chip nucleic acid assay for smart diagnosis of infectious disease. *Fundam. Res.* **2021**, *2*, 476–486. [CrossRef]
2. Cevik, S. Going Viral: A Gravity Model of Infectious Diseases and Tourism Flows. *Open Econ. Rev.* **2021**, *33*, 141–156. [CrossRef]
3. Nugent, R. Chronic Diseases in Developing Countries. *Ann. N. Y. Acad. Sci.* **2008**, *1136*, 70–79. [CrossRef]
4. Kasperson, J.X.; Kasperson, R.E. Priorities in Profile: Managing Risks in Developing Countries. In *The Social Contours of Risk*; Routledge: London, UK, 2022; pp. 172–179. [CrossRef]
5. Christaki, E. New technologies in predicting, preventing and controlling emerging infectious diseases. *Virulence* **2015**, *6*, 558–565. [CrossRef]
6. Yang, S.; E Rothman, R. PCR-based diagnostics for infectious diseases: Uses, limitations, and future applications in acute-care settings. *Lancet Infect. Dis.* **2004**, *4*, 337–348. [CrossRef]
7. Shi, A.C.; Ren, P. SARS-CoV-2 serology testing: Progress and challenges. *J. Immunol. Methods* **2021**, *494*, 113060. [CrossRef] [PubMed]
8. Peeling, R.W.; Wedderburn, C.J.; Garcia, P.J.; Boeras, D.; Fongwen, N.; Nkengasong, J.; Sall, A.; Tanuri, A.; Heymann, D.L. Serology testing in the COVID-19 pandemic response. *Lancet Infect. Dis.* **2020**, *20*, e245–e249. [CrossRef] [PubMed]
9. Magro, L.; Escadafal, C.; Garneret, P.; Jacquelin, B.; Kwasiborski, A.; Manuguerra, J.-C.; Monti, F.; Sakuntabhai, A.; Vanhomwegen, J.; Lafaye, P.; et al. Paper microfluidics for nucleic acid amplification testing (NAAT) of infectious diseases. *Lab a Chip* **2017**, *17*, 2347–2371. [CrossRef]
10. Craw, P.; Balachandran, W. Isothermal nucleic acid amplification technologies for point-of-care diagnostics: A critical review. *Lab Chip* **2012**, *12*, 2469–2486. [CrossRef]
11. Liu, D. *Molecular Detection of Human Bacterial Pathogens*; CRC Press: Boca Raton, FL, USA, 2011.
12. Kang, T.; Lu, J.; Yu, T.; Long, Y.; Liu, G. Advances in nucleic acid amplification techniques (NAATs): COVID-19 point-of-care diagnostics as an example. *Biosens. Bioelectron.* **2022**, *206*. [CrossRef]
13. Morisset, D.; Stebih, D.; Cankar, K.; Zel, J.; Gruden, K. Alternative DNA amplification methods to PCR and their application in GMO detection: A review. *Eur. Food Res. Technol.* **2008**, *227*, 1287–1297. [CrossRef]
14. Zhang, Y.; Jiang, H.-R. A review on continuous-flow microfluidic PCR in droplets: Advances, challenges and future. *Anal. Chim. Acta* **2016**, *914*, 7–16. [CrossRef] [PubMed]
15. Chang, C.-C.; Chen, C.-C.; Wei, S.-C.; Lu, H.-H.; Liang, Y.-H.; Lin, C.-W. Diagnostic Devices for Isothermal Nucleic Acid Amplification. *Sensors* **2012**, *12*, 8319–8337. [CrossRef]
16. Asiello, P.J.; Baeumner, A. Miniaturized isothermal nucleic acid amplification, a review. *Lab a Chip* **2011**, *11*, 1420–1430. [CrossRef]
17. Deiman, B.; Van Aarle, P.; Sillekens, P. Characteristics and Applications of Nucleic Acid Sequence-Based Amplification (NASBA). *Mol. Biotechnol.* **2002**, *20*, 163–180. [CrossRef]
18. Cook, N. The use of NASBA for the detection of microbial pathogens in food and environmental samples. *J. Microbiol. Methods* **2003**, *53*, 165–174. [CrossRef] [PubMed]
19. Mueller, J.D.; Pütz, B.; Höfler, H. Self-sustained sequence replication (3SR): An alternative to PCR. *Histochem. Cell Biol.* **1997**, *108*, 431–437. [CrossRef] [PubMed]
20. Fahy, E.; Kwoh, D.Y.; Gingeras, T.R. Self-sustained sequence replication (3SR): An isothermal transcription-based amplification system alternative to PCR. *Genome Res.* **1991**, *1*, 25–33. [CrossRef] [PubMed]
21. Hill, C.S. Molecular diagnostic testing for infectious diseases using TMA technology. *Expert Rev. Mol. Diagn.* **2001**, *1*, 445–455. [CrossRef] [PubMed]
22. Simmel, F.C.; Yurke, B.; Singh, H.R. Principles and Applications of Nucleic Acid Strand Displacement Reactions. *Chem. Rev.* **2019**, *119*, 6326–6369. [CrossRef]
23. Hellyer, T.J.; Nadeau, J.G. Strand displacement amplification: A versatile tool for molecular diagnostics. *Expert Rev. Mol. Diagn.* **2004**, *4*, 251–261. [CrossRef] [PubMed]
24. Motamedi, M.H.K.; Saghafinia, M.; Karami, A.; Gill, P. A review of the current isothermal amplification techniques: Applications, advantages and disadvantages. *J. Glob. Infect. Dis.* **2011**, *3*, 293. [CrossRef]
25. Ren, Y.; Cao, L.; You, M.; Ji, J.; Gong, Y.; Ren, H.; Xu, F.; Guo, H.; Hu, J.; Li, Z. "SMART" digital nucleic acid amplification technologies for lung cancer monitoring from early to advanced stages. *TrAC Trends Anal. Chem.* **2022**, *157*. [CrossRef]
26. Andresen, D.; Von Nickisch-Rosenegk, M.; Bier, F.F. Helicase-dependent amplification: Use in OnChip amplification and potential for point-of-care diagnostics. *Expert Rev. Mol. Diagn.* **2009**, *9*, 645–650. [CrossRef] [PubMed]
27. Jeong, Y.-J.; Park, K.; Kim, D.-E. Isothermal DNA amplification in vitro: The helicase-dependent amplification system. *Cell. Mol. Life Sci.* **2009**, *66*, 3325–3336. [CrossRef] [PubMed]
28. Lobato, I.M.; O'Sullivan, C.K. Recombinase polymerase amplification: Basics, applications and recent advances. *TrAC Trends Anal. Chem.* **2017**, *98*, 19–35. [CrossRef]

29. Daher, R.K.; Stewart, G.; Boissinot, M.; Bergeron, M.G. Recombinase Polymerase Amplification for Diagnostic Applications. *Clin. Chem.* **2016**, *62*, 947–958. [CrossRef]
30. Ali, M.M.; Li, F.; Zhang, Z.; Zhang, K.; Kang, D.-K.; Ankrum, J.A.; Le, X.C.; Zhao, W. Rolling circle amplification: A versatile tool for chemical biology, materials science and medicine. *Chem. Soc. Rev.* **2014**, *43*, 3324–3341. [CrossRef]
31. Zhao, W.; Ali, M.M.; Brook, M.A.; Li, Y. Rolling Circle Amplification: Applications in Nanotechnology and Biodetection with Functional Nucleic Acids. *Angew. Chem. Int. Ed.* **2008**, *47*, 6330–6337. [CrossRef]
32. Zhang, D.Y.; Brandwein, M.; Hsuih, T.; Li, H.B. Ramification Amplification: A Novel Isothermal DNA Amplification Method. *Mol. Diagn.* **2001**, *6*, 141–150. [CrossRef] [PubMed]
33. Dean, F.B.; Hosono, S.; Fang, L.; Wu, X.; Faruqi, A.F.; Bray-Ward, P.; Sun, Z.; Zong, Q.; Du, Y.; Du, J.; et al. Comprehensive human genome amplification using multiple displacement amplification. *Proc. Natl. Acad. Sci. USA* **2002**, *99*, 5261–5266. [CrossRef]
34. Lasken, R.S. Genomic DNA amplification by the multiple displacement amplification (MDA) method. *Biochem. Soc. Trans.* **2009**, *37*, 450–453. [CrossRef]
35. Notomi, T.; Mori, Y.; Tomita, N.; Kanda, H. Loop-mediated isothermal amplification (LAMP): Principle, features, and future prospects. *J. Microbiol.* **2015**, *53*, 1–5. [CrossRef]
36. Notomi, T.; Okayama, H.; Masubuchi, H.; Yonekawa, T.; Watanabe, K.; Amino, N.; Hase, T. Loop-mediated isothermal amplification of DNA. *Nucleic Acids Res.* **2000**, *28*, E63. [CrossRef]
37. Liu, L.; Yang, D.; Liu, G. Signal amplification strategies for paper-based analytical devices. *Biosens. Bioelectron.* **2019**, *136*, 60–75. [CrossRef]
38. McCalla, S.E.; Ong, C.; Sarma, A.; Opal, S.M.; Artenstein, A.W.; Tripathi, A. A Simple Method for Amplifying RNA Targets (SMART). *J. Mol. Diagn.* **2012**, *14*, 328–335. [CrossRef] [PubMed]
39. Meagher, R.J.; Priye, A.; Light, Y.K.; Huang, C.; Wang, E. Impact of primer dimers and self-amplifying hairpins on reverse transcription loop-mediated isothermal amplification detection of viral RNA. *Analyst* **2018**, *143*, 1924–1933. [CrossRef]
40. Zhang, H.; Xu, Y.; Fohlerova, Z.; Chang, H.; Iliescu, C.; Neuzil, P. LAMP-on-a-chip: Revising microfluidic platforms for loop-mediated DNA amplification. *TrAC Trends Anal. Chem.* **2019**, *113*, 44–53. [CrossRef]
41. Gong, M.M.; Sinton, D. Turning the Page: Advancing Paper-Based Microfluidics for Broad Diagnostic Application. *Chem. Rev.* **2017**, *117*, 8447–8480. [CrossRef]
42. Carrell, C.; Kava, A.; Nguyen, M.; Menger, R.; Munshi, Z.; Call, Z.; Nussbaum, M.; Henry, C. Beyond the lateral flow assay: A review of paper-based microfluidics. *Microelectron. Eng.* **2019**, *206*, 45–54. [CrossRef]
43. Fu, E.; Ramsey, S.A.; Kauffman, P.; Lutz, B.; Yager, P. Transport in two-dimensional paper networks. *Microfluid. Nanofluid.* **2010**, *10*, 29–35. [CrossRef]
44. Kaarj, K.; Akarapipad, P.; Yoon, J.-Y. Simpler, Faster, and Sensitive Zika Virus Assay Using Smartphone Detection of Loop-mediated Isothermal Amplification on Paper Microfluidic Chips. *Sci. Rep.* **2018**, *8*, 12438. [CrossRef]
45. Connelly, J.T.; Rolland, J.P.; Whitesides, G.M. "Paper Machine" for Molecular Diagnostics. *Anal. Chem.* **2015**, *87*, 7595–7601. [CrossRef]
46. Seok, Y.; Joung, H.-A.; Byun, J.-Y.; Jeon, H.-S.; Shin, S.J.; Kim, S.; Shin, Y.-B.; Han, H.S.; Kim, M.-G. A Paper-Based Device for Performing Loop-Mediated Isothermal Amplification with Real-Time Simultaneous Detection of Multiple DNA Targets. *Theranostics* **2017**, *7*, 2220–2230. [CrossRef]
47. Das, D.; Namboodiri, S. Selection of a suitable paper membrane for Loop Mediated Isothermal DNA amplification reaction (LAMP) in a point-of-care diagnostic kit—Experimental and CFD analysis. *Chem. Eng. Sci.* **2020**, *229*, 116130. [CrossRef]
48. Batule, B.S.; Seok, Y.; Kim, M.-G. Paper-based nucleic acid testing system for simple and early diagnosis of mosquito-borne RNA viruses from human serum. *Biosens. Bioelectron.* **2019**, *151*, 111998. [CrossRef]
49. Choi, J.R.; Yong, K.W.; Tang, R.; Gong, Y.; Wen, T.; Li, F.; Pingguan-Murphy, B.; Bai, D.; Xu, F. Advances and challenges of fully integrated paper-based point-of-care nucleic acid testing. *TrAC Trends Anal. Chem.* **2017**, *93*, 37–50. [CrossRef]
50. Tian, T.; Bi, Y.; Xu, X.; Zhu, Z.; Yang, C. Integrated paper-based microfluidic devices for point-of-care testing. *Anal. Methods* **2018**, *10*, 3567–3581. [CrossRef]
51. Garneret, P.; Coz, E.; Martin, E.; Manuguerra, J.-C.; Brient-Litzler, E.; Enouf, V.; Obando, D.F.G.; Olivo-Marin, J.-C.; Monti, F.; Van Der Werf, S.; et al. Performing point-of-care molecular testing for SARS-CoV-2 with RNA extraction and isothermal amplification. *PLoS ONE* **2021**, *16*, e0243712. [CrossRef]
52. Davidson, J.L.; Wang, J.; Maruthamuthu, M.K.; Dextre, A.; Pascual-Garrigos, A.; Mohan, S.; Putikam, S.V.S.; Osman, F.O.I.; McChesney, D.; Seville, J.; et al. A paper-based colorimetric molecular test for SARS-CoV-2 in saliva. *Biosens. Bioelectron. X* **2021**, *9*, 100076. [CrossRef]
53. Chowdury, M.A.; Khalid, F. Application of microfluidic paper-based analytical device (μPAD) to detect COVID-19 in energy deprived countries. *Int. J. Energy Res.* **2021**, *45*, 18275–18280. [CrossRef] [PubMed]
54. Wang, J.; Dextre, A.; Pascual-Garrigos, A.; Davidson, J.L.; Maruthamuthu, M.K.; McChesney, D.; Seville, J.; Verma, M.S. Fabrication of a paper-based colorimetric molecular test for SARS-CoV-2. *MethodsX* **2021**, *8*, 101586. [CrossRef]
55. Liu, C.; Geva, E.; Mauk, M.; Qiu, X.; Abrams, W.R.; Malamud, D.; Curtis, K.; Owen, S.M.; Bau, H.H. An isothermal amplification reactor with an integrated isolation membrane for point-of-care detection of infectious diseases. *Analyst* **2011**, *136*, 2069–2076. [CrossRef]

56. Zhang, L.; Zhang, Y.; Wang, C.; Feng, Q.; Fan, F.; Zhang, G.; Kang, X.; Qin, X.; Sun, J.; Li, Y.; et al. Integrated Microcapillary for Sample-to-Answer Nucleic Acid Pretreatment, Amplification, and Detection. *Anal. Chem.* **2014**, *86*, 10461–10466. [CrossRef]
57. Rodriguez, N.M.; Wong, W.S.; Liu, L.; Dewar, R.; Klapperich, C.M. A fully integrated paperfluidic molecular diagnostic chip for the extraction, amplification, and detection of nucleic acids from clinical samples. *Lab a Chip* **2016**, *16*, 753–763. [CrossRef]
58. Choi, J.R.; Hu, J.; Gong, Y.; Feng, S.; Abas, W.A.B.W.; Pingguan-Murphy, B.; Xu, F. An integrated lateral flow assay for effective DNA amplification and detection at the point of care. *Analyst* **2016**, *141*, 2930–2939. [CrossRef] [PubMed]
59. Choi, J.R.; Liu, Z.; Hu, J.; Tang, R.; Gong, Y.; Feng, S.; Ren, H.; Wen, T.; Yang, H.; Qu, Z.; et al. Polydimethylsiloxane-Paper Hybrid Lateral Flow Assay for Highly Sensitive Point-of-Care Nucleic Acid Testing. *Anal. Chem.* **2016**, *88*, 6254–6264. [CrossRef] [PubMed]
60. Chaumpluk, P.; Plubcharoensook, P.; Prasongsuk, S. Rapid detection of aflatoxigenic *Aspergillus* sp. in herbal specimens by a simple, bendable, paper-based lab-on-a-chip. *Biotechnol. J.* **2016**, *11*, 768–779. [CrossRef]
61. Choi, J.R.; Hu, J.; Tang, R.; Gong, Y.; Feng, S.; Ren, H.; Wen, T.; Li, X.; Abas, W.A.B.W.; Pingguan-Murphy, B.; et al. An integrated paper-based sample-to-answer biosensor for nucleic acid testing at the point of care. *Lab a Chip* **2015**, *16*, 611–621. [CrossRef]
62. Jiang, X.; Loeb, J.C.; Manzanas, C.; Lednicky, J.A.; Fan, Z.H. Valve-Enabled Sample Preparation and RNA Amplification in a Coffee Mug for Zika Virus Detection. *Angew. Chem. Int. Ed.* **2018**, *57*, 17211–17214. [CrossRef]
63. Li, B.; Zhou, X.; Liu, H.; Deng, H.; Huang, R.; Xing, D. Simultaneous Detection of Antibiotic Resistance Genes on Paper-Based Chip Using $[Ru(phen)_2dppz]^{2+}$ Turn-on Fluorescence Probe. *ACS Appl. Mater. Interfaces* **2018**, *10*, 4494–4501. [CrossRef]
64. Trieu, P.T.; Lee, N.Y. Paper-Based All-in-One Origami Microdevice for Nucleic Acid Amplification Testing for Rapid Colorimetric Identification of Live Cells for Point-of-Care Testing. *Anal. Chem.* **2019**, *91*, 11013–11022. [CrossRef] [PubMed]
65. Rofman, B.; Naddaf, R.; Bar-Dolev, M.; Gefen, T.; Ben-Assa, N.; Geva-Zatorsky, N.; Bercovici, M. Automated device for multi-stage paper-based assays enabled by an electroosmotic pumping valve. *Lab a Chip* **2022**, *22*, 4511–4520. [CrossRef]
66. Linnes, J.C.; Rodriguez, N.M.; Liu, L.; Klapperich, C.M. Polyethersulfone improves isothermal nucleic acid amplification compared to current paper-based diagnostics. *Biomed. Microdevices* **2016**, *18*, 30. [CrossRef] [PubMed]
67. Tang, R.; Xie, M.Y.; Li, M.; Cao, L.; Feng, S.; Li, Z.; Xu, F. Nitrocellulose Membrane for Paper-based Biosensor. *Appl. Mater. Today* **2021**, *26*, 101305. [CrossRef]
68. Naik, P.; Jaitpal, S.; Shetty, P.; Paul, D. An integrated one-step assay combining thermal lysis and loop-mediated isothermal DNA amplification (LAMP) in 30 min from E. coli and M. smegmatis cells on a paper substrate. *Sens. Actuators B Chem.* **2019**, *291*, 74–80. [CrossRef]
69. Suea-Ngam, A.; Choopara, I.; Li, S.; Schmelcher, M.; Somboonna, N.; Howes, P.D.; Demello, A.J. In Situ Nucleic Acid Amplification and Ultrasensitive Colorimetric Readout in a Paper-Based Analytical Device Using Silver Nanoplates. *Adv. Healthc. Mater.* **2020**, *10*, e2001755. [CrossRef]
70. Choopara, I.; Suea-Ngam, A.; Teethaisong, Y.; Howes, P.D.; Schmelcher, M.; Leelahavanichkul, A.; Thunyaharn, S.; Wongsawaeng, D.; Demello, A.J.; Dean, D.; et al. Fluorometric Paper-Based, Loop-Mediated Isothermal Amplification Devices for Quantitative Point-of-Care Detection of Methicillin-Resistant *Staphylococcus aureus* (MRSA). *ACS Sens.* **2021**, *6*, 742–751. [CrossRef] [PubMed]
71. Trinh, T.N.D.; Thai, D.A.; Lee, N.Y. Pop-up paper-based and fully integrated microdevice for point-of-care testing of vancomycin-resistant Enterococcus. *Sens. Actuators B Chem.* **2021**, *345*, 130362. [CrossRef]
72. Das, D.; Singh, T.; Ahmed, I.; Masetty, M.; Priye, A. Effects of Relative Humidity and Paper Geometry on the Imbibition Dynamics and Reactions in Lateral Flow Assays. *Langmuir* 2022. [CrossRef]
73. Wang, L.-X.; Fu, J.-J.; Zhou, Y.; Chen, G.; Fang, C.; Lu, Z.S.; Yu, L. On-chip RT-LAMP and colorimetric detection of the prostate cancer 3 biomarker with an integrated thermal and imaging box. *Talanta* **2019**, *208*, 120407. [CrossRef] [PubMed]
74. Hongwarittorrn, I.; Chaichanawongsaroj, N.; Laiwattanapaisal, W. Semi-quantitative visual detection of loop mediated isothermal amplification (LAMP)-generated DNA by distance-based measurement on a paper device. *Talanta* **2017**, *175*, 135–142. [CrossRef] [PubMed]
75. Trinh, K.T.L.; Chae, W.R.; Lee, N.Y. Recent advances in the fabrication strategies of paper-based microfluidic devices for rapid detection of bacteria and viruses. *Microchem. J.* **2022**, *180*. [CrossRef]
76. Zhu, P.; Huang, H.-L.; Zhou, C.-X.; Xu, J.; Qiao, L.-L.; Dang, C.-Y.; Pang, J.-H.; Gao, W.-F.; Yan, X.-J. Sensitive and rapid detection of Prymnesium parvum (Haptophyceae) by loop-mediated isothermal amplification combined with a lateral flow dipstick. *Aquaculture* **2019**, *505*, 199–205. [CrossRef]
77. Jawla, J.; Kumar, R.R.; Mendiratta, S.; Agarwal, R.; Singh, P.; Saxena, V.; Kumari, S.; Boby, N.; Kumar, D.; Rana, P. On-site paper-based Loop-Mediated Isothermal Amplification coupled Lateral Flow Assay for pig tissue identification targeting mitochondrial CO I gene. *J. Food Compos. Analyst* **2021**, *102*, 104036. [CrossRef]
78. Saxena, A.; Rai, P.; Mehrotra, S.; Baby, S.; Singh, S.; Srivastava, V.; Priya, S.; Sharma, S.K. Development and Clinical Validation of RT-LAMP-Based Lateral-Flow Devices and Electrochemical Sensor for Detecting Multigene Targets in SARS-CoV-2. *Int. J. Mol. Sci.* **2022**, *23*, 13105. [CrossRef]
79. Priye, A.; Ball, C.S.; Meagher, R.J. Colorimetric-Luminance Readout for Quantitative Analysis of Fluorescence Signals with a Smartphone CMOS Sensor. *Anal. Chem.* **2018**, *90*, 12385–12389. [CrossRef]
80. Priye, A.; Ugaz, V. DNA-to-Go: A Portable Smartphone-Enabled PCR Assay Platform. *arXiv* **2016**, arXiv:1606.02252.

81. Hiltunen, J.; Liedert, C.; Hiltunen, M.; Huttunen, O.-H.; Hiitola-Keinänen, J.; Aikio, S.; Harjanne, M.; Kurkinen, M.; Hakalahti, L.; Lee, L.P. Roll-to-roll fabrication of integrated PDMS–paper microfluidics for nucleic acid amplification. *Lab a Chip* **2018**, *18*, 1552–1559. [CrossRef]
82. Liu, L.; Xu, Y.; Zhong, W.; Li, L.; Li, W.; Xiao, Q. Comparison of three terminal detection methods based on loop mediated isothermal amplification (LAMP) assay for spring viremia of carp virus (SVCV). *Turk. J. Fish. Aquat. Sci.* **2019**, *19*, 805–816. [CrossRef]
83. Zhou, Q.; Pan, J.; Mo, L.; Luo, Z.; Qin, Z.; Dai, Z.; Yi, C. Fluorescent on-site detection of multiple pathogens using smartphone-based portable device with paper-based isothermal amplification chip. *Microchim. Acta* **2022**, *189*, 1–10. [CrossRef]
84. Lalkhen, A.; McCluskey, A. Clinical tests: Sensitivity and specificity. *Contin. Educ. Anaesth. Crit. Care Pain* **2008**, *8*, 221–223. [CrossRef]
85. Roy, S.; Mohd-Naim, N.F.; Safavieh, M.; Ahmed, M.U. Colorimetric Nucleic Acid Detection on Paper Microchip Using Loop Mediated Isothermal Amplification and Crystal Violet Dye. *ACS Sens.* **2017**, *2*, 1713–1720. [CrossRef]
86. Kim, J.-H.; Yoo, I.S.; An, J.H.; Kim, S. A novel paper-plastic hybrid device for the simultaneous loop-mediated isothermal amplification and detection of DNA. *Mater. Lett.* **2018**, *214*, 243–246. [CrossRef]
87. Lin, X.; Huang, X.; Urmann, K.; Xie, X.; Hoffmann, M.R. Digital Loop-Mediated Isothermal Amplification on a Commercial Membrane. *ACS Sens.* **2019**, *4*, 242–249. [CrossRef]
88. Varsha, V.; Aishwarya, S.; Murchana, S.; Naveen, G.; Ramya, M.; Rathinasabapathi, P. Correction pen based paper fluidic device for the detection of multiple gene targets of Leptospira using Loop Mediated Isothermal Amplification. *J. Microbiol. Methods* **2020**, *174*, 105962. [CrossRef]
89. Kaewphinit, T.; Arunrut, N.; Kiatpathomchai, W.; Santiwatanakul, S.; Jaratsing, P.; Chansiri, K. Detection of *Mycobacterium tuberculosis* by Using Loop-Mediated Isothermal Amplification Combined with a Lateral Flow Dipstick in Clinical Samples. *BioMed Res. Int.* **2013**, *2013*, 1–6. [CrossRef]

chemosensors

MDPI

Article

Development of a New HiBiT Biosensor Monitoring Stability of YAP/TAZ Proteins in Cells

Liqing Wu †, Anni Ge †, Yawei Hao and Xiaolong Yang *

Department of Pathology and Molecular Medicine, Queen's University, Kingston, ON K7L 3N6, Canada; yh5@queensu.ca (Y.H.)

* Correspondence: yangx@queensu.ca; Tel.: +1-613533-6000 (ext. 75998)

† These authors contributed equally to this work.

Abstract: The Hippo signaling cascade is frequently dysregulated in a variety of cancers, such as breast cancer (BC), which is one of the most commonly diagnosed malignancies in women. Among BC subtypes, triple-negative BC (TNBC) stands out due to its poor prognosis and high metastatic potential. Despite extensive research aimed at establishing treatment options, existing therapies demonstrate limited efficacy for TNBC. Recently, it has been recognized that targeting the core components of the Hippo pathway (YAP and its paralog TAZ) is a promising strategy for developing anti-cancer treatment. However, no YAP/TAZ inhibitors have been approved by the FDA as anti-TNBC treatments, and only a few compounds have been identified that directly affect YAP and TAZ activity and stability to enhance the prospect of innovative HiBiT biosensors for monitoring of YAP and TAZ in cells. Employing these biosensors, we conducted a small-scale drug screen involving 279 compounds, leading to the identification of several small molecule inhibitors (SMIs) capable of inducing YAP/TAZ degradation in diverse TNBC cell lines. It is worth noting that some drugs may indirectly affect the protein stability following prolonged treatment, and a shorter exposure can be included in the future to identify drug candidates with more direct effects. Nevertheless, our study introduces a novel approach for assessing YAP and TAZ levels, which can have significant implications for developing anti-TNBC targeted therapies.

Keywords: HiBiT; biosensor; YAP/TAZ; triple-negative breast cancer

Citation: Wu, L.; Ge, A.; Hao, Y.; Yang, X. Development of a New HiBiT Biosensor Monitoring Stability of YAP/TAZ Proteins in Cells. *Chemosensors* **2023**, *11*, 492. https://doi.org/10.3390/chemosensors11090492

Academic Editor: Chunsheng Wu

Received: 10 July 2023
Revised: 25 August 2023
Accepted: 4 September 2023
Published: 6 September 2023

1. Introduction

Cancer remains a leading cause of mortality worldwide, accounting for 10 million deaths in 2020 [1]. Among all cancer cases, breast cancer (BC) is the most common malignancy reported in women [2]. In the year 2022 alone, BC constituted approximately 25% of all newly diagnosed cancer cases, contributing to 14% of the total cancer-related deaths in women [2]. Based on the molecular features of the tumor, BC can be classified into three main subtypes: hormone receptor positive (estrogen receptor (ER+) or progesterone receptor (PR+)), human epidermal growth factor 2 (HER2) positive (HER2+), and TNBC (low or no expression of all three biomarkers) [3,4]. TNBC patients, which accounted for 15% of all breast carcinomas, often displayed high heterogeneity and earlier disease onset [3–5]. Compared to other BC subtypes, TNBC patients exhibited poorer prognoses and more aggressive pathology, with an increased risk of local and distant relapses and metastases [6–8]. Indeed, women diagnosed with TNBC usually experienced an increased possibility of distant recurrence and lower breast-specific survival relative to other types of BC, though the recurrent rate of TNBC decreased dramatically from 4 years after diagnosis [6]. Notably, different factors may predispose patients to certain BC subtypes, including lifestyle, ethnicity, and genetic mutation. Indeed, a higher risk of TNBC was observed among parous women who had several children [7]. TNBC also disproportionately occurred in African American populations with more advanced tumors and higher histologic and nuclear grade, relative to Caucasian patients [7,9]. Moreover, over 75% of BC patients carrying the *BRCA1* mutation exhibited the TNBC phenotype [5].

Given that TNBC tumors are characterized by a low or absent expression of hormonal receptors and HER2, traditional interventions like endocrine therapy or trastuzumab prove ineffective for these patients [3,5,7,10,11]. Chemotherapy (chemo) has emerged as the standard for treating TNBC, which suppresses tumor growth by facilitating the anti-tumor immune response [3]. Chemo can be administered either as a standalone treatment or in combination with Granulocyte-MQ Colony Stimulating Factor (GM-CSF), yielding a combined therapy that has demonstrated improved clinical outcomes for BC patients [3]. However, when viewed as a group, TNBC patients exhibited worse outcomes after chemo, and diverse responses to chemo had been observed among women with TNBC [5]. Neoadjuvant therapy involving pre-surgical chemo has proven effective predominantly for a subset of TNBC patients, leaving a majority with persistent residual disease post-treatment [5,11]. Despite the varied clinical outcomes after administration, chemo is still the only systemic treatment for TNBC patients [8]. On the other hand, the development of targeted therapeutic agents, such as angiogenesis inhibitor bevacizumab, may benefit women with ER- and PR- BCs [5]. Moreover, the enzyme poly (adenosine diphosphate-ribose) polymerase (PAPR) has garnered attention as a TNBC clinical target due to its involvement in DNA repair mechanisms [5]. While introducing a PAPR inhibitor alongside chemotherapy has shown augmented tumor regression, it has not translated into prolonged overall survival [5]. Furthermore, the responses to targeted agents were diverse, in part due to high heterogeneity among TNBC patients, thereby impeding effective targeted therapy [8]. Consequently, it remains important to explore more BC treatment options with improved efficacy.

The Hippo pathway was initially discovered in *Drosophila* as the regulator of cell growth [12]. Further study revealed that the Hippo signaling pathway is highly conserved in mammals, consisting of the mammalian sterile 20-like kinase (MST), large tumor suppressor (LATS) kinases, scaffold proteins Salvador homolog 1 (SAV1), and Mps One Binder kinase activator protein 1 (MOB1) [13]. In response to extracellular or internal stimuli, such as cell–cell contact and DNA damage, the Hippo signaling pathway is activated. This activation involves phosphorylation of LATS by MST with the assistance of scaffold proteins [14]. The activated LATS induces phosphorylation of the transcriptional coactivators, yes-associated protein (YAP) and its paralog Transcriptional coactivator with PDZ-binding motif (TAZ), promoting their cytoplasmic sequestration and/or proteasomal degradation [15,16]. Conversely, the suppression of this Hippo kinase cascade results in nuclear translocation of YAP and TAZ, enabling them to regulate the expression of their target genes, such as secretory protein connective tissue growth factor (*CTGF*) and cysteine-rich angiogenic inducer 61 (*CYR61*) [17–20]. Since YAP and TAZ lack DNA binding domains, they rely on interactions with transcriptional factors such as including transcriptional enhancer associate domain (TEAD) family proteins, RUNX, and SMADs to modulate gene expression [17,18]. TEAD proteins function as the major mediators of YAP/TAZ in regulating gene expression and cell proliferative [21]. Due to interdependence between YAP and TEAD, strong selection pressure has been imposed on both genes, which displayed significant coevolution [22].

Frequent observation of elevated YAP/TAZ nuclear translocation in human cancers stems from disrupted upstream Hippo kinases, resulting in the overactivation of YAP/TAZ, a condition conducive to tumorigenesis [23,24]. In non-tumor BC cells, an overexpression of YAP or TAZ promotes tumorigenic and metastatic processes, such as cell transformation and increased cell migration [25,26]. For instance, YAP overexpression in human non-transformed mammary epithelial cells induced epithelial-to-mesenchymal transition (EMT), triggered an anti-apoptosis response, and facilitated growth factor-independent growth [25]. Similarly, TAZ is an oncoprotein that is essential for sustaining self-renewing and neoplastic properties in BC stem cells (BCSCs) [23]. The prognosis of TNBC patients was discovered to be inversely correlated with the expression level of TAZ [27]. In contrast, loss of TAZ expression led to increased cell growth and anchorage-independent growth in soft agar for

184A1 and MCF7 BC cells, respectively [26,28]. Enhanced YAP/TAZ-modulated signals were also responsible for reduced early tumor suppression and TGF-β-mediated tumorigenic phenotypes in the late stage of malignancy [29]. Together, these studies suggest that the Hippo pathway plays an important role in controlling tumor development, and the disruption of YAP/TAZ activity may be a promising route for generating efficient BC therapy.

To specifically monitor the protein of interest, bioluminescence methodologies have been frequently exploited in a non-invasive manner [30]. The discovery of the luciferase enzyme has paved the way for establishing a diverse array of biochemical assays [31–33]. While firefly and Renilla luciferases are commonly utilized for assessing the biological activities of target proteins, NanoLuc—a small, bright, and stable luciferase originating from deep-sea shrimp *Oplophorus gracilirostris*—has been discovered and developed into complementary assays, which effectively detect protein–protein interactions (PPIs) and protein activity [34–37]. Referred to as the split NanoLuc luciferase reporter system, the NanoLuc complementary assay comprises two subunits: large N-terminal BiT (LgBiT; 17.6 kDa) and small C-terminal BiT (SmBiT; 1.3 kDa) [38]. The bright luminescence was detected in the presence of its substrate furimazine when the two subunits of NanoLuc (LgBiT and SmBiT) were brought together by PPIs [38]. Recently, a novel subunit, HiBiT, which is mutated from SmBiT, has been engineered to possess an exceptionally higher affinity to LgBiT, compared to the original SmBiT [39]. To use this system for quantifying protein expression levels, the HiBiT tag is first fused with a gene of interest (GOI). The resultant HiBiT-GOI fusion proteins or protein lysates expressing HiBiT-GOI are mixed with purified LgBiT. Because of the strong affinity between HiBiT and LgBiT, an active NanoLuc is generated in tandem, emitting a brilliant luminescent signal upon the introduction of its substrate [39]. The intensity of these bioluminescent signals is believed to respond to the quantities of HiBiT-GOI protein being expressed. Therefore, this HiBiT system can be used for monitoring protein levels both in vivo and in vitro.

In this research, we applied the split NanoLuc technology and generated HiBiT-tagged YAP and TAZ proteins to monitor their stability in cells (Figure 1a). After the HiBiT bioluminescent biosensors were validated in vitro and in vivo, we subjected the optimized HiBiT biosensors to a small-scale drug screening with two libraries of 279 compounds for identifying SMIs that could degrade YAP and TAZ in TNBC cells. Multiple candidates were identified to inhibit the stability of YAP and TAZ in vivo. Our data show that our established HiBiT biosensors display great potential in monitoring YAP/TAZ stability, which is an encouraging strategy to discover targeted therapies for TNBC treatment.

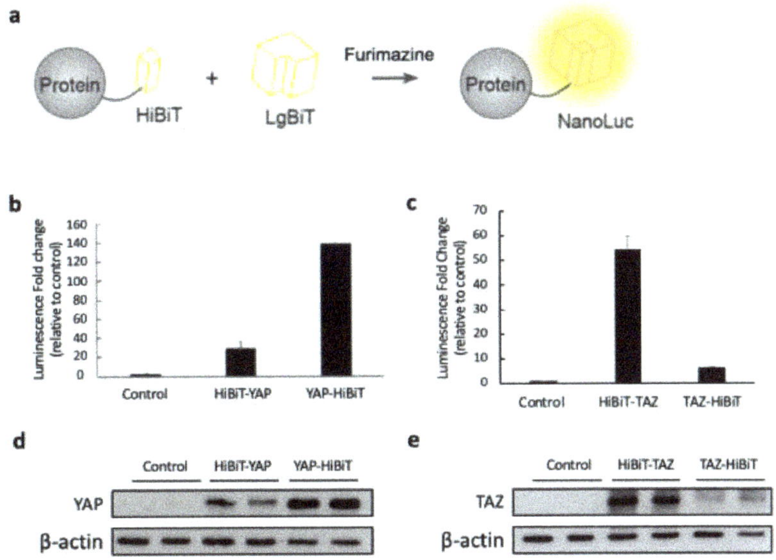

Figure 1. Establishing HiBiT-tagged YAP and TAZ biosensors monitoring protein stability. (**a**) Schematic representation of the HiBiT biosensors for detecting the YAP and TAZ protein levels. (**b**,**c**) The HiBiT sequence was attached at either the N- or C-terminus of YAP and TAZ, and each construct was subcloned into the pcDNA3.1-3xFLAG vector. These plasmids were individually transfected into the HEK293 cells in the 12-well plate. The empty pcDNA3.1 vector was transfected into the cells as a negative control. After 48 h post-transfection, protein lysates were extracted with 1× passive lysis buffer (PLB). The luciferase assay was performed by combining 2.5 μg of each lysate and 0.5 μg of purified LgBiT-His protein. The mixture was then incubated before the addition of the substrate. The luminescent fold change was determined by calculating how much the bioluminescence from the HiBiT biosensor increased compared to the empty vector control after the addition of LgBiT-His protein and NanoLuc substrate. The data are shown as mean + SD (n = 2 technical replicates) (**d**,**e**) The expression level of each construct was examined by WB using 10 μg of each cell lysate. β-actin was detected as the internal loading control.

2. Materials and Methods

2.1. Cell Culture

MDA-MB-231 (human mammary carcinoma), HEK293 (human embryonic kidney cells), and HEK293T (human embryonic kidney cells) cells were cultured in Dulbecco's Modified Eagle's Medium (DMEM; Sigma#D6429, Oakville, ON, Canada) containing 10% fetal bovine serum (FBS) and 1% penicillin/streptomycin (P/S) (Invitrogen, Burlington, ON, Canada). BT-549 (human mammary carcinoma) cells were cultured in RPMI-1640 Medium (Sigma #8758) containing 10% FBS, 1% P/S, and 0.023 U/mL of insulin. MDA-MB-231 with YAP and TAZ knockout (KO) were established as described [40] and cultured in the same medium as previously mentioned. All cells were kept at 37 °C with 5% CO_2.

2.2. Plasmid Construction

HiBiT-tagged full-length (FL) YAP and TAZ were prepared by amplifying the FL YAP and TAZ genes using PCR from cDNA and inserting into the pcDNA3.1 vector with either N- or C-terminal HiBiT (sequence: VSGWRLFKKIS). Plasmids for producing GST fusion proteins were constructed by cloning LgBiT into the BamHI/NotI sites of pGEX-4T1 vector. See supporting information for primers used in cloning (Table S1).

2.3. *LgBiT Protein Purification from Bacteria*

The BL21 *Escherichia coli* strain was transformed and utilized to purify LgBiT-His fusion protein according to the previously established protocols with minor adjustments in our lab [41]. After inoculation, when the OD_{600} reached 0.7–0.9, protein expression was induced with 0.4 mM of IPTG (isopropyl β-D-1 thiogalactopyranoside) at 37 °C incubation for 4–5 h. Bacterial cells were spun down, washed, and lysed by sonication. His-LgBiT fusion proteins were purified by Ni-NTA beads, visualized, and quantified with BSA by SDS-PAGE and Coomassie blue staining. The purified protein was stored at −80 °C.

2.4. *Protein Extraction and Western Blot Analysis*

Cells were cultured to 80–90% confluence before protein extraction. Either RIPA (50 mM of Tris HCl, 150 mM of NaCl, 1.0% (*v*/*v*) NP-40, 0.5% (*w*/*v*) Sodium Deoxycholate, 1.0 mM of EDTA, 0.1% (*w*/*v*) SDS, and 0.01% (*w*/*v*) sodium azide, pH = 7.4) with the complete EDTA-free protease inhibitor cocktail tablet (Roche, Mississauga, ON, Canada) or 1× passive lysis buffer (PLB) (Promega, Madison, WI, USA) was utilized to lyse the cells. The cell lysates were collected and centrifuged at 12,000× *g* at 4 °C for 10 min to obtain the supernatant, and the DC protein assay kit (Bio-Rad, Mississauga, ON, Canada) was used to measure the protein concentration. The protein expression levels were analyzed by Western blot (WB) using 10–30 μg of protein lysates for each sample, which was boiled with 5× loading buffer (4.2% β-Mercaptoethanol) at 95 °C for 5 min. SDS-PAGE (10–12%) was performed to resolve the proteins, which were transferred to nitrocellular membrane (Bio-Rad, Mississauga, ON, Canada). TBST buffer ((20 mM of Tris, 150 mM of NaCl, 0.1% Tween 20) containing 5% skimmed milk was used to block the membranes for one hour at room temperature. The blocked membranes were probed with primary antibodies at 4 °C overnight or for 1 h at room temperature and washed with TBST buffer before probing with secondary antibodies (1:4000 dilution) for 15 min at room temperature. The protein bands were visualized by chemiluminescence reagent (Bio-Rad). The antibodies were stripped with stripping buffer and re-probed with anti- β-actin antibody. The antibodies and corresponding dilutions are as follows: mouse monoclonal anti-FLAG (F1804, Sigma, 1:1000) antibody; rabbit polyclonal anti-YAP (sc-15407, Santa Cruz, Dallas, TX, USA, H125, 1:1000); mouse monoclonal anti-TAZ (560235, BD Biosciences, Mississauga, ON, Canada, 1:1000); mouse monoclonal anti-β-actin (A5441, Sigma, 1:1000).

2.5. *NanoGlo Luciferase Assay*

Purified His-LgBiT protein and cell lysates expressing HiBiT-tagged YAP or TAZ were mixed and incubated for at least 10 min on ice before measuring the luminescence with 1:50 diluted Nano-Glo Live Reagent (Promega) using GloMax Navigator Detection System (Promega).

2.6. *Small-Scale Drug Screen*

The drug screening with 279 compounds was performed using bioactive lipid and custom compound libraries. The experimental conditions were as follows. The MDA-MB-231 cells with YAP and TAZ KO were either transfected with the plasmid expressing YAP-HiBiT or HiBiT-TAZ in the 6-well plate. After 5 h post-transfection, the cells were trypsinized and counted for seeding into the 96-well plates with 40–50% confluency the next day. The cells were then treated with the drug candidates at a concentration of 100 μM for 48 h. After that, the bioluminescence and cell viability were measured and compared against the 1% DMSO-treated control samples for calculating the fold changes. A complete list of drug candidates can be seen in Table S2.

2.7. *Cell Viability Assay*

After treating the transfected YAP and TAZ KO MDA-MB-231 cells with different drugs, the plates and the contents were equilibrated to room temperature for approximately 30 min. A volume of CellTiter-Glo 2.0 Reagent (Promega) equal to the volume of cell culture medium per well was added and mixed for 2 min on an orbital shaker to induce cell lysis. The plate was then incubated for 10 min before examining the luminescent signal using GloMax Navigator Detection System (Promega).

2.8. Densitometric Analysis by ImageJ

After the MDA-MB-231 cells were treated with either of the 67 initial hits, cell lysates were extracted using RIPA buffer and used for WB. For protein band quantitation, the WB images were analyzed using the Gel Analyzer tool of ImageJ (NIH, Version 1.53t). The band intensity was measured for YAP, TAZ, and β-actin (loading control) and subjected to background subtraction to calculate the net signal. The protein expression for each band was normalized by determining the ratio of a net band signal over the net loading control of that lane. The YAP and TAZ expression ratios of the drug-treated samples were reported by comparing with the DMSO-treated controls (Table S4).

2.9. RNA Extraction and Quantitative Real-Time PCR

MDA-MB-231 cells were seeded in 6-well plates with 50–60% confluency the next day and then treated with either avanafil (100 μM), beta-carotene (20 μM), dipyridamole (50 μM), fluvastatin (10 μM), pterostilbene (100 μM), or XEN445 (100 μM). 48 h post-treatment, RNA was extracted using RNAZol RT reagent (200 μL/well) based on the manufacturer's instruction. A quantitative real-time PCR (qRT-PCR) was performed using 50 ng of RNA/reaction by the SuperScript III Platinum SYBR Green One-Step qRT-PCR Kit (Invitrogen). The 18S rRNA was measured as an internal control.

2.10. Statistical Analysis

For all figures, data are shown as mean + or ±SD. An ANOVA and a Student's *t*-test (two-tailed) were performed for the statistical analysis. A p-value < 0.05 was considered statistically significant.

3. Results

3.1. Design and Establishment of HiBiT Biosensors

To monitor the stability of YAP/TAZ, we established the YAP/TAZ-HiBiT biosensors in which a HiBiT peptide was fused with the N or C-terminus of a YAP or TAZ protein. The YAP/TAZ-HiBiT was expected to have a high affinity to LgBiT, forming an active NanoLuc, which could emit strong bioluminescence in the presence of its substrate furimazine (Figure 1a). HEK293T cells were transfected with the plasmids expressing either of the constructs or with an empty pcDNA3.1 vector as a negative control.

The luminescence of each HiBiT biosensor was measured by luciferase assays using purified His-tagged LgBiT from bacteria and cell lysates containing HiBiT-tagged YAP/TAZ. The combination of His-LgBiT and HiBiT-tagged YAP/TAZ displayed high luminescent increases compared to the negative control (Figure 1b,c). Interestingly, attaching the HiBiT tag to the C-terminus of YAP resulted in a stronger luminescent signal, while higher luminescence was observed when the HiBiT tag was fused to the N-terminus of TAZ (Figure 1b,c). The protein levels of the HiBiT-tagged YAP/TAZ were detected by Western blot, suggesting that the high bioluminescence of YAP-HiBiT and HiBiT-TAZ resulted from the high protein overexpression in transfected cells (Figure 1d,e).

To confirm that the substantial bioluminescent increase indeed resulted from HiBiT and LgBiT fusion protein complementation, two controls were generated by transfecting either the empty pcDNA3.1 vector or the plasmid expressing FLAG-YAP/TAZ into HEK293T cells. High protein expression of YAP/TAZ was observed in transfected cells with FLAG-YAP/TAZ or HiBiT-YAP/TAZ-expressing plasmids, whereas low protein levels were detected in the empty vector control (Figure 2a). However, the strong bioluminescent signals were only exhibited in cells transfected with HiBiT biosensor-expressing plasmids (Figure 2b). Our data indicate that the HiBiT BS were established and optimized by fusing the HiBiT to the C-terminus of YAP and the N-terminus of TAZ, which emitted strong bioluminescence after the complementation with LgBiT-His and could be useful for monitoring protein stability.

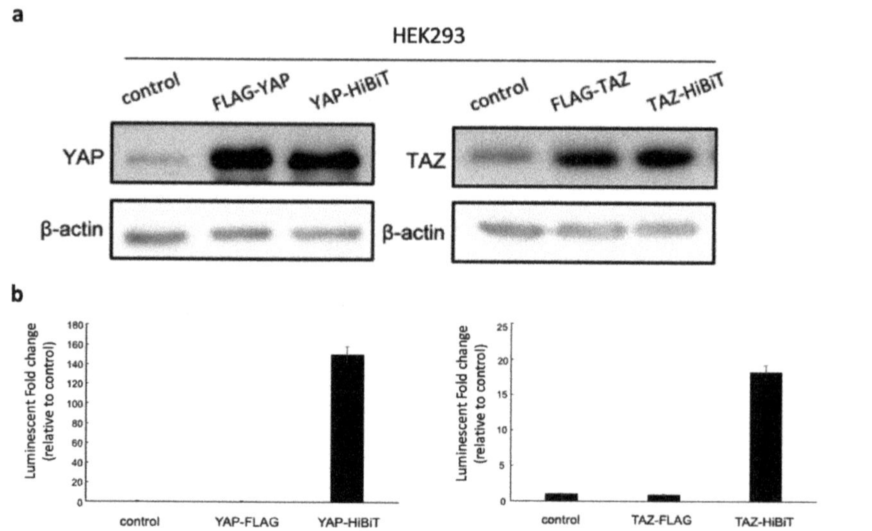

Figure 2. (**a**) The protein expression of HiBiT-tagged YAP/TAZ in mammalian cells. The HEK293 cells were either transfected with pcDNA3.1 empty vector as the control, FLAG-YAP/TAZ or HiBiT-YAP/TAZ expressing plasmids. Forty-eight hours post-transfection, cells were lysed using RIPA buffer, and 10 µg of each lysate was used for detecting YAP/TAZ levels by WB. Lysates from cells transfected with empty vector and FLAG-YAP/TAZ-expressing plasmids were included as negative and positive controls for YAP/TAZ expression. The expression of β-actin was detected as an internal control. (**b**) After the protein extraction, 5 µg of each cell lysate was combined with 1 µg of LgBiT-His protein purified from *Escherichia coli*. The mixture was incubated for 10 min before adding NanoGlo substrate for luminescent measurement. The relative luminescence of each lysate was normalized to the signal from the empty vector control to calculate the fold increase. Substantial luminescence was found exclusively in HiBiT biosensors. The data are represented as the mean + SD (n = 3 technical replicates).

3.2. *Validation of HiBiT-Tagged YAP/TAZ Biosensor*

Next, we sought to validate the HiBiT biosensors by in vitro luciferase assays. An increasing amount of plasmids expressing the optimized HiBiT-tagged biosensors was transfected into HEK293T cells. As expected, the bioluminescent signals increased as the cells were transfected with more plasmids, demonstrating that only 0.5 µg of plasmids was sufficient to produce strong luminescent signals (Figure 3a,c). An elevated expression level of YAP-HiBiT or HiBiT-TAZ coincided with the increased amount of plasmid used for transfection (Figure 3b,d). Moreover, a linear relationship was observed between the luminescent intensity and the number of transfected cells subjected to bioluminescent analysis (Figure 3e,f). These results suggest that the luminescent intensity of HiBiT biosensors was affected by the expression levels of HiBiT-tagged YAP/TAZ, indicating that our stable HiBiT biosensors could be applied for drug screening.

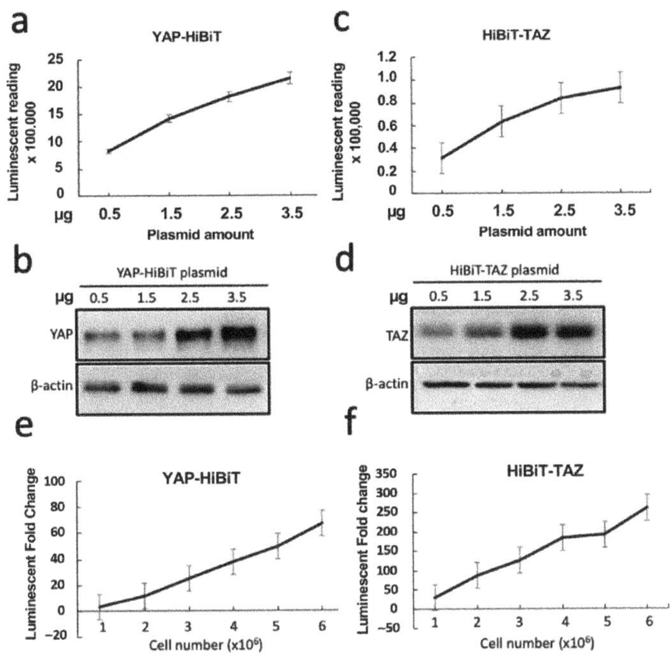

Figure 3. Validation of HiBiT BSs. (**a**–**d**) The plasmids expressing YAP-HiBiT or HiBiT-TAZ were transfected into the HEK293 cells in increasing amounts. After 48 h post-transfection, the cells were lysed with 1× PLB and subjected to luciferase assays (n = 4 biological replicates) (**a**,**c**) and WB (**b**,**d**) to compare the HiBiT BS activity and protein expression. (**e**,**f**) Detection of luminescent signal with increasing numbers of transfected HEK293 cells (n = 4 biological replicates). The cells were transfected with YAP-HiBiT- or HiBiT-TAZ-expressing plasmids in a six-well plate. After 48 h post-transfection, the transfected cells were counted and seeded into each well of a 96-well plate with increasing cell numbers, as indicated in the x-axis. The luciferase activity of each well was measured 16 h later. All bioluminescent data are represented as mean ± SD.

After we confirmed that the luminescent intensity was dependent on the protein levels of HiBiT-tagged YAP/TAZ, the HiBiT biosensors were further validated by a known YAP/TAZ inhibitor, docosahexaenoic acid (DHA), which was reported to affect the protein levels of YAP and TAZ [42]. The HEK293T cells transfected with either YAP-HiBiT or HiBiT-TAZ-expressing plasmids were treated with DHA over 4 days. The protein lysates were obtained at 0, 0.5, 1-, 2-, 3-, and 4-days post-treatment, followed by WB. Since the literature had discovered that DHA could reduce the level of β-actin in cells [43], ponceau S staining was shown instead as an indicator of equal protein loading for WB. Decreased HiBiT-tagged YAP and TAZ expression was detected by WB in a time-dependent manner after treatment (Figure 4a,b). This reduced expression of YAP/TAZ was not due to the degradation of the HiBiT tag by DHA because DHA also reduced endogenous YAP/TAZ (Figure S1), as previously reported [42]. Next, we examined the luciferase activity using the same lysates at later time points, as the most prominent changes in YAP/TAZ stability were observed after 3 and 4 days of DHA treatment. Indeed, significant reductions in the bioluminescent signal were observed on days 3 and 4 post-treatment, compared to the luminescence at day 0 (Figure 4d,e). Furthermore, to ensure that the reduced bioluminescence was not due to the direct HiBiT degradation, a control was created by directly fusing the HiBiT sequence to the C-terminus of LgBiT. The HEK293T cells transfected with LgBiT-HiBiT-expressing plasmids were treated with DHA using the same condition. Unlike previous results, DHA did not reduce the stability of LgBiT-HiBiT and its bioluminescence after prolonged treatment (Figure 4c,f). Taken together, these results demonstrate the high specificity of our HiBiT biosensors in monitoring the stability of YAP and TAZ and show that the HiBiT biosensors are applicable for the drug screening campaign to identify drugs that induce YAP/TAZ degradation.

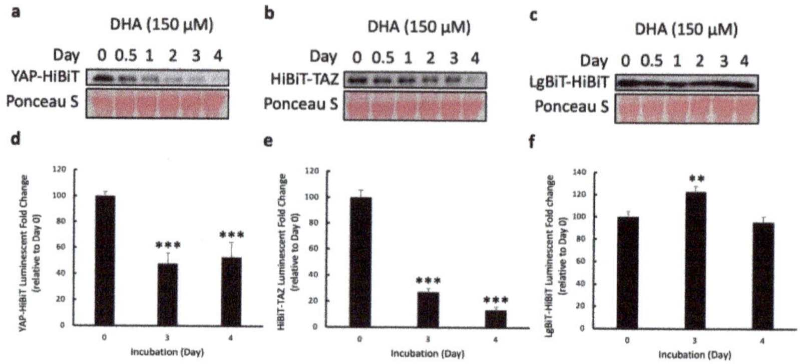

Figure 4. Validation of HiBiT BS with a known YAP/TAZ inhibitor, DHA. (**a–c**) The HEK293T cells were transfected with either HiBiT-TAZ-, YAP-HiBiT-, or LgBiT-HiBiT-expressing plasmids. After 5 h post-transfection, each transfected cell line was counted, and triplicates of 2×10^5 cells/well for each cell line were seeded into the 12-well plates. The next day, the cells were treated with 150 μM of DHA. The stability of YAP-HiBiT (**a**), HiBiT-TAZ (**b**), and LgBiT-HiBiT (**c**) was examined by WB with 10 μg of cell lysates at 0, 0.5, 1, 2, 3, 4 days post-treatment. Ponceau S staining was shown as the loading control. (**d–f**) The luciferase assays were conducted using 2.5 μg of each cell lysate at 0-, 3-, and 4-days post-treatment to assess the prolonged effect of DHA in YAP and TAZ stability. The luminescent fold change was measured by calculating how much the bioluminescence from the treated lysates changed relative to the luminescent signal at day 0 of treatment. The data are presented as mean + SD 2 independent experiments, ** $p < 0.01$, *** $p < 0.001$).

3.3. Identification and Validation of SMIs Targeting HiBiT-Tagged YAP/TAZ Biosensors in TNBC Cells

A small-scale drug screen with two libraries consisting of 279 small molecules, including the bioactive lipid library and custom compound library, was conducted using our validated HiBiT biosensors to find drugs that could degrade YAP and TAZ in TNBC cells. The YAP and TAZ KO MDA-MB231 cells expressing YAP-HiBiT or HiBiT-TAZ were treated with each drug from the libraries. The luminescence and cell viability were determined after 48 h post-treatment. After excluding the compounds that were too toxic for the cells (cell viability fold change < 0.2), a total of 67 compounds that displayed notable HiBiT signal fold changes (<0.5 or >1.8) were selected as the initial hit candidates for subsequent validations (Figure S2).

To explore the effect of the initial hits on YAP/TAZ stability, these drugs were used to treat MDA-MB-231 cells for assessing the endogenous levels of YAP and TAZ by WB after 48 h of treatment. Relative to DMSO controls, 22 out of 67 candidates greatly reduced the expression of both YAP and TAZ (expression ratio < 0.4), while eight additional drugs seemed to preferentially inhibit TAZ protein level (Figure 5; Table S4). To further verify the impact of those candidates on YAP/TAZ stability, MDA-MB-231 cells were treated with these drugs again to determine the expression levels of YAP and TAZ at different times post-treatment. Finally, six drugs were found to decrease the endogenous levels of YAP/TAZ in a time-dependent manner, including small molecules from plant extracts (pterostilbene and beta-carotene), lipid-related (fluvastatin and XEN445), and phosphodiesterase (PDE) inhibitors (avanafil and dipyridamole) (Figure 6a–f).

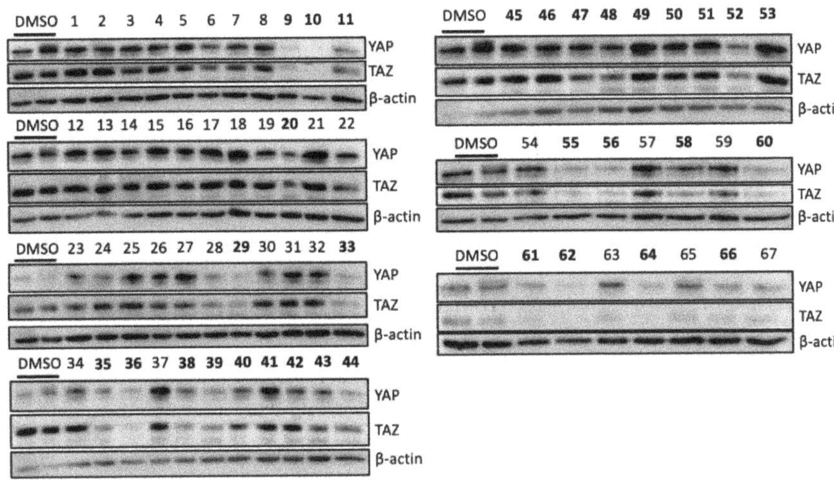

Figure 5. The effect of initial hit candidates on endogenous levels of YAP/TAZ in the TNBC cells. MDA-MB-231 cells were treated with 100 µM of 67 initial hits separately for 48 h. To compare the YAP and TAZ stability after each treatment, cells cultured with only DMSO were included as controls. Cell lysates were harvested using RIPA buffer and used for WB to assess the protein levels of YAP and TAZ. Among the initial 67 candidates, 22 compounds affected the YAP and TAZ stability in TNBC cells, and TAZ expression was specifically inhibited by additional eight drugs. The bolded numbers indicated that the corresponding drugs decreased the protein levels of both YAP and TAZ or preferentially inhibited TAZ expression.

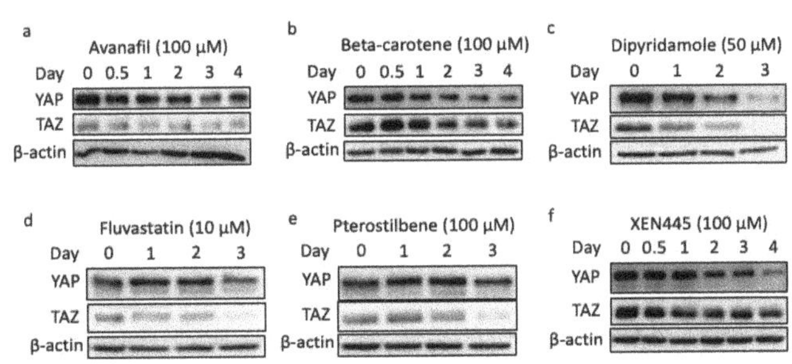

Figure 6. The YAP/TAZ stability in MDA-MB-231 cells treated with top candidates, including avanafil (**a**), beta-carotene (**b**), dipyridamole (**c**), fluvastatin (**d**), pterostilbene (**e**), and XEN445 (**f**).

To examine whether the loss of YAP/TAZ expression was due to a reduction in mRNA level rather than protein degradation, RNA was extracted from MDA-MB-231 cells that were pre-treated with either of the six candidates for 48 h. Based on our results, all of the drugs had no considerable inhibitory impact on the mRNA levels of YAP and TAZ (Figure 7). In addition, four out of the six top candidates significantly increased the level of TAZ mRNA relative to DMSO-treated controls (Figure 7). Moreover, another TNBC cell line, BT-549, was treated with the identified six drugs along with a known inhibitor, DHA, to avoid any cell line-specific effect. Consistent with our previous findings, reduced YAP and TAZ protein levels were observed in the treated cells relative to the control (Figure 8). Altogether, these data reveal multiple small molecules that affected the stability of YAP and TAZ in TNBC cells through drug screening using our HiBiT biosensors. Hence, the use of HiBiT biosensors exerts great potential for assessing the levels of YAP and TAZ proteins in vivo and identifying pre-existing compounds that regulate the stability of YAP and TAZ.

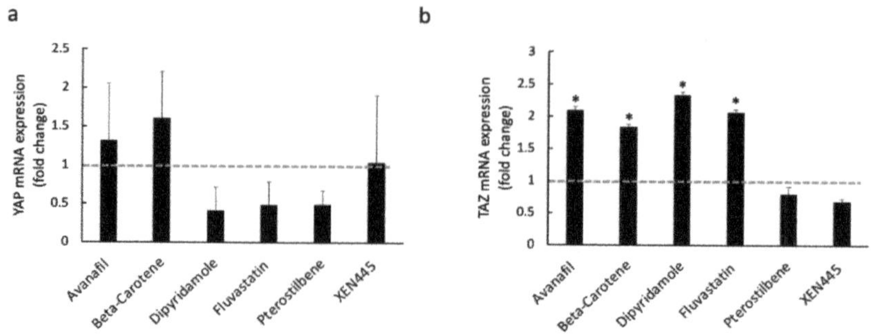

Figure 7. The mRNA levels of YAP (**a**) and TAZ (**b**) in MDA-MB-231 cells treated with either avanafil (100 μM), beta-carotene (20 μM), dipyridamole (50 μM), fluvastatin (10 μM), pterostilbene (100 μM), or XEN445 (100 μM). After 48 h post-treatment, RNA was extracted using RNAZol RT, and RT-qPCR was performed to assess the mRNA levels of YAP and TAZ. The cells treated with 1% DMSO for two days were included as the control. The mRNA expression fold change was calculated by measuring the difference between drug-treated and DMSO-treated samples. Based on ANOVA analysis of our data, the mRNA expression of YAP and TAZ is not significantly inhibited ($p \geq 0.05$) by either of the drugs in MDA-MB-231 (qRT-PCR; mean + SD, n = 2 biological replicates (2 replicates for each biological sample), *t*-test: * $p < 0.05$).

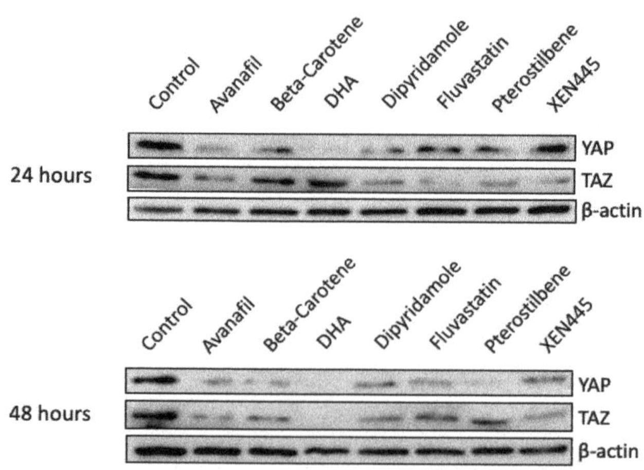

Figure 8. Validation of top candidates in BT549. Seven drugs, including avanafil (100 μM), beta-carotene (20 μM), DHA (150 μM), dipyridamole (50 μM), fluvastatin (10 μM), pterostilbene (100 μM), and XEN445 (100 μM), were used to treat another TNBC cell line to avoid cell line-specific effects. The protein lysate was collected every 24 h for a total of 2 days using RIPA buffer, and 10–30 μg of each lysate was subjected to WB. Cells treated with 1% of DMSO were included as the negative control.

4. Discussion

BC accounted for approximately 25% of all cancer cases among Canadian women in 2022 [2]. One of the BC subtypes is TNBC, which is characterized by high heterogeneity and low overall survival [7,8]. The existing medical interventions have yielded limited enhancements in the health of those affected, particularly among women diagnosed with TNBC. Adding to this challenge, TNBC patients lacking or expressing low levels of hormonal receptors and HER2, remain ineligible for the benefits of most targeted therapy drugs [3,5,7,10,11,44]. Alternatively, emerging novel therapeutics such as molecular-targeted therapy may exhibit significant efficiency in treating women with BCs [3]. Manipulating critical signaling pathways that underlie cancer progression presents a viable strategy for crafting highly targeted treatments. A pertinent example is the Hippo pathway, which is crucial not only in normal physiological functions, but also in tumorigenesis [14,43,45]. In BC cells, YAP and TAZ are often overexpressed to facilitate disease progression [23–26]. Consequently, targeting YAP and TAZ has emerged as a promising avenue for developing anti-cancer therapy. Many studies are centered around evaluating the dynamics of YAP/TAZ-TEAD interaction [17,18,21]. Currently, three drugs targeting YAP/TAZ-TEAD interaction are undergoing phase I clinical trials, along with an antisense oligonucleotide targeting YAP1 that completes the phase I clinical trial [46–49]. However, less attention is devoted to assessing the stability of YAP and TAZ, despite their potential susceptibility to degradation for developing anti-tumor treatment.

In this study, we have developed novel ultra-stable HiBiT biosensors for monitoring the levels of YAP and TAZ in cells. These differ from the conventional split NanoLuc assay that depends on the interaction of target proteins for the complementation of the SmBiT and LgBiT fragments, which do not spontaneously interact with each other [50]. Many studies have exploited this conventional approach to monitor the activity of YAP/TAZ and their binding partners, mainly TEAD [51]. Although several small molecule disruptors of YAP-TEAD PPI have been identified, most drugs indirectly regulate YAP/TAZ activity by inhibiting autopalmitoylation of TEAD, which affects protein stability [51]. To date, no small molecules have been discovered to exert a direct effect on both YAP and TAZ expression levels. Alternatively, relying on the remarkably high affinity of HiBiT to LgBiT, the levels of YAP and TAZ can be readily detected by our HiBiT biosensors through luminescent analysis upon the addition of LgBiT and their substrate. Two libraries of 276 small molecule compounds were screened using these HiBiT biosensors for identifying potential inhibitors that could degrade YAP/TAZ in TNBC cells. After validation, we uncovered six drugs, including pterostilbene, fluvastatin, beta-carotene, Avanafil, XEN445, and dipyridamole, which significantly reduced the levels of YAP and TAZ in cells.

One of the identified drugs from the screening was pterostilbene (trans-3,5-dimethoxy-4′-hydroxystilbene), which was already revealed to suppress the growth of TNBC tumors by facilitating the activation of extracellular signal-regulated kinase (ERK) 1/2 [52,53]. However, the link of this chemical to the Hippo pathway remains elusive. Our results indicate that pterostilbene can induce the degradation of downstream Hippo transducers, YAP and TAZ, although the mechanistic details require further study. Other positive hits, such as avanafil and dipyridamole, are different PDE inhibitors [54,55]. The former is frequently utilized to treat erectile dysfunction and pulmonary arterial hypertension, and dipyridamole is often prescribed for ischemic disorders [54,55]. Tumor-suppressive effects were observed in xenograft mice bearing TNBC cells treated with dipyridamole [55]. Moreover, accelerated YAP phosphorylation and inactivation can be driven by fluvastatin, which is administered for hypercholesterolemia treatment and cardiovascular disease prevention [56]. A potent and selective endothelial lipase inhibitor, XEN445, which increased plasma HDL-cholesterol concentration in mice, was also discovered in our screen [57]. While no connection is found between XEN445 and cancer therapy, there is evidence that relates lipid signaling to the Hippo pathway [58].

Due to the nature of our HiBiT biosensors, potential caveats may occur, and the major limitation of our study is that the degradation of YAP/TAZ after 48 h of treatment may be indirectly caused by drugs. Thus, a shorter treatment may be needed to identify more candidate drugs, exerting a direct effect on YAP/TAZ stability.

In summary, we have shown that our newly developed HiBiT biosensors hold substantial values in investigating the YAP/TAZ levels in cells. By applying the HiBiT biosensors, the drug screening is easily performed to identify pre-existing compounds affecting the stability of YAP/TAZ, which are key targets for anti-tumor therapy. Since no drugs targeting YAP and TAZ have been approved by FDA for TNBC treatment, our data reveal the powerful capability of HiBiT biosensors in the discovery of YAP/TAZ inhibitors to increase the likelihood of finding a successful targeted therapy for TNBC patients.

Supplementary Materials: The following supporting information can be downloaded at: https://www.mdpi.com/article/10.3390/chemosensors11090492/s1, Figure S1: Degradation of endogenous YAP/TAZ after DHA treatment; Figure S2: Small-scale screening of YAP/TAZ inhibitors with a bioactive lipid library and custom compound library; Table S1: List of primers for cloning; Table S2: List of bioactive lipid library for small-scale screen; Table S3: List of custom compound library for small-scale screen; Table S4: Quantitation of YAP and TAZ protein levels in TNBC cells treated with either of the initial hits by densitometric analysis.

Author Contributions: Conceptualization, methodology, investigation, L.W., A.G., Y.H. and X.Y.; writing—original draft preparation and editing, A.G.; supervision, project administration, and funding acquisition, X.Y. All authors have read and agreed to the published version of the manuscript.

Funding: This work was supported by grants from Canadian Institute of Health Research (CIHR) (grant number: 148629 and 186142) and Canadian Cancer Society (CCS)/Canadian Breast Cancer Foundation (CBCF) (grant number: 369676).

Institutional Review Board Statement: Not applicable.

Informed Consent Statement: Not applicable.

Data Availability Statement: Not applicable.

Acknowledgments: We would like to thank Dongsheng Tu in his help in our statistical analysis.

Conflicts of Interest: The authors declare no conflict of interest.

References

1. Cancer. Available online: https://www.who.int/news-room/fact-sheets/detail/cancer (accessed on 19 December 2022).
2. Canadian Cancer Society. Available online: https://cancer.ca/en/cancer-information/cancer-types/breast/statistics (accessed on 19 December 2022).
3. Barzaman, K.; Karami, J.; Zarei, Z.; Hosseinzadeh, A.; Kazemi, M.H.; Moradi-Kalbolandi, S.; Safari, E.; Farahmand, L. Breast cancer: Biology, biomarkers, and treatments. *Int. Immunopharmacol.* **2020**, *84*, 106535. [CrossRef] [PubMed]
4. Waks, A.G.; Winer, E.P. Breast Cancer Treatment: A Review. *JAMA* **2019**, *321*, 288–300. [CrossRef] [PubMed]
5. Foulkes, W.D.; Smith, I.E.; Reis-Filho, J.S. Triple-negative breast cancer. *N. Engl. J. Med.* **2010**, *363*, 1938–1948. [CrossRef] [PubMed]
6. Dent, R.; Trudeau, M.; Pritchard, K.I.; Hanna, W.M.; Kahn, H.K.; Sawka, C.A.; Lickley, L.A.; Rawlinson, E.; Sun, P.; Narod, S.A. Triple-negative breast cancer: Clinical features and patterns of recurrence. *Clin. Cancer Res.* **2007**, *13*, 4429–4434. [CrossRef] [PubMed]
7. Phipps, A.I.; Chlebowski, R.T.; Prentice, R.; Tiernan, A.M.; Wactawski-Wende, J.; Kuller, L.H.; Adams-Campbell, L.L.; Lane, D.; Stefanick, M.L.; Vitolins, M.; et al. Reproductive history and oral contraceptive use in relation to risk of triple-negative breast cancer. *J. Natl. Cancer Inst.* **2011**, *103*, 470–477. [CrossRef]
8. Collignon, J.; Lousberg, L.; Schroeder, H.; Jerusalem, G. Triple-negative breast cancer: Treatment challenges and solutions. *Breast Cancer* **2016**, *8*, 93–107.
9. Morris, G.J.; Naidu, S.; Topham, A.; McCue, P.; Schwartz, G.; Rosenberg, A.; Mitchell, E.P. Differences in breast carcinoma characteristics in newly diagnosed African-American and Caucasian patients: A single-institution compilation compared with the national cancer institute SEER database. *Cancer* **2007**, *110*, 876–884. [CrossRef]
10. Denkert, C.; Liedtke, C.; Tutt, A.; Minckwitz, G. Molecular alternations in triple-negative breast cancer-the road to new treatment strategies. *Lancet* **2017**, *389*, 2430–2442. [CrossRef]
11. Liedtke, C.; Mazouni, C.; Hess, K.R.; Andre, F.; Tordai, A.; Mejia, J.A.; Symmans, W.F.; Gonzalez-Angulo, A.M.; Hennessy, B.; Green, M.; et al. Response to neoadjuvant therapy and long-term survival in patients with triple-negative breast cancer. *J. Clin. Oncol.* **2008**, *26*, 1275–1281. [CrossRef]
12. Xu, T.; Wang, W.; Zhang, S.; Stewart, R.A.; Yu, W. Identifying tumor suppressors in genetic mosaics: The Drosophila lats gene encodes a putative protein kinase. *Development* **1995**, *121*, 1053–1063. [CrossRef]
13. Wu, L.; Yang, X. Targeting the Hippo Pathway for Breast Cancer Therapy. *Cancers* **2018**, *10*, 422. [CrossRef] [PubMed]
14. Halder, G.; Johnson, R.L. Hippo signaling: Growth control and beyond. *Development* **2011**, *138*, 9–22. [CrossRef] [PubMed]
15. Liu, C.; Zha, Z.; Zhou, X.; Zhang, H.; Huang, W.; Zhao, D.; Li, T.; Chan, S.W.; Lim, C.J.; Hong, W.; et al. The Hippo tumor pathway promotes TAZ degradation by phosphorylating a phosphodegron and recruiting the $SCF^{\beta\text{-TrCP}}$ E3 ligase. *J. Biol. Chem.* **2010**, *285*, 37159–37169. [CrossRef]
16. Zhao, B.; Li, L.; Tumaneng, K.; Wang, C.; Guan, K. A coordinated phosphorylating by LATS and CK1 regulates YAP stability through $SCF^{\beta\text{-TrCP}}$. *Genes. Dev.* **2010**, *24*, 72–85. [CrossRef] [PubMed]
17. Yu, F.X.; Guan, K.L. The Hippo pathway: Regulators and regulations. *Genes. Dev.* **2013**, *27*, 355–371. [CrossRef] [PubMed]
18. Holden, J.K.; Cunningham, C.N. Targeting the Hippo pathway and cancer through the TEAD family of transcription factors. *Cancers* **2018**, *10*, 81. [CrossRef] [PubMed]

19. Zhao, B.; Ye, X.; Yu, J.; Li, L.; Li, W.; Li, S.; Yu, J.; Lin, J.D.; Wang, C.; Chinnaiyan, A.M.; et al. TEAD mediates YAP-dependent gene induction and growth control. *Genes. Dev.* **2008**, *22*, 1962–1971. [CrossRef]
20. Lai, D.; Ho, K.C.; Hao, Y.; Yang, X. Taxol resistance in breast cancer cells is mediated by the hippo pathway component TAZ and its downstream transcriptional targets Cyr61 and CTGF. *Cancer Res.* **2011**, *71*, 2728–2738. [CrossRef]
21. Zhou, Z. Targeting Hippo pathway by specific interruption of YAP-TEAD interaction using cyclic YAP-like peptides. *FASEB J.* **2015**, *29*, 724–732. [CrossRef]
22. Hilman, D.; Gat, U. The evolutionary history of YAP and the Hippo/YAP pathway. *Mol. Biol. Evol.* **2011**, *28*, 2403–2417. [CrossRef]
23. Cordenonsi, M.; Zanconato, F.; Azzolin, L.; Forcato, M.; Rosato, A.; Frasson, C.; Inui, M.; Montagner, M.; Parenti, A.R.; Poletti, A.; et al. The Hippo transducer TAZ confers cancer stem cell-related traits on breast cancer cells. *Cell* **2011**, *147*, 759–772. [CrossRef] [PubMed]
24. Calvo, F.; Ege, N.; Grande-Garcia, A.; Hooper, S.; Jenkins, R.P.; Chaudhry, S.I.; Harrington, K.; Williamson, P.; Moeendarbary, E.; Charras, G.; et al. Mechanotransduction and YAP-dependent matrix remodelling is required for the generation and maintenance of cancer-associated fibroblasts. *Nat. Cell Biol.* **2013**, *15*, 637–646. [CrossRef] [PubMed]
25. Overholtzer, M.; Zhang, J.; Smolen, G.A.; Haber, D.A. Transforming properties of *YAP*, a candidate oncogene on the chromosome 11q22 amplicon. *Proc. Natl. Acad. Sci. USA* **2006**, *103*, 12405–12410. [CrossRef] [PubMed]
26. Chan, S.W.; Lim, C.J.; Guo, K.; Ng, C.P.; Lee, I.; Hunziker, W.; Zeng, Q.; Hong, W. A role for TAZ in migration, invasion, and tumorigenesis of breast cancer cells. *Cancer Res.* **2008**, *68*, 2592–2598. [CrossRef]
27. Díaz-Martín, J.; López-García, M.Á.; Romero-Pérez, L.; Atienza-Amores, M.R.; Pecero, M.L.; Castilla, M.Á.; Biscuola, M.; Santón, A.; Palacios, J. Nuclear TAZ expression associates with the triple-negative phenotype in breast cancer. *Endocr. Relat. Cancer* **2015**, *22*, 443–454. [CrossRef]
28. Zhao, D.; Zhi, X.; Zhou, Z.; Chen, C. TAZ antagonizes the WWP1-mediated KLF5 degradation and promotes breast cell proliferation and tumorigenesis. *Carcinogenesis* **2012**, *33*, 59–67. [CrossRef]
29. Hiemer, S.E.; Szymaniak, A.D.; Varelas, X. The transcriptional regulators TAZ and YAP direct transforming growth factor β-induced tumorigenic phenotypes in breast cancer cells. *J. Biol. Chem.* **2014**, *289*, 13461–13474. [CrossRef]
30. Nouri, K.; Azad, T.; Ling, M.; Van Rensburg, H.J.J.; Pipchuk, A.; Shen, H.; Hao, Y.; Zhang, J.; Yang, X. Identificatoin of Celastrol as a novel YAP-TEAD inhibitor for cancer therapy by high throughput screening with ultrasensitive YAP/TAZ-TEAD biosensors. *Cancers* **2019**, *11*, 1596. [CrossRef]
31. Azad, T.; Nouri, K.; Van Rensburg, H.J.J.; Hao, Y.; Yang, X. Monitoring Hippo signaling pathway activity using a luciferase-based large tumor suppressor (LATS) biosensor. *J. Vis. Exp.* **2018**, *139*, e58416.
32. Azad, T.; Van Rensburg, H.J.J.; Lightbody, E.D.; Neveu, B.; Champagne, A.; Ghaffari, A.; Kay, V.R.; Hao, Y.; Shen, H.; Yeung, B.; et al. A LATS biosensor screen identifies VEGFR as a regulator of the Hippo pathway in angiogenesis. *Nat. Commun.* **2018**, *9*, 1061. [CrossRef]
33. Nouri, K.; Azad, T.; Lightbody, E.; Khanal, P.; Nicol, C.; Yang, X. A kinome-wide screen using a NanoLuc LATS luminescent biosensor identifies ALK as a novel regulator of the Hippo pathway in tumorigenesis and immune evasion. *FASEB J.* **2019**, *33*, 12487–12499. [CrossRef] [PubMed]
34. Azad, T.; Tashakor, A.; Hosseinkhani, S. Split-luciferase complementary assay: Applications, recent developments, and future perspectives. *Anal. Bioanal. Chem.* **2014**, *406*, 5541–5560. [CrossRef] [PubMed]
35. Adams, S.T., Jr.; Miller, S.C. Beyond D-luciferin: Expanding the scope of bioluminescence imaging in vivo. *Curr. Opin. Chem. Biol.* **2014**, *21*, 112–120. [CrossRef] [PubMed]
36. Hall, M.P.; Unch, J.; Binkowski, B.F.; Valley, M.P.; Butler, B.L.; Wood, M.G.; Otto, P.; Zimmerman, K.; Vidugiris, G.; Machleidt, T.; et al. Engineered luciferase reporter from a deep sea shrimp utilizing a novel imidazopyrazinone substrate. *ACS Chem. Biol.* **2017**, *7*, 1848–1857. [CrossRef]
37. England, C.G.; Ehlerding, E.B.; Cai, W. NanoLuc: A small luciferase is brightening up the field of bioluminescence. *Bioconjug. Chem.* **2016**, *27*, 1175–1187. [CrossRef] [PubMed]
38. Dixon, A.S.; Schwinn, M.K.; Hall, M.P.; Zimmerman, K.; Otto, P.; Lubben, T.H.; Butter, B.L.; Binkowski, B.F.; Machleidt, T.; Kirkland, T.A.; et al. NanoLuc complementation reporter optimized for accurate measurement of protein interactions in cells. *ACS Chem. Biol.* **2016**, *11*, 400–408. [CrossRef]
39. Oh-hashi, K.; Furuta, E.; Fujimura, K.; Hirata, Y. Application of a novel HiBiT peptide tag for monitoring ATF4 protein expression in Neuro2a cells. *Biochem. Biophys. Rep.* **2017**, *12*, 40–45. [CrossRef]
40. Janse Van Rensburg, H.J.; Azad, T.; Ling, M.; Hao, Y.; Snetsinger, B.; Khanal, P.; Minassian, L.M.; Graham, C.H.; Rauh, M.J.; Yang, X. The Hippo pathway component TAZ promotes immune evasion in human cancer through PD-L1. *Cancer Res.* **2018**, *78*, 1457–1470. [CrossRef]
41. Nouri, K.; Fansa, E.K.; Amin, E.; Dvorsky, R.; Gremer, L.; Willbold, D.; Schmitt, L.; Timson, D.J.; Ahmadian, M.R. IQGAP1 interaction with RHO family proteins revisited: Kinetic and equilibrium evidence for multiple distinct binding sites. *J. Biol. Chem.* **2016**, *291*, 26364–26376. [CrossRef]
42. Zhang, K.; Chang, Y.; Shi, Z.; Han, X.; Han, Y.; Yao, Q.; Hu, Z.; Cui, H.; Zheng, L.; Han, T.; et al. ω-3 PUFAs ameliorate liver fibrosis and inhibit hepatic stellate cells proliferation and activation by promoting YAP/TAZ degradation. *Sci. Rep.* **2016**, *6*, 30029. [CrossRef]

43. Pizato, N.; Luzete, B.C.; Kiffer, L.F.M.V.; Corrêa, L.H.; Santos, I.O.; Assumpção, J.A.F.; Ito, M.K.; Magalhães, K.G. Omega-3 docosahexaenoic acid induces pyroptosis cell death in triple-negative breast cancer cells. *Sci. Rep.* **2018**, *8*, 1952. [CrossRef]
44. Canadian Cancer Society. Available online: https://cancer.ca/en/cancer-information/cancer-types/breast/treatment/targeted-therapy (accessed on 20 December 2022).
45. Ramos, A.; Camargo, F.D. The Hippo signaling pathway and stem cell biology. *Trends Cell Biol.* **2012**, *22*, 339–346. [CrossRef]
46. Study to Evaludate VT3989 in Patients with Metastatic Solid Tumors Enriched for Tumors with NF2 Gene Mutations. Available online: https://classic.clinicaltrials.gov/ct2/show/NCT04665206?term=VT3989&draw=2&rank=1 (accessed on 1 July 2023).
47. A Phase I Study of IAG933 in Patients with Advanced Mesothelioma and Other Solid Tumors. Available online: https://classic.clinicaltrials.gov/ct2/show/NCT04857372?term=IAG933&draw=2&rank=1 (accessed on 1 July 2023).
48. Oral TEAD Inhibitor Targeting the Hippo Pathway in Subjects with Advanced Solid Tumors. Available online: https://classic.clinicaltrials.gov/ct2/show/NCT05228015 (accessed on 7 July 2023).
49. A Study of ION537 in Patients with Molecularly Selected Advanced Solid Tumors. Available online: https://classic.clinicaltrials.gov/ct2/show/NCT04659096 (accessed on 7 July 2023).
50. Maity, S.; Gridnev, A.; Misra, J.R. Assays used for discovering small molecule inhibitors of YAP activity in cancers. *Cancers* **2022**, *14*, 1029. [CrossRef] [PubMed]
51. Barry, E.R.; Simov, V.; Valtingojer, I.; Venier, O. Recent therapeutic approaches to modulate the Hippo pathway in oncology and regenerative medicine. *Cells* **2021**, *10*, 2715. [CrossRef] [PubMed]
52. McCormack, D.; McFadden, D. A review of pterostilbene antioxidant activity and disease modification. *Oxid. Med. Cell. Longev.* **2013**, *2013*, 575482. [CrossRef] [PubMed]
53. Wakimoto, R.; Ono, M.; Takeshima, M.; Higuchi, T.; Nakano, S. Differential anticancer activity of pterostilbene against three subtypes of human breast cancer cells. *Anticancer Res.* **2017**, *37*, 6153–6159.
54. Pantziarka, P.; Sukhatme, V.; Crispino, S.; Bouche, G.; Meheus, L.; Sukhatme, P. Repurposing drugs in oncology (ReDO)—Selective PDE5 inhibitors as anti-cancer agents. *Ecancermedicalscience* **2018**, *12*, 824. [CrossRef]
55. Spano, D.; Marshall, J.C.; Marino, N.; De Martino, D.; Romano, A.; Scoppettuolo, M.N.; Bello, A.M.; Dato, V.D.; Navas, L.; Vita, G.D.; et al. Dipyridamole prevents triple-negative breast-cancer progression. *Clin. Exp. Metastasis* **2013**, *30*, 47–68. [CrossRef]
56. Tanaka, K.; Osada, H.; Murakami-Tonami, Y.; Horio, Y.; Hida, T.; Sekido, Y. Statin suppresses Hippo pathway-inactivated malignant mesothelioma cells and blocks the YAP/CD44 growth stimulatory axis. *Cancer Lett.* **2017**, *385*, 215–224. [CrossRef]
57. Sun, S.; Dean, R.; Jia, Q.; Zenova, A.; Zhong, J.; Grayson, C.; Xie, C.; Lindgren, A.; Samra, P.; Sojo, L.; et al. Discovery of XEN445: A potent and selective endothelial lipase inhibitor raises plasma HDL-cholesterol concentration in mice. *Bioorg. Med. Chem.* **2013**, *21*, 7724–7734. [CrossRef]
58. Han, H.; Qi, R.; Zhou, J.J.; Ta, A.P.; Yang, B.; Nakaoka, H.; Seo, G.; Guan, K.; Luo, R.; Wang, W. Regulation of the Hippo pathway by phosphatidic acid-mediated lipid-protein interaction. *Mol. Cell.* **2018**, *72*, 328–340. [CrossRef] [PubMed]

chemosensors

MDPI

Article

Trypsin-Based Chemoenzymatic Assay for Detection of Pollutants and Safety Assessment of Food Additives

Elena N. Esimbekova [1,2,*], Irina G. Torgashina [1], Elena V. Nemtseva [1,2], Anna A. Antashkevich [1], Polina Yu. Sasova [1] and Valentina A. Kratasyuk [1,2]

1 Institute of Fundamental Biology and Biotechnology, Siberian Federal University, 660041 Krasnoyarsk, Russia; torira@mail.ru (I.G.T.); enemtseva@sfu-kras.ru (E.V.N.); anneantand@gmail.com (A.A.A.); stepanovap98@gmail.com (P.Y.S.); valkrat@mail.ru (V.A.K.)

2 Laboratory of Photobiology, Institute of Biophysics of Siberian Branch of Russian Academy of Science, 660036 Krasnoyarsk, Russia

* Correspondence: esimbekova@yandex.ru; Tel.: +7-(391)-249-42-42

Abstract: Chemoenzymatic assay systems are widely used to detect toxicants in various samples, including food and environment specimens. These methods are based on the ability of various types of toxicant to specifically inhibit/activate the functions of individual enzymes or enzyme systems. The present study examines the possibility of using the proteolytic enzyme trypsin as a specific marker to detect protease inhibitors in different samples. The study shows that trypsin activity is not affected by various heavy metals, pesticides, or quinones at levels considerably greater than their maximum allowable concentrations (MACs) in water bodies. At the same time, the IC_{50} value for the food preservative potassium sorbate (E202) is 15 mg/L, which is substantially lower than its acceptable daily intake (ADI). The quenching of trypsin fluorescence in the presence of potassium sorbate suggests that inhibition could occur due to the binding of the preservative to the enzyme in the region adjacent to the active center. The trypsin was immobilized in starch gel to ensure its stability in the enzyme inhibition based assay. Single-use reagents were prepared as dry starch disks that could be stored over long periods. Their sensitivity to copper (II) chloride, potassium sorbate, and chromium (III) chloride was similar to the sensitivity of the free trypsin.

Keywords: trypsin-inhibition-based assay; detection of pollutants; food additives; chemoenzymatic assay; trypsin immobilization; protein fluorescence

Citation: Esimbekova, E.N.; Torgashina, I.G.; Nemtseva, E.V.; Antashkevich, A.A.; Sasova, P.Y.; Kratasyuk, V.A. Trypsin-Based Chemoenzymatic Assay for Detection of Pollutants and Safety Assessment of Food Additives. *Chemosensors* **2023**, *11*, 237. https://doi.org/10.3390/chemosensors11040237

Academic Editor: Xiaolong Yang

Received: 2 March 2023
Revised: 27 March 2023
Accepted: 6 April 2023
Published: 10 April 2023

1. Introduction

To maintain the health of the human population, it is critical to detect toxic substances in the environment, as there is a risk of the regular intake of various anthropogenic contaminants with food and drinks. These contaminants may include heavy metals, pesticides, phenols, and their oxidation products, quinones, which may be even more toxic than their respective polyphenols [1–5]. Another cause for concern is the use of numerous food additives, including preservatives, which are added to prevent food spoilage. The amounts of preservatives are controlled, and their concentrations are usually low: 0.01–0.2% of the weight of the product [6,7]. However, recent studies show that the consumption of increased amounts of food preservatives may produce deleterious effects through, for example, their complexing with blood proteins or the inhibition of the gene expression of the proteins involved in synthesis of coagulation factors [8,9].

Pollutants in the environment are commonly detected using biological testing systems, based on the assessment of changes in the vital functions of various test organisms [10], and chemoenzymatic methods, based on the ability of pollutants to inhibit/stimulate the activities of different enzymes or multi-enzyme systems [11]. Chemoenzymatic assay systems exhibit selective sensitivity to various classes of toxic substance and, thus, they can be used to study the effects of the mechanisms of action of toxicants on the functions of separate components of metabolic pathways [11]. However, it should be noted that prior to using enzymes as test organisms, a considerable amount of preliminary work is required, including the stabilization of the enzymes or the pretreatment of the samples to be analyzed (establishing the optimum pH for enzyme activity, ensuring the solubility of the tested compounds, removing interfering impurities) [11]. Moreover, enzymes are generally expensive reagents.

The enzymes most frequently used in these inhibition assays are hydrolases and oxidoreductases. Of the chemoenzymatic assay systems, the most widely used are the systems based on cholinesterases, which are highly sensitive to organophosphorus and carbamate pesticides [12]. The assays most commonly applied for detecting heavy meals are the assay systems based on urease, peroxidase, and glucose oxidase [11].

Protease-based chemoenzymatic assay systems are used considerably less often. The main field of application of proteases is pharmaceutics, in which drugs inhibiting protease are screened [13]. Nevertheless, a number of studies demonstrate the potential of proteases such as trypsin—an enzyme produced by the pancreas—to serve as biomarkers in the bio-diagnostics of water or soil contamination [14–16]. The proteolytic activity of trypsin can be affected by a variety of substances, which can either stimulate or inhibit its action. Substances capable of decreasing trypsin activity include, for example, specific protease inhibitors from soybeans (such as Kunitz-type inhibitors (KSTIs) and Bowman–Birk-type inhibitors (BBIs)), cadmium, etc. [17,18]. The effect of trypsin stimulation was observed in the presence of monatomic alcohol isopropanol and biogenic amine putrescine [19,20]. The data on the effect of the food preservative sodium benzoate on trypsin activity are rather contradictory. A study by Mu et al. reported the stimulation of trypsin activity by sodium benzoate as a result of the formation of a new complex and a change in the trypsin structure [21]. On the other hand, there are data showing the inhibition of trypsin by sodium benzoate [22]. The contradictions between these data suggest the importance of collecting datasets on the effects of potentially hazardous substances on proteolytic enzymes.

In the current study, we tested trypsin as the basis for the chemoenzymatic assay to detect protease inhibitors in the samples we analyzed. The effects of different types of xenobiotic—a number of salts and ions of heavy metals, pesticides, quinones, and food preservatives—on the trypsin's activity were estimated. For trypsin to be used as the basis for chemoenzymatic assays, a stable form of the enzyme should be prepared, which should retain the sensitivity of trypsin to inhibitory effects. In the present study, we propose a method involving the immobilization of trypsin in starch gel, in which the enzyme molecules are incorporated into a three-dimensional network of closely interwoven polymer chains.

2. Materials and Methods

The reagents used in this study included lyophilized trypsin from porcine pancreas, 1300 BAEE U/mg (Sigma-Aldrich, St. Louis, MO, USA), specific substrate N_{α}-Benzoyl-L-arginine ethyl ester (BAEE) (Sigma-Aldrich, St. Louis, MO, USA), hydrochloric acid (SigmaTek, Khimki, Russia), and Clark-and-Lubs buffer pH 7.6. Solvents were nitric acid (Technological Laboratory Giredmet, Khimreaktivsnab, Ufa, Russia), acetonitrile (PanReac AppliChem, Barcelona, Spain), and distilled water.

Aqueous solutions of heavy-metal salts and nitric-acid solutions of metal ions were used to study the effects of heavy metals on trypsin activity. Salts of heavy metals—copper chloride ($CuCl_2$), zinc nitrate ($Zn(NO_3)_2$), manganese chloride ($MnCl_2$), chromium chloride ($CrCl_3$), and cobalt chloride ($CoCl_2$)—were analytical-grade chemicals (OOO "Khimkraft", Kaliningrad, Russia). State-standard reference samples of ions of cadmium (Cd^{2+}), nickel (Ni^{2+}), arsenic (As^{3+}), mercury (Hg^{+}), and aluminum (Al^{3+}) (Ural Plant of Chemical Reagents, Verkhnyaya Pyshma, Russia) were used. Other chemicals whose effects on trypsin activity were studied included the following: quinones—1,4-benzoquinone and toluquinone (Sigma-Aldrich, Taufkirchen, Germany); food preservatives—sodium benzoate (Sigma-Aldrich, Amsterdam, The Netherlands) and potassium sorbate (Supelco, Bellefonte,

PA, USA); high-purity active ingredients of pesticides—fenvalerate, deltamethrin, cypermethrin, imidacloprid, glyphosate (Sigma-Aldrich, St. Louis, MO, USA), metribuzin, tebuconazole, epoxiconazole, fenoxaprop-p-ethyl (Xi'anTai Cheng Chem Co., Ltd., Yancheng, China); and glyphosate-based commercial pesticide formulation Tornado Extra (JSC Avgust, Moscow, Russia). Pesticides were dissolved in acetonitrile, except glyphosate and Tornado Extra, which were dissolved in distilled water.

Trypsin activity was determined from hydrolysis rate of its specific substrate, BAEE, by recording temporal variations in the absorbance of the reaction mixture at 253 nm, using a Shimadzu UV-2600 spectrophotometer (Shimadzu Corporation, Kyoto, Japan). The reaction mixture contained 11 U of trypsin, 450 µL of the tested or control sample, 40 µL of 0.1 M Clark-and-Lubs buffer with pH 7.6, 460 µL of 0.5-mM BAEE solution, and 40 µL of 1-mM hydrochloric acid.

The effects produced by the solvents and tested substances on trypsin activity were estimated as relative activity determined from the formula $A = (A_{exp}/A_{contr})\cdot 100\%$, where A_{exp} and A_{contr} are enzyme activities in the presence of the tested substance and in control solution, respectively. The resulting data were used as the basis for determining the values of toxicological parameters IC_{20} and IC_{50}—concentrations of the tested substances that inhibit trypsin activity by 20 and 50%, respectively.

The absorption and fluorescence spectra of trypsin were measured using a Cary 5000 spectrophotometer (Agilent Technologies, Mulgrave, Australia) and a Fluorolog-3 spectrofluorometer (Horiba, Jobin Yvon, France), respectively. Fluorescence spectra under excitation at a wavelength of 295 nm were measured in the 300–450-nm range. The spectral-slit width was 2.5 nm. The spectra were corrected for the inner-filter effect and solvent background. The absorption spectra were measured in the 200–600-nm range.

The samples for measuring fluorescence spectra had a volume of 1 mL and contained 9 U of trypsin and various concentrations of the tested substance in the buffer. The trypsin-fluorescence spectra were also studied in the presence of the commercial pesticide formulation with glyphosate as the active ingredient, Tornado Extra, whose inhibitory action was shown in our previous paper, and IC_{50} = 2.4 g/L was obtained [23].

Visualization of spatial structure of trypsin (PDB ID: 1s0q) and assessment of the distances between tryptophan residues and catalytic triad were performed using VMD 1.9.2 [24]. The solvent accessible surface areas (SASAs) of the tryptophan residues were estimated by GROMACS GMX SASA plugin with solvent-probe radius of 1.4 Å [25]. Before calculation, hydrogens were added to the trypsin crystal structure (PDB ID: 1s0q) according to CHARMM36 force-field topology.

Immobilization of trypsin was performed as follows. First, potato-starch gel was prepared by boiling starch suspension for 2 min and cooling to 25 °C. Next, the trypsin solution was mixed with potato-starch gel, and 25-µL drops were placed onto fluoropolymer film and dried at 8 °C. The resultant preparations were 4–5-mm-diameter dry disks; each disk was intended for one assay. To measure the activity of immobilized trypsin, 1 disk with immobilized enzyme, 40 µL of 1-mM hydrochloric acid, 450 µL of the tested or control sample, 40 µL of 0.1-M Clark-and-Lubs buffer with pH of 7.6, and 460 µL of 0.5-mM BAEE were added to the spectrophotometer cuvette.

The data are presented as mean values (M) with standard deviation (s). All the measurements were repeated 5 times. The significance of differences was determined by Student's *t*-test. The results were considered statistically significant at $p < 0.05$.

3. Results

3.1. The Effects of Pollutants and Food Preservatives on the Activity of Trypsin-Assay System

Enzyme-based assays have a substantial advantage over assays based on living organisms: the conditions of enzymatic assays can be varied to adjust their sensitivity to toxicants. For example, assay sensitivity can be enhanced through the pre-incubation of the enzyme in the tested solution or by altering the proportions of the components in the reaction mixture. The sensitivity of chemoenzymatic assay systems to toxic substances is largely determined by the amount of enzyme in the reaction mixture: the highest sensitivity is usually achieved at low enzyme concentrations. We varied the trypsin content and found the level that would allow its enzymatic activity to be detected while, at the same time, ensuring good reproducibility of the results, which ranged between 5 and 11 U (Figure 1). It would not have been justifiable to increase the trypsin content further, as this would have led to a decrease in the assay's sensitivity to toxicants and excessive consumption of the enzyme. A reduction in trypsin content to below 5 U resulted in a considerable decrease in the level of the detectable signal and, hence, a larger measurement error.

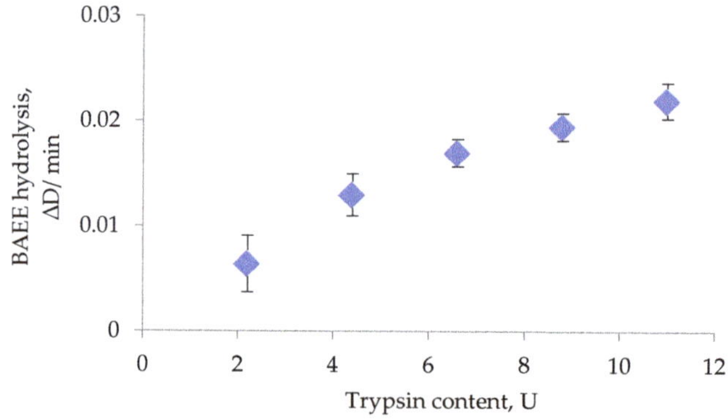

Figure 1. BAEE-hydrolysis rate versus trypsin content in the reaction mixture.

In the present study, we used trypsin amounts falling within the preliminarily determined range and analyzed the effects of a number of toxicants, such as metal salts, pesticides, quinones, and food additives on the trypsin activity. Initially, the solvents used to prepare solutions of toxicants—distilled water, nitric acid, and acetonitrile—were tested to determine their effects on the trypsin activity. The tests showed that the proportions of distilled water and acetonitrile in the reaction mixture should be no more than 45% and 3% of the total volume of the mixture, respectively. Higher percentages of these solvents inhibited the trypsin activity. The concentration of nitric acid in the reaction mixture should not exceed 0.1 mM.

The use of the optical method of signal detection imposes a number of limitations on the analysis of the effects of toxicants on enzyme activity [26]. This is particularly evident in the analysis of trypsin activity, as measurements are conducted at a wavelength of 253 nm, i.e., in the UV region, where various substances have pronounced absorption [26]. Moreover, in some of our experiments, the addition of the toxicant solutions to the reaction mixtures resulted in the production of suspended particles, turbidity, and color changes in the solution. For example, the addition of $Zn(NO_3)_2$ or more than 170 mg/L of $CuCl_2$, $MnCl_2$, $CrCl_3$, and $CoCl_2$ to the reaction mixture led to the formation of precipitate, probably due to interactions between these chemicals and the salts in the potassium-phosphate buffer. These interactions of the toxicants with the reaction-mixture components considerably limited the range of the concentrations of additives tested in the present study.

The effects of heavy metals on the activity of trypsin were studied using solutions of the heavy metals in water and in nitric acid. Concentrations of $Zn(NO_3)_2$ below 85 mg/L and $CoCl_2$ below 58.5 mg/L did not affect the trypsin's activity. The trypsin activity decreased in the presence of $CuCl_2$, with an IC_{20} value of 80 mg/L. The presence of manganese and chromium chlorides in the reaction mixture stimulated trypsin activity. The activity of trypsin exposed to 90 mg/L and 104 mg/L of $MnCl_2$ and $CrCl_3$, respectively, rose by 35%.

In the study of effects of the metals in the form of ion solutions on the trypsin activity, the range of metal concentrations was limited because of the considerable inhibitory effect of the solvent (nitric acid). Metal ions at concentrations below 0.2 mg/L did not affect the trypsin activity. An additional process—the 5-min incubation of trypsin in solutions of metal salts or ions—did not lead to the emergence of an inhibitory effect within the tested range of concentrations.

Most of the tested pesticides were poorly soluble in water, and they were dissolved in acetonitrile. However, the addition of high concentrations of pesticides dissolved in acetonitrile to the reaction mixture often caused the turbidity of the solution and the production of suspended particles, which rendered the analysis impossible. Thus, we determined the highest possible concentrations of the tested pesticides: up to 9 mg/L for metribuzin, tebuconazole, fenoxaprop-P-ethyl, and epoxiconazole; and up to 10 mg/L for deltamethrin, fenvalerate, cypermethrin, and imidacloprid. At these concentrations of pesticides, no changes in trypsin activity were observed. The additional process—enzyme incubation in the tested pesticide solution—increased the trypsin sensitivity to metribuzin and fenoxaprop-P-ethyl only (Figure 2). In the presence of the water-soluble glyphosate, pesticide-concentration dependence was obtained without enzyme pre-incubation (Figure 3).

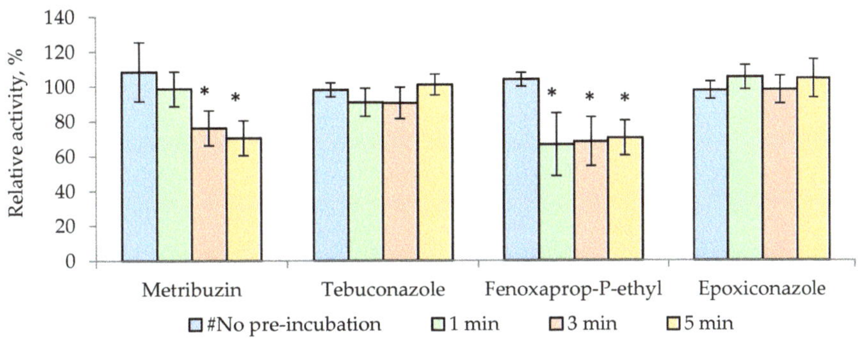

Figure 2. Dependence of relative activity of trypsin on the time of incubation of the enzyme in solutions with pesticide concentration of 9 mg/L. A 20% inhibition of trypsin activity was achieved by pre-incubation of trypsin in metribuzin and fenoxaprop-P-ethyl solutions for 3 and 1 min, respectively. #No pre-incubation denotes that the pesticide solution was added to the reaction mixture immediately before signal detection. * Values of $p < 0.05$ when comparing the relative activity values obtained without pre-incubation of trypsin in the pesticide solution.

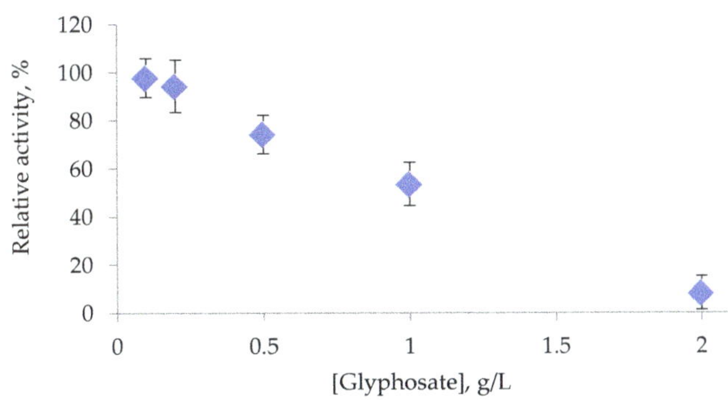

Figure 3. Relative activity of trypsin in the presence of glyphosate. The amount of trypsin in the reaction mixture was 11 U. The values of IC_{20} and IC_{50} for glyphosate were determined as 0.5 and 1 g/L, respectively.

The highest concentrations of the quinones—toluquinone and benzoquinone—in the reaction mixture that made it possible to detect changes in its absorbance were 6 and 16 mg/L, respectively. Benzoquinone at a concentration of 16 mg/L inhibited the trypsin activity by 20%. The decrease in the amount of trypsin to 5.5 U only insignificantly increased the sensitivity of the trypsin-assay system to benzoquinone and toluquinone (Figure 4).

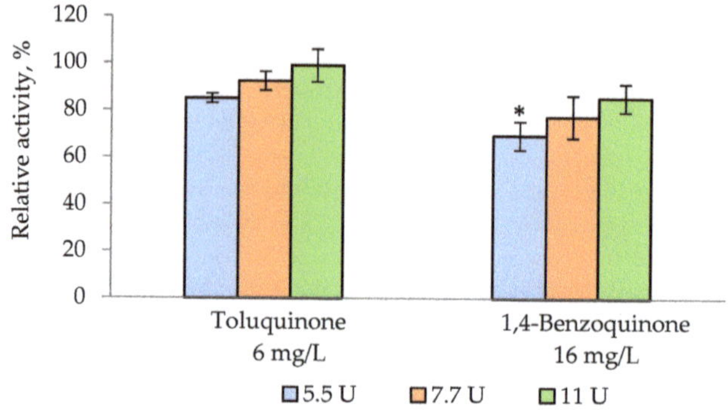

Figure 4. Dependence of the relative activity of trypsin in the presence of quinones on the amount of the enzyme. A decrease in the amount of trypsin in the reaction mixture led to an insubstantial increase in the sensitivity of the trypsin-assay system to benzoquinone and toluquinone. * Values of $p < 0.05$ when comparing the relative activity values obtained in the presence of 11 U of trypsin.

The highest tested concentration of the food preservative sodium benzoate, 144 mg/L, did not affect the trypsin activity. At the same time, another food preservative—potassium sorbate—considerably inhibited the trypsin activity; the IC_{50} value for the potassium sorbate was 15 mg/L, and it was substantially lower than its acceptable daily intake (ADI), which, for drinks, for example, is 300 mg/L [7].

Thus, the analysis of different types of xenobiotic revealed the selective sensitivity of trypsin to toxicants and food preservatives. Table 1 lists the IC_{20} and IC_{50} values for some toxicants and food preservatives.

Table 1. IC_{20} (mg/L) and IC_{50} (mg/L) values determined from the effects of toxicants and food preservatives on trypsin-assay system * #.

Inhibitors	Concentrations Studied (mg/L)	IC_{20} (mg/L)	IC_{50} (mg/L)	ADI RUS (mg/L)
Copper(II) chloride	<80	80	–	0.001–0.002
Glyphosate	100–2000	500	1000	0.02–5.0
Metribuzin &	<9	9	–	0.1–0.2
Fenoxaprop-P-ethyl &	<9	9	–	0.0003–0.04
1,4-Benzoquinone	<16	16	–	0.001–0.2
Potassium sorbate	<15	10	15	200–300

"–" No inhibitory effect of the toxicant was detected in the tested concentration range. * For the other additives (zinc nitrate, cobalt chloride; ions of cadmium, nickel, arsenic, mercury, and aluminum; toluquinone, sodium benzoate; pesticides: tebuconazole, fenoxaprop-P-ethyl, deltamethrin, fenvalerate, cypermethrin, and imidacloprid) either no effect was found or enzyme activation was observed (manganese chloride, chromium chloride) at studied concentrations. # Trypsin content in all cases was 11 U, except for the test with 1,4-benzoquinone, on which trypsin content was 7.7 U. & The pre-incubation for 3 min was performed for two toxicants marked by & and no pre-incubations—in the other assays in the table.

3.2. *Interaction of Food Preservatives and Some Toxicants with Trypsin*

One of the possible reasons for the variations in the enzyme activity is a change in the enzyme structure caused by its interactions with the toxicants. To reveal changes in the tertiary structure of the trypsin, we measured the fluorescence spectra in the presence of the toxicants that had a pronounced effect on the trypsin activity. These were the pesticide formulation Tornado Extra (the effect of which was studied in our previous work [23]), the salts $CuCl_2$, $MnCl_2$, and $CrCl_3$, and potassium sorbate. The toxicants were preliminarily dissolved in water. The protein spectra in the presence of a toxicant were compared with the spectra of the control sample, to which a corresponding amount of water was added. In each treatment, the trypsin-to-toxicant concentration ratio was close to the ratio obtained under the maximal inhibitory or stimulatory effect. The fluorescence spectra were measured at an excitation wavelength of 295 nm, at which only the tryptophan residues of proteins were excited. Highly sensitive to microenvironment polarity, tryptophan fluorophores respond to variations in microenvironment parameters caused by disturbances in the protein structure or direct contact with bound additives via changes in fluorescence intensity and the position of the spectral maximum.

Figure 5 shows the fluorescence spectra of the trypsin in the presence of different concentrations of various additives. The trypsin exhibited fluorescence typical for this protein, with a maximum of about 330 nm, which did not change in the experiments. The addition of the pesticide formulations only slightly perturbed the intensity of the trypsin fluorescence, without producing concentration-dependent effects (an example is shown in Figure 5a). This can be attributed to the high level of background emissions of the complex mixtures, which were difficult to take into account and subtract completely. The salts $CrCl_3$, $MnCl_2$, and $CuCl_2$ produced no effect on the trypsin fluorescence (the example with $CuCl_2$ is shown in Figure 5b), implying no change in the protein structure after their addition. The only additive that caused a change in the fluorescence spectrum was potassium sorbate (Figure 5c). This food preservative is very soluble in water and has favorable spectral properties, which made it possible to detect clearly a gradual decrease in protein fluorescence with an increase in its concentration.

The Stern–Volmer plot for the quenching of the trypsin fluorescence by the potassium sorbate is shown in Figure 5d. The corresponding Stern–Volmer constant K_{SV} was calculated as 85.99 ± 3.42 M^{-1}. Taking into account that the average fluorescence lifetime of trypsin in the buffer was measured as 2.51 ns [27], we estimated the bimolecular quenching constant k_q as 3.4×10^{10} $M^{-1}s^{-1}$. This value is greater than the diffusion-controlled limit of k_q, which is about 1×10^{10} $M^{-1}s^{-1}$ [28]. Thus, we concluded that both the dynamic mechanism of the quenching due to the collision with the quencher and the static mechanism of the quenching due to the binding of the potassium sorbate by the trypsin contributed to the total effect. To distinguish between the contributions of these two quenching types, a detailed study of the fluorescence lifetimes of trypsin in the presence of potassium sorbate and under varying experimental temperatures should be performed, which will be the subject of the future research.

To assess the possibility of a relationship between the inhibition of the trypsin activity and the quenching of its fluorescence by the potassium sorbate, we considered the relative positions of the catalytic triad and the tryptophan residues in the trypsin globe (Figure 6). Of the four tryptophans, Trp215 was the closest to the active site of the enzyme. It was located at 5.16–6.43 Å from the triad residues. Thus, the interaction of the potassium sorbate with the surface area adjacent to the site of the Trp215 localization could have influenced the trypsin activity. The tryptophan residues of trypsin have different solvent accessible surface areas (SASAs), which means that they are not equally available for quenching by additives. The following SASAs were estimated for the tryptophans ($Å^2$): Trp51—0, Trp141—3.3, Trp215—56.3, Trp237—37.9. This indicated that only the last two were able to make contact with the quenchers. Therefore, we suggest that Trp215 and Trp237 could have been responsible for the observed protein-fluorescence quenching with the potassium sorbate, and that one of the two was located close to the active site of the enzyme. Although the crystal structure of trypsin does not reflect protein dynamics in the solution, which could cause variation in the estimated distances, the relative positions of the amino acids and their accessibility to the solvent are unlikely to change considerably because of structural fluctuations. A previous study led to the conclusion that trypsin has nearly identical molecular architectures in solution and in three crystal forms [29].

Thus, the fluorescence spectroscopy revealed direct intermolecular contact between the trypsin and the additive, affecting its activity for potassium sorbate only. It can be concluded that the inhibitory effects of the $CuCl_2$ and the pesticides or the stimulatory effects of the $CrCl_3$ and the $MnCl_2$ resulted from other mechanisms, such as the effects on the ligands and enzyme substrates, or the disruption of the enzyme–substrate interaction.

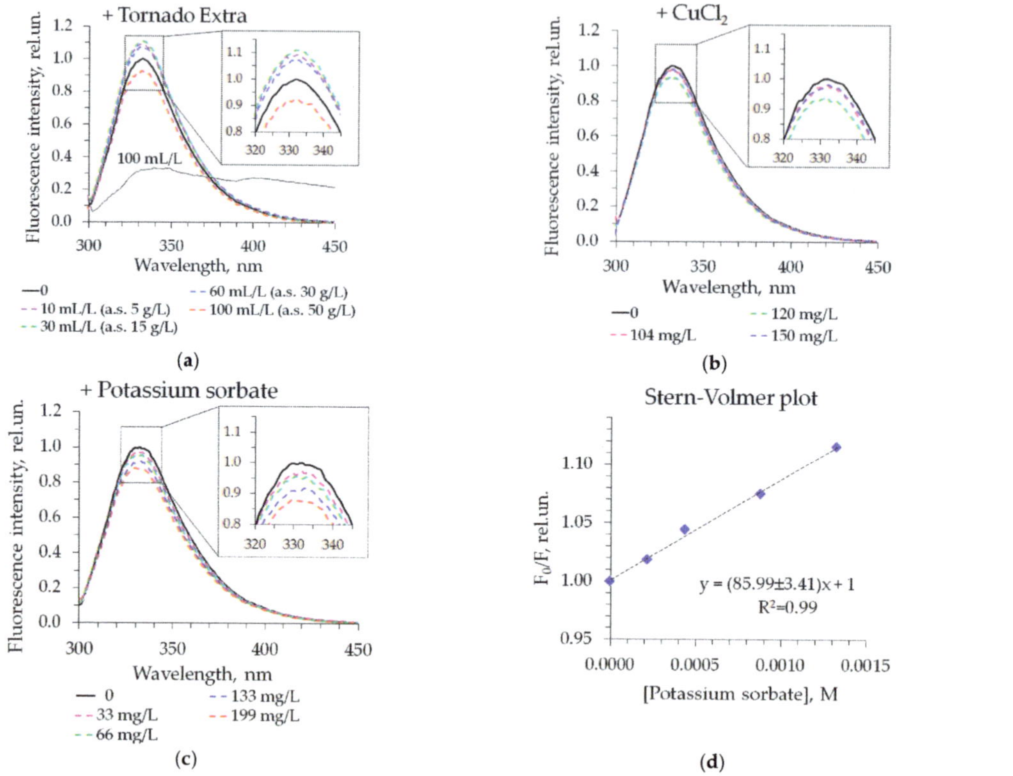

Figure 5. Fluorescence spectra of trypsin under 295-nm excitation in the presence (colored dashed lines) and in the absence (thick black lines) of additives: (**a**) Tornado Extra pesticide formulation, (**b**) $CuCl_2$, (**c**) potassium sorbate. Inserts show the intensity change near the spectral maximum. Protein content was 9 U. Thin black line in panel (**a**) refers to the fluorescence spectrum of the additive without trypsin at indicated level under 295-nm excitation. Panel (**d**) shows the Stern–Volmer plot for quenching of trypsin fluorescence by potassium sorbate. Equation for the linear approximation and an adjusted coefficient of determination R^2 are indicated.

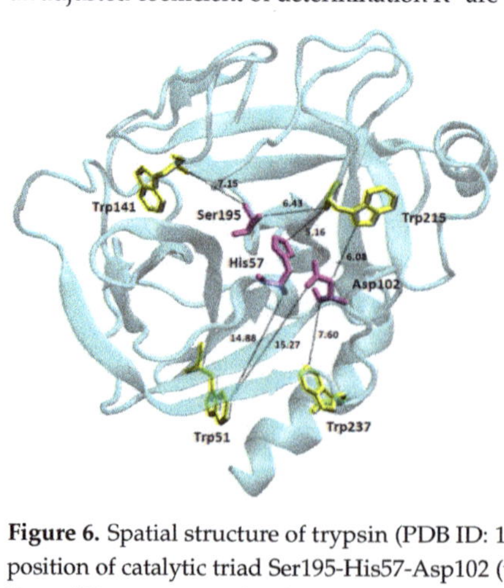

Figure 6. Spatial structure of trypsin (PDB ID: 1s0q) shown by ribbon representation with indicated position of catalytic triad Ser195-His57-Asp102 (violet sticks) and tryptophan residues (yellow sticks). Dotted lines connect the closest atoms of the triad and tryptophans; corresponding distances are indicated in Å.

3.3. Constructing the Biological Module for Chemosensor by Immobilizing Trypsin in Starch Gel

The trypsin-based assay can only be used as an indicator to detect protease inhibitors if the stable form of the enzyme is available. It is a challenging task to prepare a stable enzyme that retains the necessary sensitivity to toxicants. In the present study, trypsin was immobilized into starch-polymer gel for the first time. The resultant preparations were 4–5-mm-diameter dry starch disks; one disk was intended for one assay.

When developing a reagent based on immobilized trypsin, we studied the effects of the potato-starch concentration and drying temperature on the trypsin activity (Figure 7). The activities of the reagents immobilized into 2.5, 3, and 3.5% starch gels did not differ considerably. At the same time, the reagents prepared using the 2.5% starch gel were brittle, and the 3.5% gel was too viscous to measure accurately the sizes of the drops that were dried to make the starch disks. Therefore, the 3% starch gel was chosen as the matrix for the trypsin immobilization. No significant differences in the activities of trypsin immobilized in the starch gel were observed between the disks dried at 8 °C and 20 °C.

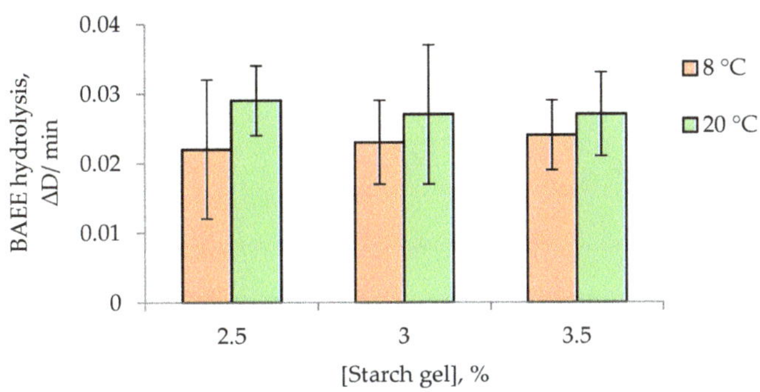

Figure 7. The rate of BAEE hydrolysis by trypsin versus concentration of potato-starch gel. Trypsin content was 11 U in the starch disks.

The single-use disks prepared in the present study contained 11 U of immobilized trypsin (the same amount as the solution), and disks containing increased amounts of trypsin—22 and 29 U—were also prepared. The activity of the disks with immobilized trypsin was about 100% compared to the activity of the free forms with the same amounts of trypsin. The immobilized trypsin reagents stored at 8 °C for 4 months retained 80% of their activity.

The sensitivity of the trypsin immobilized in starch gel to the toxicants was assessed in experiments with glyphosate as the active ingredient of the commercial pesticide Tornado Extra, $CuCl_2$, and potassium sorbate.

As the trypsin immobilized in starch gel had the form of dry starch disks, for the enzymatic reaction to occur, it was necessary to enable the diffusion of the substrate and inhibitor to the immobilized trypsin. This was achieved by causing the disk swell while it was pre-incubated in the tested solution, before the addition of the substrate. The disks with immobilized trypsin were incubated in the pesticide solution for 2 and 5 min. The results of the measurements of the trypsin activity in the disks with the lower trypsin contents (11 and 22 U) had poorer reproducibility (Figure 8). In addition, the incubation of the disks containing different amounts of trypsin in the glyphosate solution for 5 min did not result in any significant differences in the degree of inhibition of enzyme activity.

The sensitivity of the trypsin immobilized in starch gel to $CuCl_2$ and potassium sorbate was comparable to the sensitivity of the free enzyme. The IC_{20} and IC_{50} values for the potassium sorbate were 12 and 20 mg/L, respectively. Furthermore, the activity of the immobilized trypsin, as with that of the free enzyme, was stimulated in the presence of $CrCl_3$.

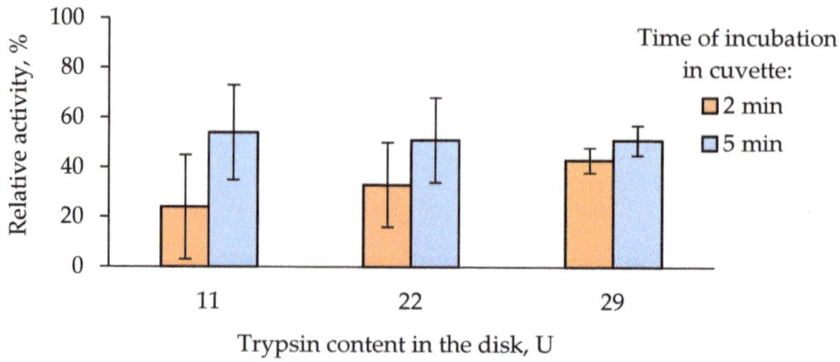

Figure 8. Relative activity of trypsin immobilized in starch gel in the presence of 5 g/L of glyphosate as the active ingredient of commercial pesticide Tornado Extra. Five g/L of glyphosate was determined as the IC_{50} value for trypsin immobilized in starch gel; this value was obtained through pre-incubation of the immobilized enzyme in the pesticide solution for 5 min.

4. Discussion

Chemoenzymatic assay systems are widely used to detect various chemical and biological components for clinical, food, and environmental monitoring [30]. Sensors based on enzyme activation or inhibition are effective tools because of their high sensitivity and simplicity. The methods used to detect analytes using chemoenzymatic assay systems can be either specific, i.e., sensitive to certain substances, or integrated, with the combination of substances contained in the tested sample affecting the assay system. The method chosen to detect analytes should correspond to the aims of the study, and the choice of the method determines the choice of the chemoenzymatic assay system. A good example of a specific detection method is provided by chemoenzymatic assays based on cholinesterases used to detect organophosphorus compounds and carbamate pesticides [11,30]. Integrated methods include assay systems based on enzyme systems conjugated with bacterial luciferase: the enzymes in these systems are sensitive to a wide range of toxicants, and the systems can be used to assess the overall toxicity of samples [11,31].

In the current study, we examined the possibility of using trypsin as the basis for a chemoenzymatic assay system for detecting toxic substances in aqueous media. The preparation of the solutions of the tested toxicants was a challenging task, which placed serious limitations on the investigations of the trypsin's sensitivity to the toxicants. For instance, most pesticides are soluble in acetonitrile, but the maximum volume of this solvent that could be added to the reaction mixture without affecting trypsin activity was only 3% of the reaction mixture volume. Moreover, in most of the experiments, when the toxicant solutions were prepared and when they interacted with the trypsin reaction mixture, the solution absorbance increased considerably, preventing the performance of accurate assays. However, certain adjustments to the procedure, such as decreasing the amount of the enzyme in the reaction mixture and/or increasing the time of enzyme incubation in the toxicant solution, enhanced the trypsin sensitivity to some of the toxicants.

Nevertheless, despite the measures taken to increase the sensitivity of the trypsin-based chemoenzymatic assay to the toxicants, the study showed that the trypsin activity was inhibited by some of the large number of tested substances only when their concentrations in the solutions were very high—considerably greater than their maximum allowable concentrations (MACs). Of all the tested heavy metals, the trypsin activity was slightly affected by $CuCl_2$, which inhibited it, and $MnCl_2$ and $CrCl_3$, which stimulated its activity. We assume that the distinctions between the effects caused by the four salts of the heavy metals originated from the properties of cations. This is for two reasons. Firstly, the content of chloride anion in the samples with the highest salt concentrations was in the narrow range of 0.9–1.2 mM, but the effects differed significantly. Secondly, a recent study of trypsin inhibition by anions [32] demonstrated that the effect of chloride ions on this enzyme was rather weak (at a 1-M concentration, it suppressed trypsin activity by 25%).

Among the pesticides, the metribuzin and fenoxaprop-P-ethyl herbicides produced insignificant inhibitory effects on the trypsin activity, but their IC_{20} values (9 mg/L) were considerably greater than their MACs for water bodies—0.1 and 0.0003 mg/L, respectively. The glyphosate herbicide substantially inhibited the trypsin activity. However, again, the IC_{20} and IC_{50} values for the glyphosate—0.5 and 1 g/L—were much higher than its MAC in water bodies (0.02 mg/L). These results indicate that the trypsin-based chemoenzymatic assay is not sensitive to different types of pesticide. This method can only detect very high levels of pesticides in samples. Similar results were obtained in the experiments with the quinones. The trypsin activity was insignificantly inhibited by the benzoquinone at concentrations that were higher than its MAC by a factor of 160.

The study of the trypsin fluorescence in the presence of different concentrations of pesticide formulations indicated that protein tertiary structure did not change and the components of the formulations did not interact with the tryptophan residues of the trypsin. The same results were obtained for the salts $CrCl_3$, $MnCl_2$, and $CuCl_2$.

Of particular interest was the study of the effects of the food preservatives on the trypsin activity. Food preservatives are not toxic substances, and they are commonly used to prevent the spoilage of food during storage. Trypsin was found to be highly sensitive to potassium sorbate: the IC_{50} value was 15 mg/L, which was noticeably lower than the ADI established for the potassium sorbate—300 mg/L. In the presence of this additive, the quenching of the trypsin fluorescence was detected with a high bimolecular-quenching constant, which indicates the binding of the sorbate with the trypsin in the area near the tryptophan and the formation of a non-fluorescent complex. In a previous study, the strong interaction of potassium benzoate with trypsin was revealed by a number of spectroscopic techniques [20]; however, this food preservative did not substantially change trypsin's activity in either that study [20] or our current experiments.

Thus, the present study demonstrates that the trypsin-based chemoenzymatic assay is a promising tool to estimate the amounts of the food preservative potassium sorbate in food samples in order to prevent its excessive intake, which may cause health problems.

To use trypsin in chemoenzymatic assays, it is necessary to prepare its stable form. Trypsin immobilization has been carried out using a wide range of methods, such as covalent binding on magnetic nanoparticles [33] and surface adsorption using silicon dioxide [34], chitosan [35], alginate [36], and other carriers. The immobilization of enzymes usually results in a decrease in their activity [37–39]. For example, the activity of covalent immobilized trypsin onto poly[(methyl methacrylate)-co-(ethyl acrylate)-co-(acrylic acid)] latex particles was 51.2% of the free trypsin activity [39]. A number of studies showed, however, that immobilized proteolytic enzymes, such as trypsin, were even more effective than their free forms [40–43]. For instance, the catalytic activity of trypsin immobilized on cyclodextrin-modified gold nanospheres was 100–120% [40]. Other studies reported that compared with the free enzyme, immobilized trypsin exhibited greater resistance against thermal inactivation and denaturants and showed good durability for multiple recycling [32,33]. The potential of immobilized trypsin for use in large-scale proteomics studies and practical applications has been shown [33,34].

Starch gel is a promising carrier for trypsin immobilization. Indeed, natural starch, which is a renewable, biocompatible, biodegradable, and relatively inexpensive material, is a convenient biopolymer for biomedical and pharmaceutical applications [44,45]. Starch is composed of two main components, amylose and amylopectin, and their ratio is determined by the starch source and pretreatment [46]. Corn, rice, wheat, and potato starches contain about 70–80% of amylopectin and 20–30% of amylose. Starch granules are not soluble in cold water, but in large amounts of water at temperatures of >40 °C, they are gelatinized. Starches extracted from different sources have different gelatinization temperatures. Potato starch has the lowest gelatinization temperature (55–65 °C), making it convenient for immobilizing enzymes. The process of immobilizing enzymes into starch gel is very simple:

starch gel is mixed with the enzyme solution, pipetted, and dried. Starch gel was previously used as a carrier to prepare highly active reagents based on butyrylcholinesterase [47] and a bioluminescence enzyme system of luminescent bacteria [48], which exhibited a longer storage time and stability under varying physical and chemical environmental conditions (temperature, pH, and the ionic strengths of solutions) [44,49,50].

Despite the advantages of the integration of enzymes into polymeric matrices, trypsin has not been immobilized into starch gel until now. Based on our previous studies [50], here, we attempted to immobilize trypsin into starch gel and, at the same time, preserve its activity and sensitivity to toxicants. In the current study, we prepared single-use dry starch disks with trypsin to simplify the enzyme-inhibition assay. The trypsin was immobilized into the 3% starch gel, which was convenient to pipette and enabled the preparation of mechanically strong, dry, single-use reagents.

The sensitivity of trypsin immobilized in starch gel differs depending on the time it spends in contact with the toxicant solution. During its incubation in the solution, the starch disk with the immobilized trypsin swelled, and, probably due to the quickening of the diffusion processes, the enzyme became more accessible to the inhibitor and the substrate. Thus, by varying the incubation time, it is possible to control the sensitivity of chemosensors based on trypsin immobilized in starch gel to toxicants. The sensitivity of the immobilized trypsin to $CuCl_2$, potassium sorbate, and $CrCl_3$ was comparable to the sensitivity of the free trypsin.

To conclude, we must note that certain difficulties may be encountered when chemosensors are used to analyze real environmental/food samples rather than laboratory samples, as accurate interpretations of results cannot be achieved without taking into account the properties and effects of natural components of the tested media [31,51].

5. Conclusions

It is unlikely that chemoenzymatic assays based on the measurement of trypsin activity will be used to detect toxicants such as metals (copper, chromium, manganese, zinc, cobalt, cadmium, nickel, arsenic, mercury, and aluminum), pesticides (fenvalerate, deltamethrin, cypermethrin, imidacloprid, glyphosate, metribuzin, tebuconazole, epoxiconazole, and fenoxaprop-p-ethyl), or quinones (benzoquinone and toluquinone) in environmental or food samples. These toxicants can produce inhibitory/stimulatory effects on trypsin activity only if their concentrations in tested samples are considerably higher than their MACs. However, chemoenzymatic assays with trypsin may be used to assess the safety of food additives. The present study demonstrated a substantial inhibitory effect of the food preservative potassium sorbate on trypsin activity, which was probably associated with the binding of this additive to the enzyme at the active center. The trypsin immobilized in the single-use dry starch disks retained its high activity during storage and showed sensitivity to toxicants similar to that of the free enzyme. Thus, the present study demonstrates that trypsin immobilized in starch gel can potentially be used to design a chemosensor for assessing the safety of food additives.

Author Contributions: Conceptualization, E.N.E.; formal analysis, I.G.T., E.V.N. and A.A.A.; funding acquisition, E.N.E.; investigation, I.G.T., E.V.N., A.A.A. and P.Y.S.; project administration, E.N.E.; supervision, E.N.E. and V.A.K.; visualization, I.G.T., A.A.A. and P.Y.S.; writing—original draft, I.G.T.; writing—review and editing, E.N.E., E.V.N. and V.A.K. All authors have read and agreed to the published version of the manuscript.

Funding: The reported study was funded by Strategic Project "Gastronomic RnD Park" of the Development Program of the Federal State Educational Institution of Higher Education, "Siberian Federal University", for 2021–2030 and by State Assignment of the Ministry of Science and Higher Education of the Russian Federation (project no. 0287-2021-0020).

Institutional Review Board Statement: Not applicable.

Data Availability Statement: Not applicable.

Acknowledgments: The authors are grateful to A.A. Deeva and L.A. Sukovatyi (Siberian Federal University, Krasnoyarsk, Russia) for their help in analyzing the structure of the trypsin.

Conflicts of Interest: The authors declare that they have no known competing financial interest or personal relationship that could have appeared to influence the work reported in this paper. The funders had no role in the design of the study; in the collection, analyses, or interpretation of data; in the writing of the manuscript; or in the decision to publish the results.

References

1. Häder, D.P.; Banaszak, A.T.; Villafañe, V.E.; Narvarte, M.E.; González, R.A.; Helbling, E.W. Anthropogenic pollution of aquatic ecosystems: Emerging problems with global implications. *Sci. Total Environ.* **2020**, *713*, 136586. [CrossRef] [PubMed]
2. Stom, D.I. Influence of polyphenols and quinones on aquatic plants and their blocking of SH-groups. *Acta Hydrochim. Hydrobiol.* **1977**, *5*, 291–298. [CrossRef]
3. Kalyabina, V.P.; Esimbekova, E.N.; Kopylova, K.V.; Kratasyuk, V.A. Pesticides: Formulants, distribution pathways and effects on human health—A review. *Toxicol. Rep.* **2021**, *8*, 1179–1192. [CrossRef] [PubMed]
4. Tchounwou, P.B.; Yedjou, C.G.; Patlolla, A.K.; Sutton, D.J. Heavy Metal Toxicity and the Environment. *Exp. Suppl.* **2012**, *101*, 133–164. [CrossRef] [PubMed]
5. Kudryasheva, N.S.; Esimbekova, E.N.; Remmel, N.N.; Kratasyuk, V.A. Effect of quinones and phenols on a triple enzymic bioluminescent system with protease. *Luminescence* **2003**, *18*, 224–228. [CrossRef] [PubMed]
6. Lennerz, B.S.; Vafai, S.B.; Delaney, N.F.; Clish, C.B.; Deik, A.A.; Pierce, K.A.; Ludwig, D.L.; Mootha, V.K. Effects of sodium benzoate, a widely used food preservative, on glucose homeostasis and metabolic profiles in humans. *Mol. Genet. Metab.* **2015**, *114*, 73–79. [CrossRef]
7. Electronic Fund of Regulatory, Technical and Regulatory Information of the Consortium "Kodeks". Available online: https://docs.cntd.ru/document/902359401 (accessed on 6 February 2023).
8. Mohammadzadeh-Aghdash, H.; Akbari, N.; Esazadeh, K.; Dolatabad, J.E.N. Molecular and technical aspects on the interaction of serum albumin with multifunctional food preservatives. *Food Chem.* **2019**, *293*, 491–498. [CrossRef]
9. Sambu, S.; Hemaram, U.; Murugan, R.; Alsofi, A.A. Toxicological and Teratogenic Effect of Various Food Additives: An Updated Review. *BioMed Res. Int.* **2022**, *2022*, 6829409. [CrossRef]
10. Yi, X.; Gao, Z.; Liu, L.; Zhu, Q.; Hu, G.; Zhou, X. Acute toxicity assessment of drinking water source with luminescent bacteria: Impact of environmental conditions and a case study in Luoma Lake, East China. *Front. Environ. Sci. Eng.* **2020**, *14*, 109. [CrossRef]
11. Esimbekova, E.N.; Torgashina, I.G.; Kalyabina, V.P.; Kratasyuk, V.A. Enzymatic Biotesting: Scientific Basis and Application. *Contemp. Probl. Ecol.* **2021**, *14*, 290–304. [CrossRef]
12. Amine, A.; Arduini, F.; Moscone, D.; Palleschi, G. Recent advances in biosensors based on enzyme inhibition. *Biosens. Bioelectron.* **2016**, *76*, 180–194. [CrossRef]
13. Batt, A.R.; St Germain, C.P.; Gokey, T.; Guliaev, A.B.; Baird, T., Jr. Engineering trypsin for inhibitor resistance. *Protein Sci.* **2015**, *24*, 1463–1474. [CrossRef] [PubMed]
14. Caruso, G.; De Pasquale, F.; Mita, D.G.; Micale, V. Digestive enzymatic patterns as possible biomarkers of endocrine disruption in the red mullet (*Mullus barbatus*): A preliminary investigation. *Mar. Pollut. Bull.* **2016**, *105*, 37–42. [CrossRef]
15. Hani, Y.M.I.; Turies, C.; Palluel, O.; Delahaut, L.; Gaillet, V.; Bado-Nilles, A.; Porcher, J.M.; Geffard, A.; Dedourge-Geffard, O. Effects of chronic exposure to cadmium and temperature, alone or combined, on the threespine stickleback (*Gasterosteus aculeatus*): Interest of digestive enzymes as biomarkers. *Aquat. Toxicol.* **2018**, *199*, 252–262. [CrossRef]
16. Renella, G.; Mench, M.; van der Lelie, D.; Pietramellara, G.; Ascher, J.; Ceccherini, M.T.; Landi, L.; Nannipieri, P. Hydrolase activity, microbial biomass and community structure in long-term Cd-contaminated soils. *Soil. Biol. Biochem.* **2004**, *36*, 443–451. [CrossRef]
17. Savage, G.P.; Morrison, S.C. Trypsin inhibitors. In *Encyclopedia of Food Sciences and Nutrition*, 2nd ed.; Caballero, B., Ed.; Academic Press: Cambridge, MA, USA, 2003; pp. 5878–5884. [CrossRef]
18. Shimada, H.; Funakoshi, T.; Waalkes, M.P. Acute, nontoxic cadmium exposure inhibits pancreatic protease activities in the mouse. *Toxicol. Sci.* **2000**, *53*, 474–480. [CrossRef] [PubMed]
19. Hu, X.; Yu, Z.; Liu, R. Spectroscopic investigations on the interactions between isopropanol and trypsin at molecular level. *Spectrochim. Acta A Mol. Biomol. Spectrosc.* **2013**, *108*, 50–54. [CrossRef]
20. Momeni, L.; Shareghi, B.; Sabouryc, A.A.; Farhadian, S.; Reisi, F. A spectroscopic and thermal stability study on the interaction between putrescine and bovine trypsin. *Int. J. Biol. Macromol.* **2017**, *94*, 145–153. [CrossRef] [PubMed]
21. Mu, Y.; Lin, J.; Liu, R. Interaction of sodium benzoate with trypsin by spectroscopic techniques. *Spectrochim. Acta A Mol. Biomol. Spectrosc.* **2011**, *83*, 130–135. [CrossRef]
22. Esimbekova, E.N.; Asanova, A.A.; Deeva, A.A.; Kratasyuk, V.A. Inhibition effect of food preservatives on endoproteinases. *Food Chem.* **2017**, *235*, 294–297. [CrossRef]

23. Esimbekova, E.N.; Kalyabina, V.P.; Kopylova, K.V.; Lonshakova-Mukina, V.I.; Antashkevich, A.A.; Torgashina, I.G.; Lukyanenko, K.A.; Nemtseva, E.V.; Kratasyuk, V.A. The Effects of Commercial Pesticide Formulations on the Function of In Vitro and In Vivo Assay Systems. A Comparative Analysis. *Chemosensors* **2022**, *10*, 328. [CrossRef]
24. Humphrey, W.; Dalke, A.; Schulten, K. VMD: Visual molecular dynamics. *J. Mol. Graph* **1996**, *14*, 33–38. [CrossRef] [PubMed]
25. Van Der Spoel, D.; Lindahl, E.; Hess, B.; Groenhof, G.; Mark, A.E.; Berendsen, H.J. GROMACS: Fast, flexible, and free. *J. Comput. Chem.* **2005**, *26*, 1701–1718. [CrossRef] [PubMed]
26. Esimbekova, E.N.; Kalyabina, V.P.; Kopylova, K.V.; Lonshakova-Mukina, V.I.; Antashkevich, A.A.; Torgashina, I.G.; Lukyanenko, K.A.; Nemtseva, E.V.; Kratasyuk, V.A. Enzyme Inhibition-Based Assay to Estimate the Contribution of Formulants to the Effect of Commercial Pesticide Formulations. *Int. J. Mol. Sci.* **2023**, *24*, 2268. [CrossRef]
27. Song, W.; Yu, Z.; Hu, X.; Liu, R. Dissection of the binding of hydrogen peroxide to trypsin using spectroscopic methods and molecular modeling. *Spectrochim. Acta A Mol. Biomol. Spectrosc.* **2015**, *137*, 286–293. [CrossRef]
28. Lakowicz, J. *Principles of Fluorescence Spectroscopy*, 3rd ed.; Springer Science & Business Media: New York, NY, USA, 2006. Available online: https://link.springer.com/book/10.1007/978-0-387-46312-4 (accessed on 5 December 2007).
29. Berndt, K.D.; Güntert, P.; Orbons, L.P.; Wüthrich, K. Determination of a high-quality nuclear magnetic resonance solution structure of the bovine pancreatic trypsin inhibitor and comparison with three crystal structures. *J. Mol. Biol.* **1992**, *227*, 757–775. [CrossRef] [PubMed]
30. Amine, A.; Mohammadi, H.; Bourais, I.; Palleschi, G. Enzyme inhibition-based biosensors for food safety and environmental monitoring. *Biosens. Bioelectron.* **2006**, *21*, 1405–1423. [CrossRef] [PubMed]
31. Esimbekova, E.N.; Kalyabina, V.P.; Kopylova, K.V.; Torgashina, I.G.; Kratasyuk, V.A. Design of bioluminescent biosensors for assessing contamination of complex matrices. *Talanta* **2021**, *233*, 122509. [CrossRef]
32. Dušeková, E.; Garajová, K.; Yavaşer, R.; Tomková, M.; Sedláková, D.; Dzurillová, V.; Kulik, N.; Fadaei, F.; Shaposhnikova, A.; Minofar, B.; et al. Modulation of global stability, ligand binding and catalytic properties of trypsin by anions. *Biophys. Chem.* **2022**, *288*, 106856. [CrossRef]
33. Sahin, S.; Ozmen, I. Covalent immobilization of trypsin on polyvinyl alcohol-coated magnetic nanoparticles activated with glutaraldehyde. *J. Pharm. Biomed. Anal.* **2020**, *184*, 113195. [CrossRef]
34. Sun, X.; Cai, X.; Wang, R.Q.; Xiao, J. Immobilized trypsin on hydrophobic cellulose decorated nanoparticles shows good stability and reusability for protein digestion. *Anal. Biochem.* **2015**, *477*, 21–27. [CrossRef]
35. Sun, J.; Yang, L.; Jiang, M.; Shi, Y.; Xu, B.; Ma, H.L. Stability and activity of immobilized trypsin on carboxymethyl chitosan-functionalized magnetic nanoparticles cross-linked with carbodiimide and glutaraldehyde. *J. Chromatogr. B Anal. Technol. Biomed. Life Sci.* **2017**, *1054*, 57–63. [CrossRef] [PubMed]
36. Ganachaud, C.; Bernin, D.; Isaksson, D.; Holmberg, K. An anomalous behavior of trypsin immobilized in alginate network. *Appl. Microbiol. Biotechnol.* **2013**, *97*, 4403–4414. [CrossRef] [PubMed]
37. Du, Y.; Zhong, L.; Wang, Z.; Feng, Y.; Cui, J.; Jia, S. Immobilization of multienzymes: Problems and solutions. In *Biocatalyst Immobilization*; Ferreira, M.L., Ed.; Academic Press: Cambridge, MA, USA, 2022; pp. 317–340. [CrossRef]
38. Daglioglu, C.; Zihnioglu, F. Covalent immobilization of trypsin on glutaraldehyde-activated silica for protein fragmentation. *Artif. Cells Blood Substit. Immobil. Biotechnol.* **2012**, *40*, 378–384. [CrossRef] [PubMed]
39. Kang, K.; Kan, C.; Yeung, A.; Liu, D. The properties of covalently immobilized trypsin on soap-free P(MMA-EA-AA) latex particles. *Macromol. Biosci.* **2005**, *5*, 344–351. [CrossRef]
40. Villalonga, R.; Fragoso, A.; Cao, R.; Ortiz, P.D.; Villalonga, M.L.; Damiao, A.E. Supramolecular-mediated Immobilization of Trypsin on Cyclodextrin-modified Gold Nanospheres. *Supramol. Chem.* **2005**, *17*, 387–391. [CrossRef]
41. Aslani, E.; Abri, A.; Pazhang, M. Immobilization of trypsin onto Fe_3O_4@SiO_2-NH_2 and study of its activity and stability. *Colloids. Surf. B Biointerfaces* **2018**, *170*, 553–562. [CrossRef]
42. De Oliveira, J.M.; Fernandes, P.; Benevides, R.G.; de Assis, S.A. Production, characterization, and immobilization of protease from the yeast *Rhodotorula oryzicola*. *Biotechnol. Appl. Biochem.* **2021**, *68*, 1033–1043. [CrossRef]
43. Mahmood, M.S.; Asghar, H.; Riaz, S.; Shaukat, I.; Zeeshan, N.; Gul, R.; Ashraf, N.M.; Saleem, M. Expression and immobilization of trypsin-like domain of serine protease from Pseudomonas aeruginosa for improved stability and catalytic activity. *Proteins* **2022**, *90*, 1425–1433. [CrossRef]
44. Chakraborty, R.; Kalita, P.; Sen, S. Natural Starch in Biomedical and Food Industry: Perception and Overview. *Curr. Drug Discov. Technol.* **2019**, *16*, 355–367. [CrossRef]
45. Gamage, A.; Thiviya, P.; Mani, S.; Ponnusamy, P.G.; Manamperi, A.; Evon, P.; Merah, O.; Madhujith, T. Environmental Properties and Applications of Biodegradable Starch-Based Nanocomposites. *Polymers* **2022**, *14*, 4578. [CrossRef]
46. French, D. Chemical and Physical Properties of Starch. *J. Anim. Sci.* **1973**, *37*, 1048–1061. [CrossRef]
47. Lonshakova-Mukina, V.I.; Esimbekova, E.N.; Kratasyuk, V.A. Multicomponent butyrylcholinesterase preparation for enzyme inhibition-based assay of organophosphorus pesticides. *Catalysts* **2022**, *12*, 643. [CrossRef]
48. Lonshakova-Mukina, V.; Esimbekova, E.; Kratasyuk, V. Impact of enzyme stabilizers on the characteristics of biomodules for bioluminescent biosensors. *Sens. Actuators B Chem.* **2015**, *213*, 244–247. [CrossRef]
49. Lonshakova-Mukina, V.; Esimbekova, E.; Kratasyuk, V. Thermal inactivation of butyrylcholinesterase in starch and gelatin gels. *Catalysts* **2021**, *11*, 492. [CrossRef]

50. Bezrukikh, A.E.; Esimbekova, E.N.; Nemtseva, E.V.; Kratasyuk, V.A.; Shimomura, O. Gelatin and starch as stabilizers for the coupled enzyme system of luminous bacteria NADH:FMN-oxidoreductase-luciferase. *Anal. Bioanal. Chem.* **2014**, *406*, 5743–5747. [CrossRef] [PubMed]
51. Jisui Tan, I.; Liu, L.; Li, F.; Chen, Z.; Chen, G.Y.; Fang, F.; Guo, J.; He, M.; Zhou, X. Screening of Endocrine Disrupting Potential of Surface Waters via an Affinity-Based Biosensor in a Rural Community in the Yellow River Basin, China. *Environ. Sci. Technol.* **2022**, *56*, 14350–14360. [CrossRef]

chemosensors

MDPI

Article

Design of Peptide Ligand for Lactoferrin and Study of Its Binding Specificity

Tatiana Zimina [1,2,*], Nikita Sitkov [1,2,*], Vladimir Karasev [1,2], Yury Skorik [3,4], Alexey Kolobov [2,5], Alexander Kolobov [5], Nikolay Bunenkov [6] and Viktor Luchinin [1]

1 Department of Micro and Nanoelectronics, Saint Petersburg Electrotechnical University "LETI", 197022 Saint Petersburg, Russia
2 Centre for Digital Telecommunication Technologies, Saint Petersburg Electrotechnical University "LETI", 197022 Saint Petersburg, Russia
3 Almazov National Medical Research Centre, 197341 Saint Petersburg, Russia
4 Institute of Macromolecular Compounds, Russian Academy of Sciences, 199004 Saint Petersburg, Russia
5 Laboratory of Peptide Chemistry, Institute of Human Hygiene, Occupational Pathology and Ecology, 188663 Saint Petersburg, Russia
6 Department of Bone Marrow Transplantation, Raisa Gorbacheva Research Institute of Children Oncology, Hematology and Transplantation of Pavlov First Saint Petersburg State Medical University, 197022 Saint Petersburg, Russia
* Correspondence: tmzimina@gmail.com (T.Z.); sitkov93@yandex.ru (N.S.)

Abstract: The in silico modelling of peptides complementary to lactoferrin was carried out using the Protein 3D software package and replication of the natural bonding site between pneumococcal surface protein (PSP) and lactoferrin (LF). The modeling was based on analysis of the conjugated ion–hydrogen bond systems between these proteins (CIHBS). The oligopeptide EEVAPQAQAKIAE-LENQVHRLE was proposed via computer modelling and synthesized using the solid phase synthesis technique, purified, and analyzed with MS and HPLC methods to confirm >95% purity. The peptide was then studied by capillary electrophoresis (CE). The CE experiments demonstrated the split of peptide zone in the presence of LF, due to complex formation and subsequent mobility change of the system peptide-protein. The reference experiments with homomyeloperoxidase and myoglobin did not show binding with LETI-11.

Keywords: peptide aptamers; biosensors; lactoferrin; electrophoresis; ligands

Citation: Zimina, T.; Sitkov, N.; Karasev, V.; Skorik, Y.; Kolobov, A.; Kolobov, A.; Bunenkov, N.; Luchinin, V. Design of Peptide Ligand for Lactoferrin and Study of Its Binding Specificity. *Chemosensors* **2023**, *11*, 162. https://doi.org/10.3390/chemosensors11030162

Academic Editor: Xiaolong Yang

Received: 3 February 2023
Revised: 18 February 2023
Accepted: 24 February 2023
Published: 27 February 2023

1. Introduction

The development of multiparametric diagnostics of non-communicable diseases (NCD) is becoming increasingly actual [1,2]. In fact, the prevalence of such diseases has begun to exceed the level of pandemic virus diseases, such as influenza and COVID-19, while their diversity requires special, complex, multifunctional diagnostics with the participation of many specialists [3]. A promising approach towards solving this problem is the application of multiparametric biosensors for detection of protein diagnostic marker-arrays and development of computer expert systems on this basis [4–6].

The main problem of such a diagnostic approach is the achievement of the selective recognition and binding of target marker protein (TMP). Antibodies are widely used as bioselective elements for binding TMPs in biosensor systems since they are natural protein molecules with high specificity and affinity for the target proteins. Antibodies have a number of limitations, namely, low stability at high temperatures, limited service life, high production cost, and the complexity of their chemical modification to increase the sensitivity and selectivity of the method [7]. These limitations could be minimized by using aptamers—synthetic oligonucleotides or oligopeptides with spatial complementarity to a certain region of the TMPs [8–11].

Originally, aptamers were produced by in vitro directed selection from combinatorial libraries of oligonucleotides. Aptamer isolation technology is called SELEX (systematic evolution of ligands by exponential enrichment) and is a multistage process [12].

Peptide aptamers are oligopeptides (consisting of 10–30 amino acids) that form spatial structures with a high degree of affinity and specificity of attachment to the TMP molecule. Peptide aptamers have dissociation constants that are comparable to, and sometimes even better than, those of antibodies [13]. Therefore, they are extremely promising for the development of biosensors for the detection of protein structures.

The number and diversity of approaches to designing peptide aptamers are ever-growing and their capabilities are improving every year. Modern in vitro methods are based on the improvement of the SELEX technology and are generally very expensive and laborious [12,13]. Alternatives to these methods are approaches based on the in silico atomic-molecular design of peptide aptamers and the analysis of their interactions with target proteins. The main directions of these approaches are molecular docking and non-stationary dynamic modeling [14–16].

Molecular docking is performed by virtual docking of the ligand to the binding site of the target protein and its movement to determine the location and conformation that will be most beneficial for selective binding. Proteins can possess several binding sites, and the use of docking is also one of the methods for assessing the quality of binding of a peptide aptamer to a target protein [14]. Non-stationary dynamic modeling allows one to form a detailed course of the aptamer-protein binding process [17–19]. Based on these methods, various computational services have been developed to search for binding sites for peptide aptamers with target proteins, such as pepATTRACT, FlexPepDock, HADDOCK [14–16], PEP-SiteFinder, and others [17–19].

The Engineering Center for Microtechnology and Diagnostics (CMID) of St. Petersburg Electrotechnical University "LETI" has developed an approach for atomic and molecular modeling of protein structures based on the model of a protein molecular vector machine (MVM) and the concept of Conjugated Ionic-Hydrogen Bond Systems (CIHBS) [20–22].

It is known from electronics that the features of functional electronic devices are determined by the choice of charge transfer principles and architectonics, and design methods. In biomolecular systems, processes associated with the transfer of charges (electrons and protons) also occur. Supramolecular structures of biosystems are similar to hierarchically organized electronic devices of ultra-high complexity [23], and their structure and properties should also be related to the principles of charge transfer [24]. As part of the development of the problem of charge transfer in biomolecular structures, the concept of CIHBS was proposed, which acts as the basis for their construction, as well as charge transfer channels design. Hydrogen atoms are taken as inputs, and lone pairs of electrons are taken as outputs in such molecular models, where molecular valves, signal delay elements, and charge sources are distinguished. The main conditions for the implementation of the model of charge transfer along with the CIHBS are the symmetry and subunit structure.

The atomic-molecular design of the peptide aptamers in silico is realized based on spatial analysis of the areas of possible interaction between the ligand and the marker. For the visualization of protein structures, the Protein 3D software developed earlier was used [21]. The software makes it possible to predict the amino acid sequence for the synthesis of peptide aptamers with spatial complementarity to target proteins, based on the contact region of neighboring globular proteins. Using this program, peptide sequences have been earlier proposed for a number of proteins that can be used as bioselective elements in analytical microsystems [4,5,25].

Thus, the considered methods of atomic-molecular design of bioselective elements in silico make it possible to create highly specific peptide recognition elements that can be integrated into multimodal microanalytical systems capable of implementing express-multiparametric detection, thus increasing the performance of the clinical analysis.

At present time, many groups of proteins have been identified as markers of various diseases, such as cardiovascular, oncological, lung diseases, diabetes, pancreatitis, etc., and described in data banks [26–29]. Lactoferrin (LF) is one of the important targets for diagnosing acute inflammatory processes and infections [30,31]. The important role of LF in maintaining the oxidative balance of the body is also noted [32]. LF is present in neutrophil granules, which tend to degrade under ischemic stress, releasing LF as well as myeloperoxidase (MPO). Concentration of LF correlates with concentration of MPO, which in turn is the marker of poor outcome of organ transplantation. It has also been reported that LF has an important role in supporting physiologic homeostasis during development of various pathologies due to its immunomodulatory nature, explained by the ability to monitor the immune status of an organism and act accordingly [33]. The functional role of LF in physiological homeostasis of an organism during development of disease and associated pathology is described in detail in the review [34]. The review data confirm an importance of LF as a prognostic marker of acute disease, when it fulfills its anti-inflammatory roles via different cell receptors and activation of various cell signaling pathways [34]. These features particularly explain an interest for clinical monitoring of LF levels during surgery, transplantation, acute viral or bacterial infections, etc. [35]. Normally, LF content in blood is reported as 59–345 ng/mL, while in pathology it could be about 10–500 μg/mL [36,37]. With such a scatter of values in the human blood, the dynamics of changes in this parameter are usually monitored [34,35].

Analytical strategies for determination of LF are described in a recent review [36]. In clinical practice, immunoassays (radial immunodiffusion (RID), enzyme-linked immunosorbent assay (ELISA), instrumental methods (reversed-phase high performance liquid chromatography (RP-HPLC)), and capillary electrophoresis (CE) are usually used [36]. Microfluidic devices using various detection techniques (fluorescence-based biosensor, sensor based on oligonucleotide aptamers with conjugated FITC, electrochemical immunobiosensor) are described [36]. A label-free surface plasmon resonance (SPR) immunoassay was used for detecting LF from human milk in the range of up to 1000 ng/mL [36].

It is important to note that LF is observed in blood in oligomer forms and as various complexes, for example, myeloperoxidase-lactoferrin-ceruloplasmin (MPO-LF-CP) [33,38,39]. This tendency for forming protein associations outputs further questions regarding selective determination via spatial recognition. For example, in [33], the interaction between this complex and peptides obtained via proteolysis of ceruloplasmin (CP), in which the peptides displaced CP bound to LF, is described. Thus, the problem of interaction site geometry and special location in the complex needs to be analyzed. It will be completed below in Section 3 of this article.

As shown in recent reports for the detection of LF using biosensor systems, antibodies or oligonucleotide aptamers are most often used as a biorecognition elements [36,37]. Peptide aptamers can be comparable to them in selectivity, and their synthesis is much cheaper. Therefore, the search for a peptide sequence capable of selectively recognizing and binding LF, and subsequently being used for its rapid detection in a biosensor format seems to be an actual and important enough task. However, the ultimate goal of the effort is development and implementation of high dimension/density multiparametric miniature devise for realization of system diagnostics. The implementing of such diagnostic systems needs a number of additional problems to be solved as well, besides in silico modelling of complementary ligands (aptamers) for selective binding of target proteins. These include design of high dimension/density miniature devises, and development of affordable registration principles [40], which are in the agenda for future work.

Presented in this article are results of the development and testing of a peptide to bind to a protein of the acute phase of inflammation, which contributes to regulating the functions of immunocompetent cells—lactoferrin (LF) [36,37].

2. Materials and Methods

2.1. Peptide Modelling

The analysis of the spatial structure of globular marker proteins capable of forming oligomers was carried out using the Protein 3D software (ETU, St. Petersburg) [21]. This software is designed to visualize supramolecular structures (primarily proteins) presented in the Protein Data Bank in PDB format [26]. As one of the forms of representation of the structure (rendering), the program implements the conjugated ionic-hydrogen bonds systems (CIHBS). The software allows the selection of continuous CIHBS analysis. Using this program, complementary regions of the subunit contact area were identified, and their prospects were assessed for the formation of stable systems of CIHBS, enabling the amino acid sequences for the creation of aptamer peptides on their basis to be proposed. Where necessary, sequence optimization could be performed to exclude fluorescent amino acid side chains interfering with fluorescent detection [4]. Thus, as an additional criterion in the process of selecting protein fragments suitable for use as aptamer templates, the analysis of CIHBS in the selected structures was used.

The concept of CIHBS is described earlier in the most complete form in [22]. The essence of this concept consists of the following statements:

(1) Supramolecular biostructures are built on the basis of the principle of continuity of type systems:

$$\ldots HQ_1\text{-}\overset{\overset{\textstyle R}{|}}{R} = X_1 \ldots Z_1 \ldots HQ_2\text{-}R = X_2 \ldots HQ_2\text{-}R = X_2 \ldots \qquad \ldots HQ_n\text{-}\overset{\overset{\textstyle R}{|}}{R} = X_n \ldots Z_m \ldots . \tag{1}$$

where: R—Zi—simple and Qi–R = Xi—resonance groups, containing instead of R, Z, Q, X, atoms of organogenic elements (C, N, O, P, S).

(2) These systems are utilized in biostructures as channels for energy and charge transfer. Formation of CIHBS in the area of subunit contact considerably promotes it stabilization.

2.2. Peptide Synthesis

Synthesis of the peptide was carried out by solid state method, in situ, using synthesizer "Applied Biosystems 430A (Applied Biosystems, Waltham, MA, USA)" and Nα-Boc-protected amino acid derivatives. The following types of trifunctional derivatives of amino acids were used: Boc-(Mts)Arg-OH, Boc-(Dnp)-His-OH, Boc-(OBzl)-Glu-OH, Boc-(ClZ)-Lyz-OH, etc. The synthesis was carried out using 0.1 mM of corresponding aminoacyl polymers with initial capacity of 0.5–0.8 mM/g. The polymer was 1% copolymer of styrene-divinylbenzene. Deblocking was performed using trifluoroacetic acid (TFA) for 1 min, twice. Neutralization at first condensation was made by adding a threefold excess of diisopropylethylamine (DIPEA) directly into the reaction chamber at the stage of bonding of amino acid residue [41]. Secondary condensation was made after the washing of peptydil polymer with 10% solution of DIPEA in dimethylformamide. Addition of amino acid residues was carried out by the method of 1-hydroxibenzotriazole ethers, using 2–5 fold excess of reagents and ninhydrin and bromophenol tests for control over the completeness of the reaction.

On completion of the peptide chain growth process, the product was treated with DMF:EDT:DIPEA (7:2:1 *v*/*v*) for 30 min and washed with DMF and DCM. Then washed with TFA twice for 1 min, washed with DCM and DEE, removed from the reactor, and dried.

The separation of peptide from polymer and removing of the side protective groups was carried out using liquid HF according to Sn1/Sn2 mechanisms under the presence of scavengers. The isolated products were cleaned using semi-preparative reversed-phase HPLC (Waters Prep Nova-Pak HR C-18, 6 μ, 60 Å, 19 × 300 mm^2) under gradient of acetonitrile. Detection occurred at 220 nm. The isolated fraction was analyzed with MS and HPLC methods showing the purity of >95%.

The samples of proteins and synthesized peptide LETI-11 are described in Table 1. The model proteins, MPO and MB, were selected for the following reasons: MPO as a protein, as its content often correlates with LF content in blood, forming complexes with LF in blood, and which was isolated for us, specifically for this work, from human leukocytes, by our scientific partners at The Institute of Experimental Medicine RAS; MB was selected as an iron-binding protein, which can be present in blood as a marker of myocardial infarction, and of appropriate *pI* and MW values. BSA could be a candidate, but its *pI* value was too close to LETI-11, which was not convenient for electrokinetic experiments.

Table 1. Proteins and peptide investigated and their properties.

No.	Sample	MW, Da	*pI*	Reference
1	Myoglobin from equine heart (MB); (Sigma-Aldrich, St. Louis, MO, USA)	117,199	6.8	[42,43]
2	Myeloperoxidase from human leucocytes in monomeric form (homoMPO); (IEM RAS, St. Petersburg, Russia)	75,000	5.62	[44–46]
3	Lactoferrin from human milk (LF) (IEM RAS, St. Petersburg)	80,000	8–9	[47–49]
4	Peptide (LETI-11): EEVAQAQAKIAELENQVHRLE (SPB ETU—IOPHS, St. Petersburg)	2345	5.6	[50]

2.3. Capillary Electrophoresis

The measurements were carried out using a Capel 105M capillary electrophoresis system (Lumex, Russia) [51] equipped with quartz capillaries with an internal diameter of 75 μm, a total capillary length of 60 cm, and the effective length of 50 cm. The experiments were carried out at temperature of 30 °C. The registration of the zones of the studied substances was carried out using a spectrophotometric detector at a wavelength of 200 nm. The wavelength was chosen due to the high level of absorbance of the peptide at the range of 200–220 nm [52,53].

Capillaries were prepared in two ways—A and B. Preparation of the capillary A procedure: washed with a solution of 1M HCl for 10 min, washing with deionized water for 10 min. Preparation of capillary B procedure: washing with deionized water for 5 min, washing with a solution of 1M HCl for 10 min, washing with deionized water for 10 min, washing with 0.5 M NaOH solution for 15 min, and finally washing with deionized water for 10 min. The capillaries were treated before every run.

The proteins and peptide were dissolved in 0.01 M Tris-HCl buffer solution with pH 7.2. The selection of chemical composition and pH of the buffer solution was determined by the goal of approaching native conditions for proteins.

3. Results and Discussion

3.1. Modelling of Peptide Ligand for Lactoferrin

Lactoferrin (LF) is a polyfunctional protein from the transferrin family [33,47–49,54]. It is a globular glycoprotein, and the human form contains 691 amino acids (AA). It consists of two domains linked by a covalent bond (analogues of the subunit). As shown in Figure 1, LF contains both α-helical regions (red fragments) and β-folds regions (green fragments).

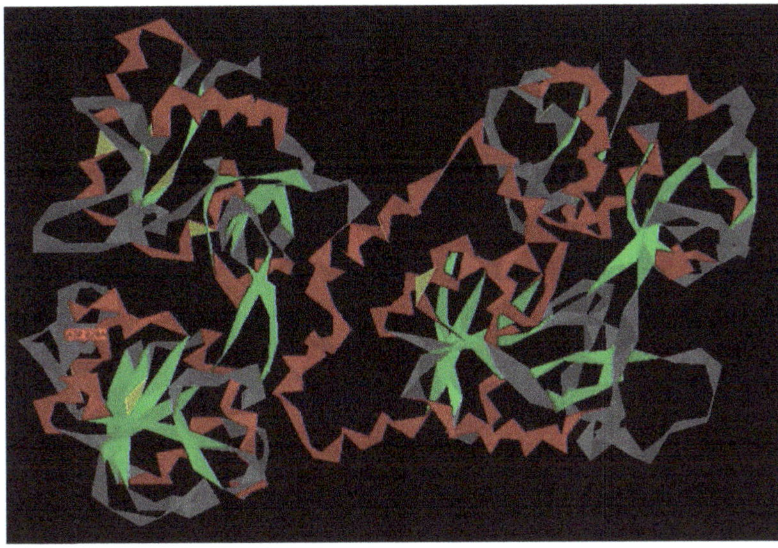

Figure 1. General view of human LF as a paper model (file 1B0L at PDB) obtained by rendering using Protein 3D software. α-Spiral parts (red); β-structures (green); left (N-lobe); and right (C-lobe).

LF is widely present in various secretory fluids such as milk, saliva, tears, and nasal secretions, as well as in blood. This protein is one of the components of the body's immune system, takes part in the system of nonspecific humoral immunity, regulates the functions of immunocompetent cells, and is an acute phase protein of inflammation. LF interacts with DNA and RNA, polysaccharides, heparin, and exhibits some of its biological functions in the form of complexes with these ligands.

However, for the purposes of this work, i.e., creating ligands that provide quantitative determination of this protein in tissues and biological fluids, these complexes are not very suitable, since they can also form complexes with other biomolecules.

Numerous structures of this protein are known, studied by X-ray diffraction analysis (XRD), some of which are presented in RSCB Protein Data Bank [26]. In our previous studies [4,25], we applied an approach that uses two principles: complementarity of the subunit contact area (to create ligands, amino acid sequences from the subunit contact area were used) and formation of systems of conjugated ionic-hydrogen bonds (CIHBS) to select optimal areas in the contact area. However, in the case of LF, the first principle does not work, since both parts of the protein are linked by a covalent bond and the formation of a complex with a ligand from the subunit contact area is impossible. For this reason, it was necessary to look for other complexes in which the patterns of protein-protein recognition are manifested.

One of the possible options used in this work is the formation of the LF complex with microbial proteins. For example, a complex described in [49] in which the LF complex with pneumococcal surface protein (PSP) was obtained. The authors of this work consider the formation of this complex, 2PMS, as an example of the manifestation of the immune antimicrobial properties of LF [26,49].

The entire protein was not used as a model, only the N-lobe of LF was used. The resulting complex, which crystallizes as a dimer, was examined by X-ray diffraction analysis with a resolution of 2.9 Å. This, of course, is not a very strong resolution, but when comparing the two subunits of the complex with other structures, as well as with the primary structures of homologous proteins, this complex was taken as the basis for the development of the ligand. As before, we used the Protein 3D software package [21].

As shown in Figure 2, the PSP molecule consists of three α-helical fragments, and not all of it participates in the formation of the complex, rather only one fragment participates.

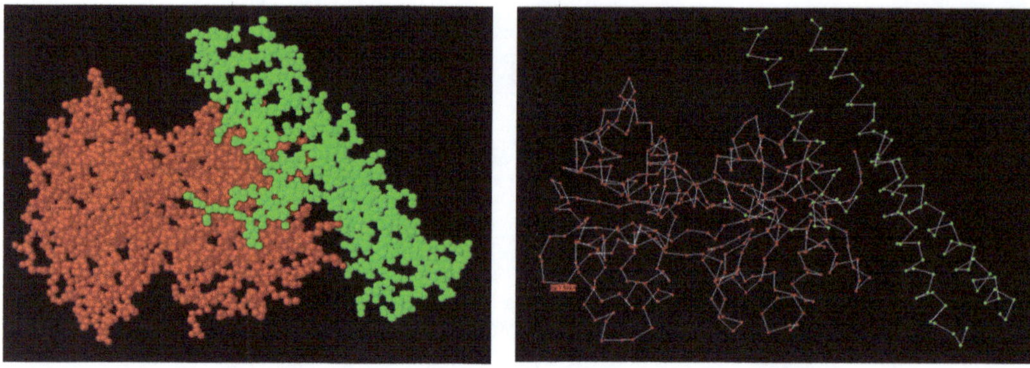

(**a**) (**b**)

Figure 2. General view of complex of N-lobe of LF (red) with pneumococcus surface protein (PSP—green) by rendering atoms (**a**); complex of N-lobe of LF with PSP by rendering CA-skeleton (**b**).

For the purposes of a more detailed analysis, we have reduced the number of amino acids in the file output by Protein 3D down to 60 amino acids in LF and to 20 amino acids in PSP. At the same time, it is possible to compare the A–C and B–D complexes (Figure 3a,b). Figure 3 shows that both variants are close in structure, although an identical arrangement cannot be obtained.

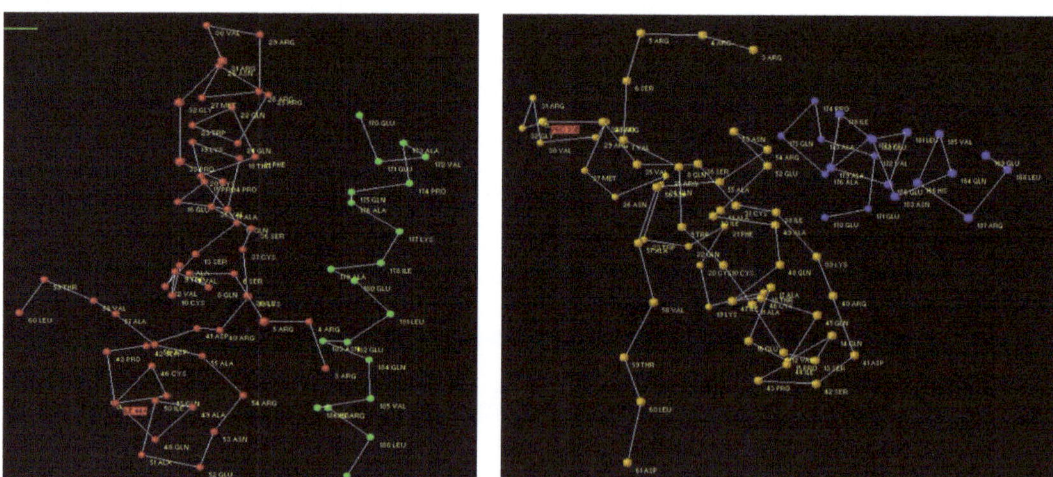

(**a**) (**b**)

Figure 3. Fragments of LF–PSP complex used for analysis: complex between A (red) and C (green) subunits (**a**), complex between B (yellow) and D (blue) subunits (**b**).

Further analysis was carried out separately for each of the two variants using CIHBS. Figure 4 shows CIHBS during the formation of hydrogen bonds between His 186, Glu 182 of PSP and Gln 45, Arg 40 of LF. It is easy to verify that the composition obtained for both variants is identical.

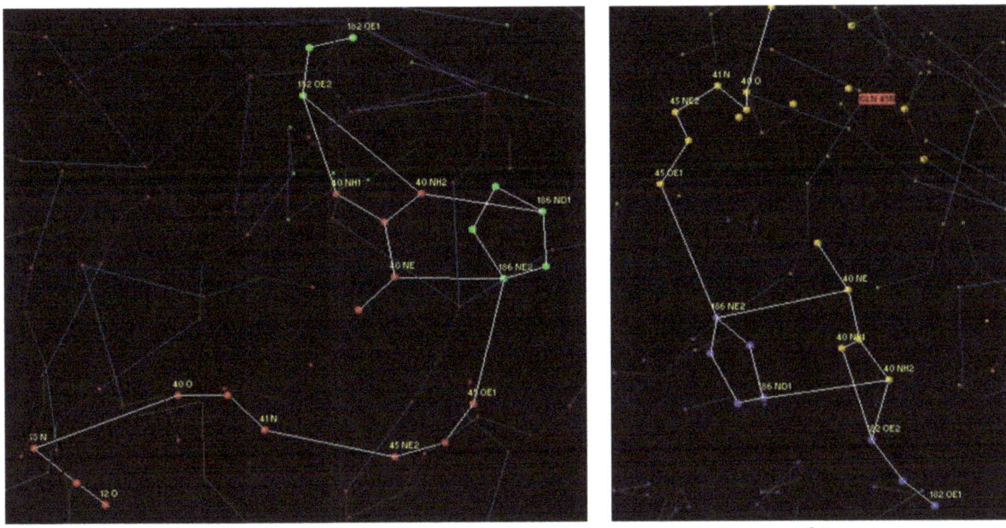

(**a**) (**b**)

Figure 4. Comparative analysis of CIHBS in complex between A (red) and C (green) subunits (**a**), complex between B (yellow) and D (blue) subunits (**b**).

Further analysis was made for Gln 175 of PSP and side chains Arg 25 and Gln 24 of LF, which form lengthy bonds inside LF. The structure for A–B and C–D was identical. Then analysis was completed for Arg 187 and Gln 183 of PSP and system of HN-C=O-groups of LF. For both cases, structures of CIHBS were identical. In the next step, in complex A–C, amino acids Glu 180, Gln 184 of PSP and Lys 39 of LF were analyzed, analogical amino acids were found in the B-D complex. Bonds were formed between Glu 171 of PSP and Gln 14 of LF. However, in this case, in the A–C complex, the bond spread through Thr 18 into the protein molecule. While in complex B–D, the bond was broken as a result of structural differences. Finally, a coupled hydrogen bond was formed between Glu 189 of PSP and Arg 54 of LF, which was identical for both complexes.

Figure 5 shows the structure of the PSP protein fragment for which the isolation of CIHBS was carried out. The names of the amino acids that are involved in the formation of CIHBS with LF are placed in white rectangles. It can be seen that almost all amino acids capable of forming H-bonds participate in the formation of the complex. The exceptions are Glu 170 and Lys 177. The latter forms an intramolecular CIHBS.

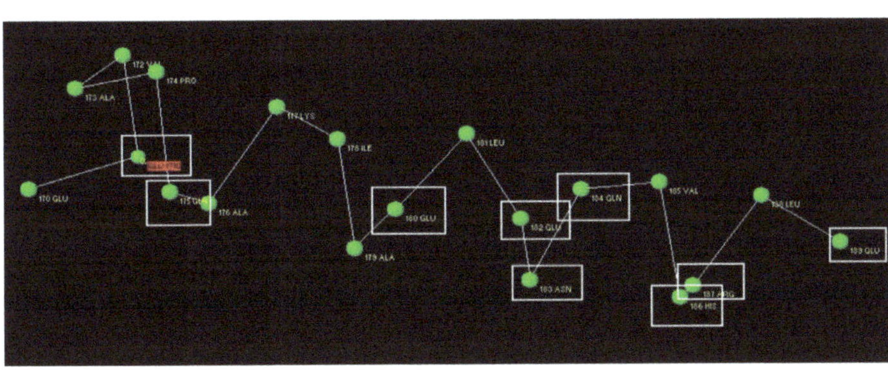

Figure 5. Fragment of PSP protein, forming a CIHBS link with N-lobe of LF.

The same amino acids are highlighted in bold on the proposed amino acid sequence to create a ligand for LG (Table 2). The peptide contains 40.91% of hydrophobic groups, 22.73% of neutral groups, 22.73% of acidic groups, and 13.64% of basic groups.

Table 2. The isolated fragment to form the ligand (peptide named LETI-11) to bind to lactoferrin.

170	**176**	**180**	**184**	**189**
GLU **GLU** VAL ALA PRO	**GLN** ALA LYS	ILE ALA **GLU** LEU	**GLU ASN GLN** VAL **HIS**	**ARG** LEU **GLU**

3.2. Study of Peptide LETI-11 and Its Selective Binding of Lactoferrin

The peptide was studied with MS Voyager-DE BioSpectrometry Workstation (Applied Biosystems, Waltham, MA, USA) (Figure 6).

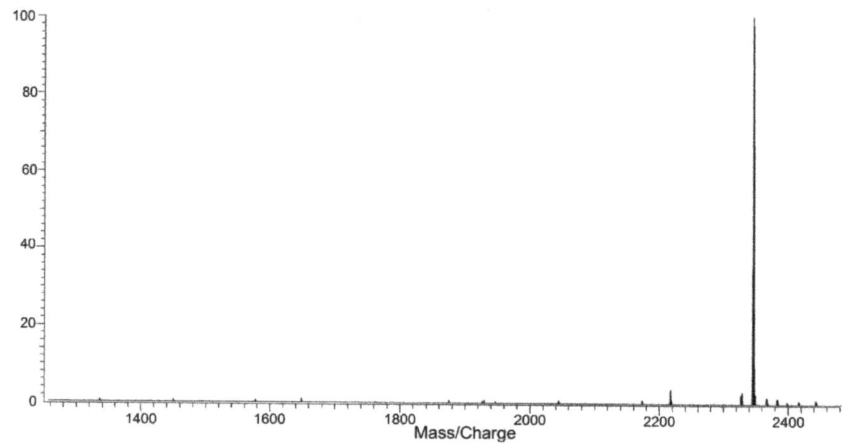

Figure 6. Mass-spectrometry data on the synthesized peptide LETI-11 (Mw = 2344.63); MS Voyager-DE BioSpectrometry Workstation (Applied Biosystems, Waltham, MA, USA).

The peptide LETI-11 was further analyzed by HPLC using chromatograph Gilson (France). The results of HPLC analysis are presented in Figure 7.

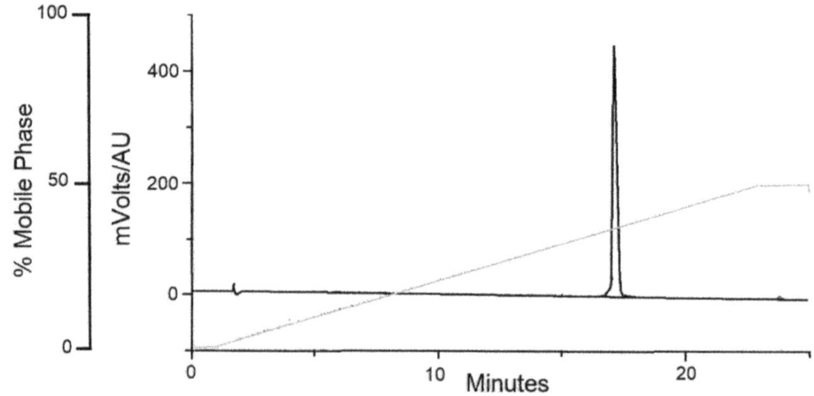

Figure 7. HPLC analysis of peptide LETI-11 sample. Column: DeltaPak C-18, 5 µm, 100 Å, 3.9 × 150 mm^2, gradient (0.1–50)% TFA in 20 min. Detector UV 220 nm.

The capillary electrophoresis system was used to study the interactions of peptide LETI-11 with target protein LF and two reference proteins—myeloperoxidase (MPO) and myoglobin (MGB). At the first stage, the capillary system was calibrated with a series of peptide LETI-11 solutions of different concentrations within a range of 0.2–3.6 μM, using capillary prepared as procedure A. The electric field strength was 400 B.cm^{-1}. A sample electropherogram of LETI-11 of 3.6 μM concentration in Tris 0.01 M buffer pH 7.2 is presented in Figure 8a. The elution time was (5.2 ± 0.3) min. A relationship of zone areas at electropherograms versus sample concentration is presented in Figure 8b.

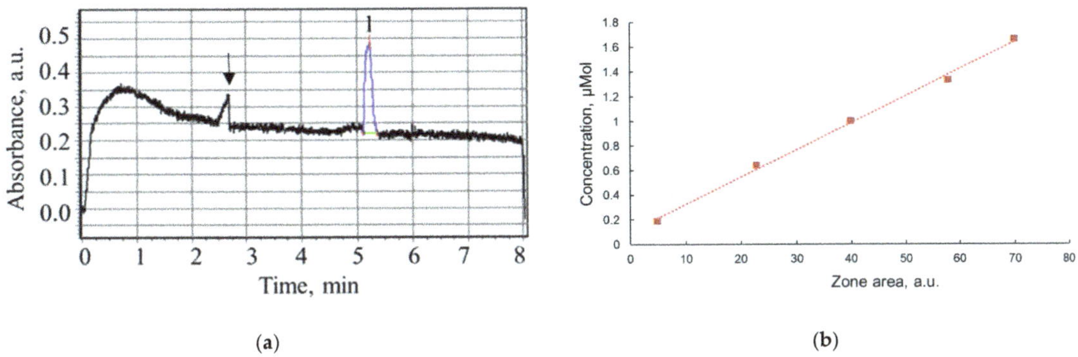

Figure 8. Electropherogram of peptide LETI-11 (1), concentration 3.6 μM, Tris 0.01 M buffer pH 7.2, U = 20 kV, capillary length 50 cm, detection wavelength 200 nm. The leading peak reflects the system zone indicating the buffer velocity. Capillary treatment type A, the arrow shows the systemic peak corresponding to the electroosmotic flow velocity (**a**); relationship of LETI-11 zone areas, S (in arbitrary units) versus molar peptide LETI-11 concentrations (**b**).

Further experiments were carried out for LF (c = 3.6 μM) and the mixture of peptide LETI-11 and LF in the following molar ratios: 1:1; 1.5:1; 2:1; 2.5:1. The zone of LF did not appear in the detector area, due to the protein charge and polarity values that kept it stuck in the inlet volume (Figure 9a). The experiments of LF and LETI-11 mixtures demonstrated the split of the LETI-11 zone into 2 zones: 2 (5.2 ± 0.3 min) and 3 (7.8 ± 0.3 min), which could be attributed to the binding of LETI-11 to LF and subsequent reduction of the mobility of the complex compared to the LETI-11. Under selected conditions of CE, closer to physiological values in blood, the LF molecule is anionic and does not migrate, but after binding to a cationic peptide, the complex moves (Figure 9b).

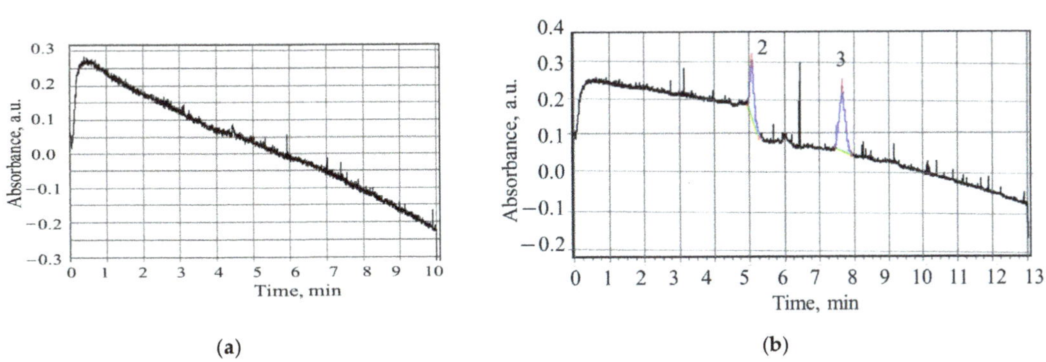

Figure 9. Electropherograms of the LF (**a**) and the mixture of LETI-11 and LF at molar proportion of 1:1 (**b**). The conditions are as in Figure 8. Zone 2 (free LETI-11); zone 3 (complex of LETI-11 and LF). Capillary pretreatment type A.

The relationship of the zone areas presented in Figure 10 shows the inflection point at 1:1 molar concentration relationship of the components ($C_{LETI-11}$:C_{LF}), which is due to the excess of peptide LETI-11 molar content over protein molar content, which remained constant. Figure 10 shows that the dependence of the area of the zone 2 (see Figure 8, i.e., the zone corresponding to the position of the LETI-11 zone without the addition of LF) versus content in mixture is linear. The dependence of the area of the zone 3 versus the molar ratio of the reaction components has a logarithmic character. This indicates the concentration dependence of the kinetics and equilibrium of the reaction. Based on the data obtained, it can be concluded that there is a statistical distribution of the reaction components and, as a result, a partial migration of LF molecules, and a partial capture of the peptide ligand by the protein.

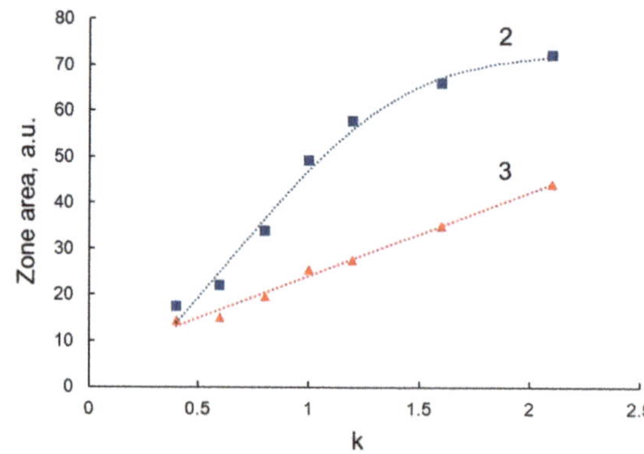

Figure 10. Relationship of sample areas of the split zones of LETI-11 (2) and complex of LETI-11 and LF (3) during electrophoresis under the presence of LF in solution, and the sum of the areas versus the ratio of components, k.

The reference experiments were performed with myeloperoxidase monomer (homoMPO) and myoglobin (MB). In Figure 11a, the electropherogram of LETI-11 and homoMPO is presented, showing the separate migration of the zones. In Figure 11b the electrophoregram of LETI-11 and myoglobin is presented, showing their separate migration.

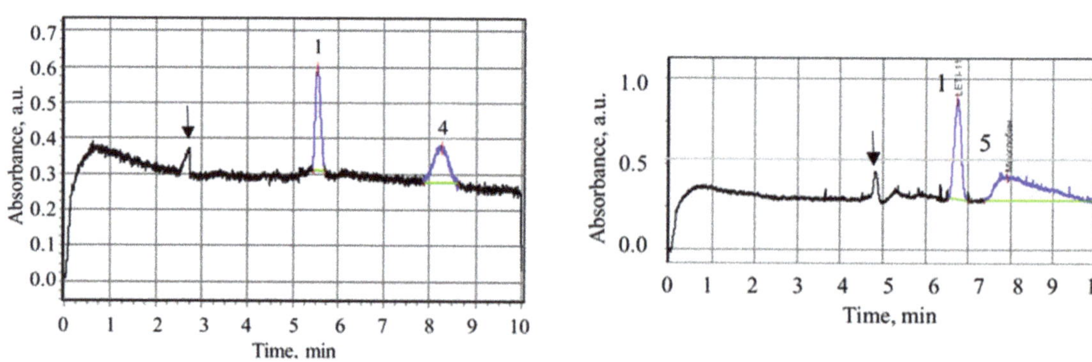

(a) (b)

Figure 11. Electropherogram of the mixture of (1) LETI-11 (1.5 µM), 5.5 min and (4) homoMPO (2.2 µM), 8.2 min, capillary pretreatment type A (**a**) and (1) LETI-11 (2.0 µM), 6.7 min and (5) MB (4.2 µM), 8 min, capillary pretreatment type B (**b**), pH = 7.2, 0.01 M tris-HCl buffer solution; the arrows show the systemic peak corresponding to the electroosmotic flow velocity.

The results demonstrate the lack of interaction between peptide LETI-11 and reference proteins MPO and MB, which manifested itself in separate migration of corresponding zones. The range of concentrations investigated was: LETI-11 2.0 μM—MB (0.5–4.2 μM) and LETI-11 2.0 μM—MPO (0.5–2 μM).

Under conditions of type B preparation of capillary, which provide higher polarity of capillary wall, the higher migration times of all components were observed, and the transport effect for LF by LETI-11 has not been demonstrated.

As an additional test, we made a theoretical assessment of the peptide LETI-11 binding to human serum albumin (HAS), a protein present in the blood at a very high concentration. We performed molecular docking using the open-source service pepATTRACT [55,56], from which we selected the model of the complex of LETI-11 and HAS with the best docking score. Further, this model was transferred to the Protein 3D software, in which the absence of CIHBS was found at the site of the supposed docking contact, when in contrast they were observed in the case of lactoferrin.

4. Conclusions

In this work, the computer design of peptide architecture providing the complementary binding to lactoferrin (LF) for use in a biosensor system is presented. Modeling was carried out using the Protein 3D program and the natural architectonics of the interaction of LF and pneumococcal surface protein (PSP). The selection of the target protein was based on consideration that the formation of this complex is an example of the immune antimicrobial properties of LF [37]. The peptide sequence EEVAPQAQAKIAELENQVHRLE was proposed, which should demonstrate spatial complementarity to LF surface fragment. The peptide was synthesized and analyzed for purity and composition, which confirmed more than 95% purity. The interaction between LF and peptide (LETI-11) was studied using capillary electrophoresis at detection wavelength of 200 nm, where molar extinction coefficients of peptides are high. Cleavage of the LETI-11 peptide zone in the presence of LF in solution under mild conditions, close to physiological conditions, was observed. At the same time, no cleavage was observed in reference experiments with homomyeloperoxidase and myoglobin. The results indicate the interaction between LF and LETI-11.

Author Contributions: Conceptualization, T.Z., N.S., V.K. and V.L.; methodology, T.Z., N.S., V.K., A.K. (Alexey Kolobov), A.K. (Alexander Kolobov), Y.S., N.B. and V.L.; software, V.K.; validation, T.Z., N.S., A.K. (Alexander Kolobov), N.B. and Y.S.; formal analysis, N.B., Y.S. and V.L.; investigation, T.Z., N.S., V.K., A.K. (Alexey Kolobov), A.K. (Alexander Kolobov) and Y.S.; resources, T.Z., N.S., Y.S., A.K. (Alexey Kolobov), A.K. (Alexander Kolobov), N.B. and V.L.; data curation, T.Z. and V.K.; writing—original draft preparation, T.Z. and N.S.; writing—review and editing, T.Z. and N.S.; visualization, T.Z., N.S., V.K. and Y.S.; supervision, T.Z. and V.L.; project administration, T.Z. and V.L.; funding acquisition, T.Z. and N.S. All authors have read and agreed to the published version of the manuscript.

Funding: The reported study was funded by the Russian Science Foundation, project number 21-79-20219.

Institutional Review Board Statement: Not applicable.

Informed Consent Statement: Not applicable.

Data Availability Statement: Not applicable.

Acknowledgments: Authors acknowledge with thanks Sokolov V. A. from the Institute of Experimental Medicine, Russian Academy of Sciences, St. Petersburg for giving samples of proteins (homomyeloperoxydase and lactoferrin). The authors are also grateful to the Department of Mathematics and Natural Sciences of the Almazov National Medical Research Centre for providing equipment for capillary electrophoresis and to M.V. Baranova for carrying out the electrophoretic experiment.

Conflicts of Interest: The authors declare no conflict of interest.

References

1. Bhosale, S.D.; Moulder, R.; Venäläinen, M.S.; Koskinen, J.S.; Pitkänen, N.; Juonala, M.T.; Kähönen, M.A.P.; Lehtimäki, T.J.; Viikari, J.S.A.; Elo, L.L.; et al. Serum Proteomic Profiling to Identify Biomarkers of Premature Carotid Atherosclerosis. *Sci. Rep.* **2018**, *8*, 9209. [CrossRef] [PubMed]
2. Emilsson, V.; Gudmundsdottir, V.; Gudjonsson, A.; Jonmundsson, T.; Jonsson, B.G.; Karim, M.A.; Ilkov, M.; Staley, J.R.; Gudmundsson, E.F.; Launer, L.J.; et al. Coding and Regulatory Variants Are Associated with Serum Protein Levels and Disease. *Nat. Commun.* **2022**, *13*, 481. [CrossRef] [PubMed]
3. WHO. Noncommunicable Diseases. Key Facts. Available online: https://www.who.int/news-room/fact-sheets/detail/noncommunicable-diseases (accessed on 30 November 2022).
4. Sitkov, N.O.; Zimina, T.M.; Soloviev, A.V.; Lemozerskii, V.E.; Karasev, V.A. Development of Peptide Aptamers—3d Complementary Ligands for Selective Binding of Target Biomarkers of Diseases in Multiparametric Sensor Systems. In Proceedings of the 2019 IEEE Conference of Russian Young Researchers in Electrical and Electronic Engineering (EIConRus), St. Petersburg, Russia, 28–30 January 2019. [CrossRef]
5. Molvin, J.; Pareek, M.; Jujic, A.; Melander, O.; Råstam, L.; Lindblad, U.; Daka, B.; Leósdóttir, M.; Nilsson, P.M.; Olsen, M.H.; et al. Using a Targeted Proteomics Chip to Explore Pathophysiological Pathways for Incident Diabetes—The Malmö Preventive Project. *Sci. Rep.* **2019**, *9*, 272. [CrossRef] [PubMed]
6. Sitkov, N.; Zimina, T.; Kolobov, A.; Karasev, V.; Romanov, A.; Luchinin, V.; Kaplun, D. Toward Development of a Label-Free Detection Technique for Microfluidic Fluorometric Peptide-Based Biosensor Systems. *Micromachines* **2021**, *12*, 691. [CrossRef]
7. Neeli, I.; Radic, M. Current Challenges and Limitations in Antibody-Based Detection of Citrullinated Histones. *Front. Immunol.* **2016**, *7*, 528. [CrossRef] [PubMed]
8. Zhang, Y.; Lai, B.; Juhas, M. Recent Advances in Aptamer Discovery and Applications. *Molecules* **2019**, *24*, 941. [CrossRef]
9. Fathil, M.F.M.; Md Arshad, M.K.; Gopinath, S.C.B.; Hashim, U.; Adzhri, R.; Ayub, R.M.; Ruslinda, A.R.; Nuzaihan, M.N.M.; Azman, A.H.; Zaki, M.; et al. Diagnostics on Acute Myocardial Infarction: Cardiac Troponin Biomarkers. *Biosens. Bioelectron.* **2015**, *70*, 209–220. [CrossRef] [PubMed]
10. Ali, M.H.; Elsherbiny, M.E.; Emara, M. Updates on Aptamer Research. *Int. J. Mol. Sci.* **2019**, *20*, 2511. [CrossRef]
11. Li, J.; Tan, S.; Chen, X.; Zhang, C.-Y.; Zhang, Y. Peptide Aptamers with Biological and Therapeutic Applications. *Curr. Med. Chem.* **2011**, *18*, 4215–4222. [CrossRef]
12. Gold, L. SELEX: How It Happened and Where It Will Go. *J. Mol. Evol.* **2015**, *81*, 140–143. [CrossRef]
13. Dausse, E.; Barré, A.; Aimé, A.; Groppi, A.; Rico, A.; Ainali, C.; Salgado, G.; Palau, W.; Daguerre, E.; Nikolski, M.; et al. Aptamer Selection by Direct Microfluidic Recovery and Surface Plasmon Resonance Evaluation. *Biosens. Bioelectron.* **2016**, *80*, 418–425. [CrossRef]
14. Morris, G.M.; Lim-Wilby, M. Molecular Docking. *Mol. Modeling Proteins* **2008**, *443*, 365–382. [CrossRef]
15. Lensink, M.F.; Velankar, S.; Wodak, S.J. Modeling Protein-Protein and Protein-Peptide Complexes: CAPRI 6th Edition. *Proteins Struct. Funct. Bioinform.* **2016**, *85*, 359–377. [CrossRef]
16. Agrawal, P.; Singh, H.; Srivastava, H.K.; Singh, S.; Kishore, G.; Raghava, G.P.S. Benchmarking of Different Molecular Docking Methods for Protein-Peptide Docking. *BMC Bioinform.* **2019**, *19*, 105–124. [CrossRef]
17. Rhinehardt, K.L.; Mohan, R.V.; Srinivas, G. Computational Modeling of Peptide–Aptamer Binding. *Comput. Pept.* **2014**, 313–333. [CrossRef]
18. Karplus, M.; McCammon, J.A. Molecular Dynamics Simulations of Biomolecules. *Nat. Struct. Biol.* **2002**, *9*, 646–652. [CrossRef]
19. Senn, H.M.; Thiel, W. QM/MM Methods for Biomolecular Systems. *Angew. Chem. Int. Ed.* **2009**, *48*, 1198–1229. [CrossRef] [PubMed]
20. Karasev, V. A Model of Molecular Vector Machine of Proteins. *Biosystems* **2019**, *180*, 7–18. [CrossRef] [PubMed]
21. Visualizer of Supramolecular Biostructures «Protein 3d». Available online: http://protein-3d.ru/index_e.html (accessed on 30 November 2022).
22. Karasev, V.A.; Stefanov, V.E. Topological Nature of the Genetic Code. *J. Theor. Biol.* **2001**, *209*, 303–317. [CrossRef]
23. Pokrovskii, V.N. *The Mesoscopic Theory of Polymer Dynamics*; Springer: Dordrecht, The Netherlands, 2010. [CrossRef]
24. Karasev, V.A.; Luchinin, V.V.; Stefanov, V.E. Topological Coding: Towards New Materials for Molecular Electronics. *Adv. Funct. Mater.* **2002**, *12*, 461–469. [CrossRef]
25. Sitkov, N.O.; Zimina, T.M.; Karasev, V.A.; Lemozerskii, V.E.; Kolobov, A.A. Design of Peptide Ligands (Aptamers) for Determination of Myeloperoxidase Level in Blood Using Biochips. In Proceedings of the 2020 IEEE Conference of Russian Young Researchers in Electrical and Electronic Engineering (EIConRus), St. Petersburg, Russia, 27–30 January 2020. [CrossRef]
26. RCSB Protein Data. Available online: https://www.rcsb.org/ (accessed on 30 November 2022).
27. Misra, S.; Kumar, A.; Kumar, P.; Yadav, A.K.; Mohania, D.; Pandit, A.K.; Prasad, K.; Vibha, D. Blood-Based Protein Biomarkers for Stroke Differentiation: A Systematic Review. *PROTEOMICS—Clin. Appl.* **2017**, *11*, 1700007. [CrossRef] [PubMed]
28. Barro, C.; Zetterberg, H. The Blood Biomarkers Puzzle—A Review of Protein Biomarkers in Neurodegenerative Diseases. *J. Neurosci. Methods* **2021**, *361*, 109281. [CrossRef]
29. Ananthan, K.; Lyon, A.R. The Role of Biomarkers in Cardio-Oncology. *J. Cardiovasc. Transl. Res.* **2020**, *13*, 431–450. [CrossRef]

30. Sokolov, A.V.; Zakahrova, E.T.; Kostevich, V.A.; Samygina, V.R.; Vasilyev, V.B. Lactoferrin, Myeloperoxidase, and Ceruloplasmin: Complementary Gearwheels Cranking Physiological and Pathological Processes. *BioMetals* **2014**, *27*, 815–828. [CrossRef] [PubMed]
31. Lepanto, M.S.; Rosa, L.; Paesano, R.; Valenti, P.; Cutone, A. Lactoferrin in Aseptic and Septic Inflammation. *Molecules* **2019**, *24*, 1323. [CrossRef]
32. Kruzel, M.L.; Zimecki, M.; Actor, J.K. Lactoferrin in a Context of Inflammation-Induced Pathology. *Front. Immunol.* **2017**, *8*, 1438. [CrossRef]
33. Sokolov, A.V.; Dubrovskaya, N.M.; Kostevich, V.A.; Vasilev, D.S.; Voynova, I.V.; Zakharova, E.T.; Runova, O.L.; Semak, I.V.; Budevich, A.I.; Nalivaeva, N.N.; et al. Lactoferrin Induces Erythropoietin Synthesis and Rescues Cognitive Functions in the Offspring of Rats Subjected to Prenatal Hypoxia. *Nutrients* **2022**, *14*, 1399. [CrossRef] [PubMed]
34. Skarżyńska, E.; Żytyńska-Daniluk, J.; Lisowska-Myjak, B. Correlations between Ceruloplasmin, Lactoferrin and Myeloperoxidase in Meconium. *J. Trace Elem. Med. Biol.* **2017**, *43*, 58–62. [CrossRef]
35. Grudinin, N.V.; Karev, V.E.; Bunenkov, N.S.; Komok, V.V.; Kostevitch, V.V.; Gorbunov, N.P.; Sokolov, A.V.; Karpov, A.A.; Lepik, K.V.; Shvetsov, A.N.; et al. Histological features and blood plasma changes after heterotopic heart transplantation. *Cell. Ther. Transplant.* **2018**, *7*, 59–60.
36. Zhang, Y.; Lu, C.; Zhang, J. Lactoferrin and Its Detection Methods: A Review. *Nutrients* **2021**, *13*, 2492. [CrossRef] [PubMed]
37. Monnard, C.; Vernet, M. Détermination de la lactoferrine plasmatique par une méthode immunoenzymatique "double sandwich" en phase hétérogène [Determination of plasma lactoferrin by the "double sandwich" immunoenzymatic method in heterogeneous phase]. *Pathol. -Biol.* **1988**, *36*, 941–944. [PubMed]
38. Mantel, C.; Miyazawa, K.; Broxmeyer, H.E. Physical Characteristics and Polymerization During Iron Saturation of Lactoferrin, A Myelopoietic Regulatory Molecule With Suppressor Activity. *Adv. Exp. Med. Biol.* **1994**, 121–132. [CrossRef]
39. Bagby, G.J.; Bennett, R. Feedback Regulation of Granulopoiesis: Polymerization of Lactoferrin Abrogates Its Ability to Inhibit CSA Production. *Blood* **1982**, *60*, 108–112. [CrossRef] [PubMed]
40. Sitkov, N.; Zimina, T.; Kolobov, A.; Sevostyanov, E.; Trushlyakova, V.; Luchinin, V.; Krasichkov, A.; Markelov, O.; Galagudza, M.; Kaplun, D. Study of the Fabrication Technology of Hybrid Microfluidic Biochips for Label-Free Detection of Proteins. *Micromachines* **2021**, *13*, 20. [CrossRef] [PubMed]
41. Jaradat, D.M. Thirteen Decades of Peptide Synthesis: Key Developments in Solid Phase Peptide Synthesis and Amide Bond Formation Utilized in Peptide Ligation. *Amino Acids* **2017**, *50*, 39–68. [CrossRef] [PubMed]
42. Zaia, J.; Annan, R.S.; Biemann, K. The Correct Molecular Weight of Myoglobin, a Common Calibrant for Mass Spectrometry. *Rapid Commun. Mass Spectrom.* **1992**, *6*, 32–36. [CrossRef]
43. Myoglobin from Equine Heart. Available online: https://www.sigmaaldrich.com/RU/en/product/sigma/m9267 (accessed on 30 November 2022).
44. Davies, M.J. Myeloperoxidase-Derived Oxidation: Mechanisms of Biological Damage and Its Prevention. *J. Clin. Biochem. Nutr.* **2010**, *48*, 8–19. [CrossRef]
45. Malle, E.; Furtmüller, P.G.; Sattler, W.; Obinger, C. Myeloperoxidase: A Target for New Drug Development? *Br. J. Pharmacol.* **2007**, *152*, 838–854. [CrossRef]
46. Mathy-Hartert, M.; Bourgeois, E.; Grülke, S.; Deby-Dupont, G.; Caudron, I.; Deby, C.; Lamy, M.; Serteyn, D. Purification of myeloperoxidase from equine polymorphonuclear leucocytes. *Can. J. Vet. Res.* **1998**, *62*, 127. [PubMed]
47. Li, J.; Ding, X.; Chen, Y.; Song, B.; Zhao, S.; Wang, Z. Determination of Bovine Lactoferrin in Infant Formula by Capillary Electrophoresis with Ultraviolet Detection. *J. Chromatogr. A* **2012**, *1244*, 178–183. [CrossRef]
48. Shimazaki, K.-I.; Kawaguchi, A.; Sato, T.; Ueda, Y.; Tomimura, T.; Shimamura, S. Analysis of Human and Bovine Milk Lactoferrins by Rotofor and Chromatofocusing. *Int. J. Biochem.* **1993**, *25*, 1653–1658. [CrossRef]
49. Senkovich, O.; Cook, W.J.; Mirza, S.; Hollingshead, S.K.; Protasevich, I.I.; Briles, D.E.; Chattopadhyay, D. Structure of a Complex of Human Lactoferrin N-Lobe with Pneumococcal Surface Protein A Provides Insight into Microbial Defense Mechanism. *J. Mol. Biol.* **2007**, *370*, 701–713. [CrossRef]
50. Isoelectric Point Calculator. Available online: https://www.bachem.com/knowledge-center/peptide-calculator/ (accessed on 30 November 2022).
51. Andreeva, A.; Budenkova, E.; Babich, O.; Sukhikh, S.; Ulrikh, E.; Ivanova, S.; Prosekov, A.; Dolganyuk, V. Production, Purification, and Study of the Amino Acid Composition of Microalgae Proteins. *Molecules* **2021**, *26*, 2767. [CrossRef] [PubMed]
52. Gill, S.C.; von Hippel, P.H. Calculation of Protein Extinction Coefficients from Amino Acid Sequence Data. *Anal. Biochem.* **1989**, *182*, 319–326. [CrossRef] [PubMed]
53. Saraiva, M.A. Interpretation of α-Synuclein UV Absorption Spectra in the Peptide Bond and the Aromatic Regions. *J. Photochem. Photobiol. B Biol.* **2020**, *212*, 112022. [CrossRef] [PubMed]
54. Sun, X.-L.; Baker, H.M.; Shewry, S.C.; Jameson, G.B.; Baker, E.N. Structure of Recombinant Human Lactoferrin Expressed in Aspergillus Awamori. *Acta Crystallogr. Sect. D Biol. Crystallogr.* **1999**, *55*, 403–407. [CrossRef] [PubMed]

55. Schindler, C.E.M.; de Vries, S.J.; Zacharias, M. Fully Blind Peptide-Protein Docking with PepATTRACT. *Structure* **2015**, *23*, 1507–1515. [CrossRef]
56. De Vries, S.J.; Rey, J.; Schindler, C.E.M.; Zacharias, M.; Tuffery, P. The PepATTRACT Web Server for Blind, Large-Scale Peptide–Protein Docking. *Nucleic Acids Res.* **2017**, *45*, W361–W364. [CrossRef]

chemosensors

MDPI

Article

Highly Efficient, Non-Covalent Functionalization of CVD-Graphene via Novel Pyrene-Based Supporter Construct

Misbah Shahzadi [1,†], Sobia Nisar [2,†], Deok-Kee Kim [2,*], Nasir Sarwar [3], Aamir Rasheed [4], Waqas Ahmad [5], Amir Muhammad Afzal [6], Muhammad Imran [7], Mohammed Ali Assiri [7], Zafar M. Shahzad [3,5,*] and Ghulam Dastgeer [8,*]

1 Department of Chemistry, University of Agriculture, Faisalabad 38040, Pakistan
2 Department of Electrical Engineering, Sejong University, Seoul 05006, Republic of Korea
3 Department of Textile Engineering, University of Engineering & Technology, Lahore, Faisalabad Campus, Faisalabad 38000, Pakistan
4 Department of Chemistry, University of Ulsan, Ulsan 44610, Republic of Korea
5 Department of Chemical and Polymer Engineering, University of Engineering & Technology, Lahore, Faisalabad Campus, Faisalabad 38000, Pakistan
6 Department of Physics, Riphah International University, Lahore Campus, 13-Km Raiwind Road Lahore, Lahore 54000, Pakistan
7 Department of Chemistry, Faculty of Science, King Khalid University, P.O. Box 9004, Abha 61413, Saudi Arabia
8 Department of Physics and Astronomy, Sejong University, Seoul 05006, Republic of Korea
* Correspondence: deokkeekim@sejong.ac.kr (D.-K.K.); mshahzad@skku.edu (Z.M.S.); gdastgeer@sejong.ac.kr (G.D.)
† These authors contributed equally to this work.

Abstract: Ultra-thin two-dimensional (2D) materials have attained huge interest for biosensing applications because of their strong electrostatic coupling with target molecules such as spike proteins and DNA. One such 2D material is graphene, which is extremely thin and flexible and has a strong non-covalent interaction with the supporting constructs needed to detect biomolecules. This work aimed to develop a way to efficiently functionalize the surface of 2D material using a pyrene-based supporter construct to detect the target protein. For this purpose, high-quality, pristine graphene was grown via the chemical vapor deposition (CVD) method and transferred over the Si/SiO_2 substrate for its functionalization using our engineered pyrene–lysine-based supporter construct (PLB). The construct was synthesized using the solid-phase peptide synthesis (SPPS) method and utilized to functionalize the graphene-channel-based field-effect transistor (FET) device via non-covalent $\pi-\pi$ stacking interaction. The optimum concentration of the functionalized PLB was evaluated via atomic force microscopy (AFM), Raman spectroscopy, and real-time electrical measurements. The characterization techniques successfully provide an overview of the effect of the concentration of PLB used for functionalization. Moreover, the performance was tested and compared in terms of the percentage response of the device generated after the detection of various concentrations of the streptavidin protein. This research could be useful in determining how to functionalize any 2D material by designing a supporter construct without material degradation and owing to over-stacking or bypassing surface screening effects.

Keywords: protein; graphene FET; device functionalization; biosensing

Citation: Shahzadi, M.; Nisar, S.; Kim, D.-K.; Sarwar, N.; Rasheed, A.; Ahmad, W.; Afzal, A.M.; Imran, M.; Assiri, M.A.; Shahzad, Z.M.; et al. Highly Efficient, Non-Covalent Functionalization of CVD-Graphene via Novel Pyrene-Based Supporter Construct. *Chemosensors* **2023**, *11*, 83. https://doi.org/10.3390/chemosensors11020083

Academic Editor: Xiaolong Yang

Received: 27 December 2022
Revised: 17 January 2023
Accepted: 19 January 2023
Published: 21 January 2023

1. Introduction

Two-dimensional materials such as graphene are being functionalized through various routes to explore their structural and morphological features for utilization in catalysis, energy-harvesting materials, and biosensing [1–5]. The biosensors based on graphene or semiconductor 2D materials can sense surface changes from their surroundings and offer the best sensing environment for ultrasensitive and low-noise detection.

This makes graphene-based FET technology extremely appealing for uses involving delicate immunological diagnostics. As a matter of fact, covalent functionalization constitutes the sp^2 structure of graphene lattices, thus resulting in defects and loss of the electronic properties during functionalization [1,6]. In contrast, non-covalent functionalization is largely preferred as it does not alter the structure and electronic properties of graphene while inducing new groups onto the surface, possibly due to forces such as $\pi-\pi$ interactions, hydrogen bonding, and van der Waals forces. Hence, it leads to enhanced dispersibility, biocompatibility, reactivity, binding capacity, and sensing properties [7–11]. However, the main challenge in functionalizing graphene and graphene-like material is determining the optimum concentration of the receptor so as to avoid the loss of material due to over-stacking and also avoiding various screening effects, such as Debye screening length. [12–15] We tried to cover this gap by engineering a short-peptide-containing pyrene moiety and utilizing it for the functionalization of CVD-grown good-quality pristine graphene. For the fabrication of the graphene-based FET biosensor, the p-doped silicon substrates were coated with 300 nm thick silicon oxide (SiO_2) and washed with acetone and isopropanol and dried with nitrogen gas (N_2) to ensure absolute cleanliness. The CVD-grown graphene was transferred onto our working substrate (Si/SiO_2) and analyzed under Raman spectroscopy to analyze the properties of graphene after its functionalization using various concentrations of our PLB construct. Our engineered construct had a pyrene molecule at the terminal end so that it could make stable $\pi-\pi$ stacking interactions with the hexagonal structure of graphene rings. Pyrene has a larger number of free electrons as compared to graphene. Hence, it shared its free electrons with the graphene and made non-covalent ($\pi-\pi$) bonding interactions, as shown in Figure 1a. This study is backed by the AFM measurements of the height profile of the substrate before and after functionalization. In addition to the above, we patterned the electrodes using photolithography and a thermal metal deposition chamber on the same substate and fabricated a back-gated FET device for detailed electrical characterization. The 1 nM concentration was found to be ideal for channel functionalization with our designed construct of PLB.

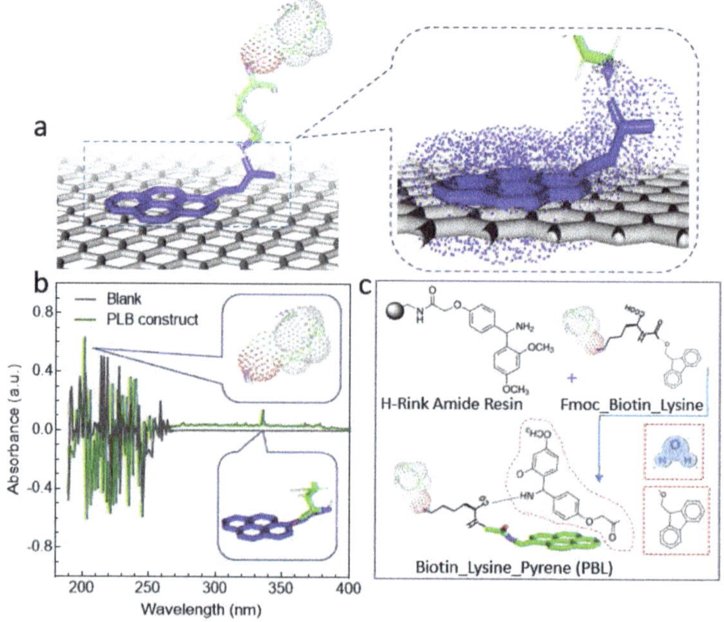

Figure 1. A schematic demonstration of functionalizing graphene with our engineered pyrene-based construct. (**a**) Pyrene interacts with graphene rings by sharing its free electrons. (**b**) The synthesized

PLB construct was tested (via middle/long-wave UV spectra) for the successful conjugation of pyrene and the presence of biotin. Sharp clear peaks at 202 nm and 335 nm confirm the presence of biotin and pyrene. (**c**) The reaction mechanism for our engineered PLB construct. The synthesis reaction constitutes a series of deprotection and coupling steps. The Fmoc was removed via deprotection using piperidine while coupling was assisted by coupling solution containing DIC and Oxyma.

2. Material and Methods

2.1. The Growth of High-Quality Pristine Graphene (PG)

Usually, substrates are functionalized using various two-dimensional (2D) materials [16–19], and their properties, which reflect various binding events, are monitored. In this regard, a 2D sp2-bonded carbon allotrope, graphene, has attracted enormous interest over the past decade due to its unique properties, such as ultrahigh electron mobility, uniform broadband optical absorption, and high tensile strength. Hence, monolayer, large-size, and high-quality pristine graphene was grown via the chemical vapor deposition (CVD) method. This method was selected since CVD-derived graphene is polycrystalline and good-quality [20–23]. The copper-foil nook was shaped by twisting a 25 mm thick copper foil (Nilaco Corporation, Tokyo, Nilaco Bldg., Japan) and afterward creasing the three leftover sides. Methane flow was kept at 1 sccm at 50 mTorr, and the temperature remained at 1035 °C followed by the 2.0 h ramping process. At temperatures above 1035 °C, large-domain graphene growth was observed on the copper-foil enclosure. The procedure is described in agreement with the literature [24–28].

Graphene grew on both the inside and outside of the Cu enclosure. The back-end graphene was etched via oxygen plasma etching. The graphene was coated with PMMA using a spin coater at 4000 PRM for 1 min. After transferring, the Si/SiO_2 substrate was dried in an oven at 80 °C for 1 h Finally, the substrate was dried by a nitrogen blow followed by washing using acetone for PMMA removal and methanol for acetone removal. The graphene layer was transferred to our working substrate (Si/SiO_2) via the wet transfer method, using iron (III) chloride as an etchant. The transferred monolayer graphene was characterized using Raman spectra analysis, showing an excellent 2D/G ratio with a negligible defect peak, as shown in Figure 2a.

2.2. The Device Fabrication for Electrical Measurements

After transferring the CVD-grown graphene over the Si/SiO_2 substrates, the substrates were coated with polymers of EL-9 and photoresist (PR) and baked over the hot plate at 170 °C and 100 °C, respectively. The photolithography technique was utilized to draw the electrode pattern over the graphene. After photolithography, all the substrates were loaded in the thermal evaporator chamber for the deposition of the Cr/Au electrodes. In a high vacuum of 5×10^{-6} torr, 5 nm thick Cr and 50 nm thick gold films were deposed to draw the source–drain electrodes. After metal deposition, the substrates were washed with acetone and methanol for the liftoff process and dried with N_2. In order to measure the electrical properties of the device, these source and drain electrodes were used to apply biasing voltage and measure the output current, respectively. After the deposition of electrodes, the final devices were annealed at 200 °C for 2.0 h under a suitable pressure, in the presence of argon. The electrical properties of our graphene-channel-based field-effect transistor were measured using a three-probe station. The inner probes were connected to a voltage meter, while the two outer probes were connected to source and drain electrodes drawn over the graphene channel. The gate voltage and biasing voltages were controlled via a source meter (Keitley K-2400), and the output current was recorded using a pico-ammeter (Keithley K-6485). The voltage drop across the inner probes was measured as current flowed between the outer probes. All the electrical and real-time measurements were performed in the ambient environment and at room temperature. The resistivity of the material being tested determines how the current and voltage values relate to one another.

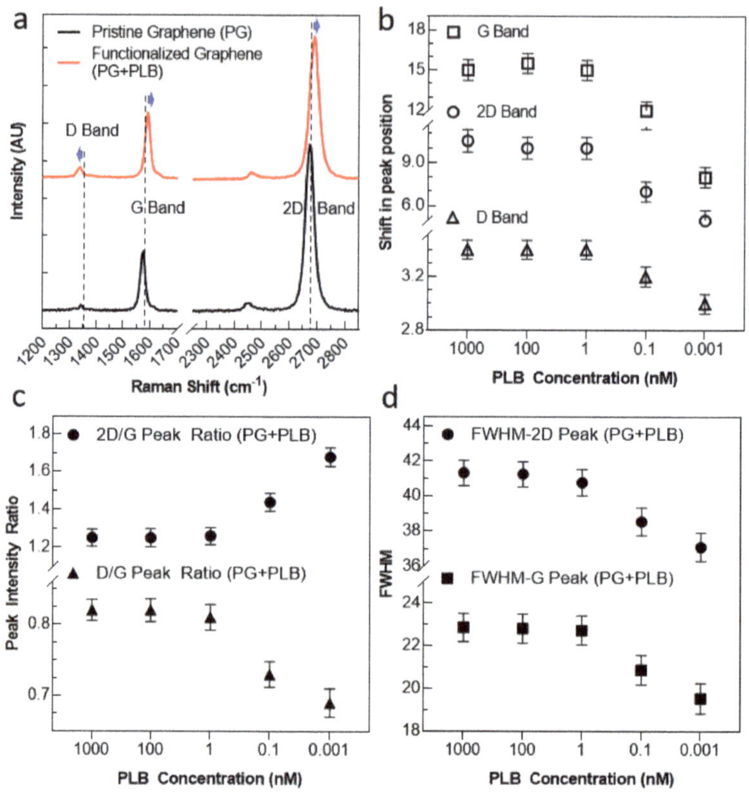

Figure 2. The Raman spectra analysis of our substrate before and after functionalization using various PLB concentrations. (**a**) Pyrene interacts with graphene rings by sharing its free electrons. (**b**) The synthesized PLB construct was tested (via middle/long-wave UV. (**c**) The peak ratios were plotted as a function of PLB molecule concentration. (**d**) The FWHM for 2D peak and G peak are illustrated as a function of PLB molecule concentration.

2.3. The PLB Synthesis

In this work, we utilized a self-engineered pyrene–lysine–biotin (PLB) construct for the functionalization of our substrate material (graphene). Using Fmoc chemistry-based H-Rink amide resin, a manual solid-phase peptide synthesizer (SPPS) was used to synthesize the PLB construct. The resin, because of the 0.14 mmol synthesis scale, was expanded in DMF for 30 min and deprotected utilizing 20% piperidine in N,N-dimethylformamide at 90 °C ± 5 for 50 s followed by 75 °C ± 5 for 15 s. The principal wash was conducted with DMF and the secondary was carried out using dichloromethane in between each deprotecting and coupling step. For synthesis, 0.2 M Fmoc-Lys (Biotin) (prepared according to provider protocol) was added to activated amino acid (DIPEA), using DMF as a solvent in the presence of 0.5 M DIC and 1 M Oxyma at 90 °C ± 5 for 110 s followed by 75 °C ± 5 for 15 s. Repeating the deprotection step, the 1-pyrene butyric acid was used for the coupling reaction in the circumstances referenced previously. The finished product was dried for 1 h in a desiccator. Using a trifluoroacetic-acid-containing cleavage solution, the product was cleaved from the solid support and was achieved by mixing it with triisopropylsilane (2.5%) and water (2.5%) at room temperature for 2 h. The confirmation of coupling and deprotection was made at each phase using the ninhydrin test. A schematic of this reaction is presented in Figure 1c. Before utilizing this construct for functionalization, the presence of pyrene was confirmed by measuring absorbance at 335 nm, as shown in Figure 1b.

3. Results and Discussion

In our experimental investigation, a pyrene-based construct was engineered and used to functionalize high-quality pristine graphene. In order to find out the best conditions of functionalization, avoiding over-stacking and loss of material, solutions (with the PLB construct) with various concentration were prepared and drop-cast to functionalize the channel material. After washing and drying, the final graphene-based FET device was characterized using various techniques, such as Raman spectra, AFM, and electrical measurements, and the results are presented below.

3.1. The Raman Spectra Analysis

To evaluate the efficiency of functionalization and find the optimum concentration for functionalization, the substrate, containing various concentrations of PLB, was drop-cased and washed–dried for analysis. The substrate was subjected to Raman spectroscopy (under 532 nm excitation) to ensure the quality of graphene and assembly of our PLB onto the graphene surface. As the first step, the Raman spectra of bare graphene were recorded. Two clear peaks (G and 2D peaks) were observed (black line in Figure 2a). The G peak corresponds to the lattice vibration mode, and the 2D peak stems from second-order Raman scattering. There was no D peak seen corresponding to any defects while transferring graphene to the Si/SiO_2 substrate. After confirming the presence of defect-free pristine graphene, we functionalized the channel using our PLB construct. In spectra (red line in Figure 2a), a clear right shift in the G peak and 2D peak position can be seen, which indicates p-doping by the π–π interaction between the pyrene molecule and graphene. To further validate this observation, we measured the spectra at various devices and plotted the peak positions, as shown in Supplementary Figure S1, to confirm this shifting upon functionalization. Moreover, the layer-dependent ratio of the intensity of the 2D band to the G band (2D/G) fell from 1.71 to 1.26, as shown in Supplementary Figure S2a. Adding to this, a D peak (at ~1342 cm^{-1}) was also observed and shifted downwards upon functionalization. This D peak appears to be due to the introduction of disorder arising from the orbital hybridization of the molecule with the graphene plane [29–31]. Further, an increase in FWHM in the G and 2D peaks can be seen, depicting the loss in crystallinity of graphene on the surface after its functionalization, as shown in Supplementary Figure S2b.

Further, the Raman spectra peak property analysis was carried out by varying the concentration of the PLB construct from 0.001 nM to 1000 nM, as shown in Figure 2b–d. From the peak property analysis, a clear shift in all peak positions can be seen from concentrations of 0.001 nM to 1 nM, whereas a marginal shift in peak position was observed upon the further increase in PLB concentration, indicating the saturation of the surface at ~1 nM PLB. The trend seems consistent with all the peak positions (Figure 2b). In order to validate the data, the testing was performed using three different substrates ($N = 3$), and its mean was plotted along with the standard deviation to avoid errors in measurement. Moreover, a similar leveling off of variation trend was observed with the peak intensity ratio (Figure 2c) and FWHM of the main and second resonance peaks (Figure 2d). Hence, through the Raman spectra investigation, a concentration of around 1 nM of our PLB construct was considered sufficient to fully functionalize the substrate, without causing unwanted doping and loss of material.

3.2. The AFM Analysis

Furthermore, the substrate was analyzed using atomic force microscope (AFM, I-Nexus Co., Ltd., Seoul, Republic of Korea) measurements to investigate surface conditions. Since the capacitance (C) of a graphene FET is inversely proportional to the distance (d) between the graphene and the charged biomolecule, the thickness of the receptor is an important parameter for determining the electrostatic coupling strength between the graphene and the charged biomolecules, when the graphene is functionalized.

$$C = \frac{\alpha\beta A}{d} \tag{1}$$

where C is the capacitance, β is the absolute permittivity, d is the thickness of the PLB construct (receptor), A is the area of the graphene channel, and α is the relative permittivity.

Hence, the graphene surface was functionalized with a buffer solution containing various concentrations of PLB to confirm the effect of the receptor thickness on the graphene biosensor. The self-assembled PLB construct's AFM profile on the graphene surface at various concentrations is depicted in Figure 3. For the self-assembly property study, the lyophilized powder of pure PLB was dissolved in distilled water, Tris buffer (20 mM), and NaCl (5 mM) to create solutions of 0.001 nM, 100 nM, and 1000 nM. Both the procedure and the buffer were chosen in accordance with the literature. Before the AFM analysis, the substrate was triple-washed with the same buffer and dried. Based on the interaction, the pyrene residue of the PLB construct interacted with graphene, enabling the construct to be uniformly assembled on the graphene surface without leaving any voids. The substrate with a high concentration of PLB showed stacking of material after the surface saturated, as shown in Figure 3c,d. It can be observed that the average height of the surface containing bare graphene was ~0.4 nm while the height increased up to ~1.5 nm upon functionalization with the PLB construct, which can be attributed to the formation of a monolayer PLB construct on the graphene surface. Moreover, it can be observed that, when using a low concentration of PLB (0.001 nM, Figure 3a), there were a lot of empty spaces still vacant for functionalization. However, after increasing the concentration to 1 nM, most of the surfaces became occupied, leaving fewer vacant positions, Figure 3b. It is noteworthy that by further increasing PLB concentration, the height increased, forming rough lumps without an ordered orientation, which indicated clear stacking of material onto the surface, as seen in Figure 3d.

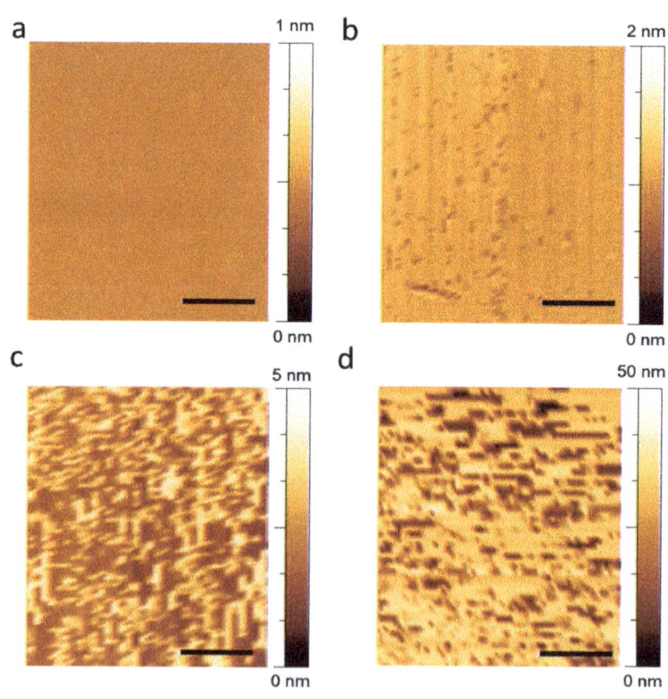

Figure 3. The atomic force microscopic image of our graphene channel substrate after functionalizing using PLB construct at increasing concentrations. (**a**) The AFM image of graphene surface containing 0.001 nM PLB construct for functionalization. (**b**–**d**) The AFM image of graphene-channel-based substrate with 1 nM, 100 nM, and 1000 nM PLB, respectively. The inset scale bar in each figure is equivalent to 2 µm.

3.3. The Electrical Measurements

To further validate the results, we made an FET device by impeding source and drain electrodes and subjected the device to detailed electrical measurements (see Figures S3 and S4). The linear I_{DS}-V_{DS} curves at zero gate voltage are presented in Figure 4a, exhibiting the drain characteristics of the device after functionalization. Initially, the pristine graphene FET was measured without any treatment of the PLB molecules by increasing the source–drain voltage from −2 V to 2V. In pristine condition (black line in Figure 4a), a straight line passing through the origin means the device has ohmic behavior. It can be observed that the output current started to rise after increasing the concentration of the PLB supporter construct over the graphene surface. Further, after functionalizing with a higher concentration of PLB, a dominant increase in current was observed, saturating at ~1 nM PLB concentration. This saturation in current shows a reasonable functionalization of the graphene surface. Further, the transfer curves of the graphene FET were measured by increasing the back gate voltage from −30 to 30 V. A sufficient variation in Dirac point voltage can also be observed before and after functionalization of the graphene FET, as shown in Figure 4b. For the pristine graphene, the Dirac point existed at zero voltage, which shifted to around 10 V after functionalization with the PLB. The scanning electron microscopeic (SEM) image of the final device is illustrated in the Figure 4b inset. All the measurements were conducted using three different devices, and the mean/standard deviation is plotted in Figure 4c,d.

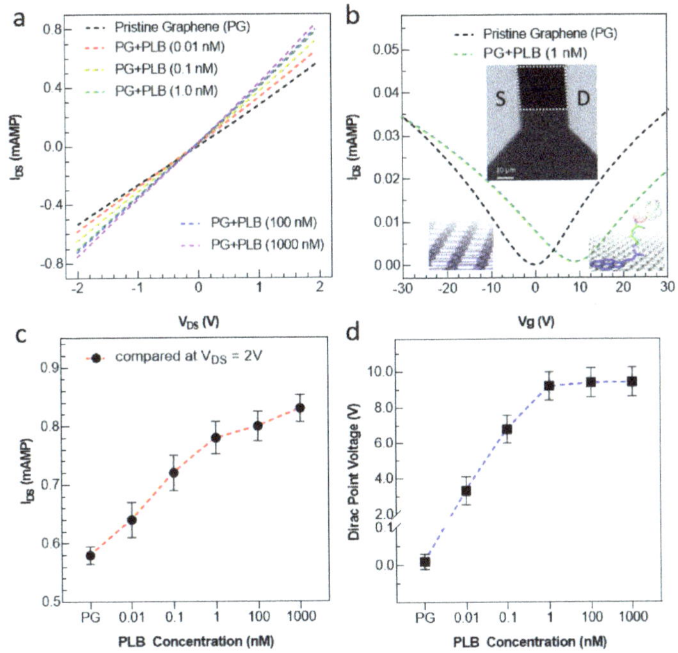

Figure 4. Electrical property analysis of the device before and after functionalizing with our PLB construct at various concentrations. (**a**) The output characteristics (at V_g = 0 V) of the device before and after functionalization. The current increases upon functionalization due to transfer of free electrons from pyrene to graphene. (**b**) The transfer characteristic of the device before and after functionalization. The Dirac point lies around 0 V, showing perfect growth and transfer procedure without any serious external doping. After functionalization, a clear shift in DIRAC point voltage was observed, owing to the p-doping of graphene upon functionalization. The SEM image of the final graphene FET device is shown in the inset with a scale bar of 10 μm. The white dotted rectangular region shows the graphene channel area. (**c,d**) A relative increase in current at (V_{DS} = 2 V) and Dirac point voltage V_{DS} = 0.01 V) was achieved by increasing the concentration of PLB.

3.4. The Real-Time Response

In testing the performance of the device in real time, a reduction in the current (I_{DS}) from its baseline was noted (for three different devices) concerning time, as seen Figure 5. For a comparison without error, the normalized level of current was calculated using Equation (2).

$$Normalized\ I_{DS} = \frac{I - Io}{Io} \quad (2)$$

where I is the current at any time (t) while Io is the current at the time (t = 0). The real-time electrical measurements were performed at a steady bias voltage of 0.01 V. Three different devices with uncovered graphene channels were equilibrated with buffer, and their ongoing current profiles with PLB-covered graphene channels of various concentrations (0.001 nM (red line), 1 nM (green line), and 100 nM (blue line)) are presented in Figure 5a. None of the devices showed any considerable current change during buffer application, so we accepted this as a baseline for functionalization. When the solution containing 0.01 nM PLB was drop-cast onto the channel, the level of current increased but the trend leveled off within a short time of 2 min. This short time of functionalization is due to the perfection of our engineered construct, which is free from any kind of surface screening effects and other issues, such as steric hindrance. After the current was maintained, the device was washed using the same buffer, and the level of current remained the same before and after washing, as shown in Figure 5a (in red line). To improve the level of current, we increased the PLB concentration to 1 nM with the second device and noted the current profile. It was observed that the level of current improved, and after washing, a negligible drop in current was observed, as shown in Figure 5a (green line). However, when we further increased the PLB concentration on our third device, the level of current improved a little bit, and after washing, it dropped sharply, indicating the stacking of PLB molecules due to saturation of the sensor surface at that high concentration (100 nM), as shown in Figure 5a (blue line). All the real-time measurements were repeated using three different devices, and the results are presented in Figure 5b. Since our construct was biotinylated and its presence was confirmed using middle/long-wave UV spectra (Figure 1b), the device was subjected to a solution containing streptavidin. The source–drain current decreased after the streptavidin addition because some of the PLB molecules' charges were occupied by streptavidin molecules. As a result, a small number of charges were transferred from the PLB to the graphene surface, which caused a decrease in current [5,9,10]. The time in which device current is saturated after capturing the streptavidin is known as the response time. The real-time measurements exhibited that our device current was saturated within ~30 s after adding the target protein (streptavidin) solution, which means that graphene devices can successfully detect the target protein (streptavidin) within ~30 s, as shown in Figure 5c. The comparison of the graphene FET with other biosensor devices is illustrated in Figure S5. It is worth noting that the level of current after streptavidin application and adding another washing step (washing-2) was almost the same. Moreover, the specificity of the device was tested by applying a similar concentration of nearly similar molecular weight protein, i.e., bovine serum albumin (BSA), which gave no considerable response as compared to the target protein (see Figure S6). Followed by initial trials of concentration, a fixed concentration of 1 nM streptavidin was used to compare the devices functionalized using three various concentrations of the PLB construct, the results of which are presented in Figure 5d.

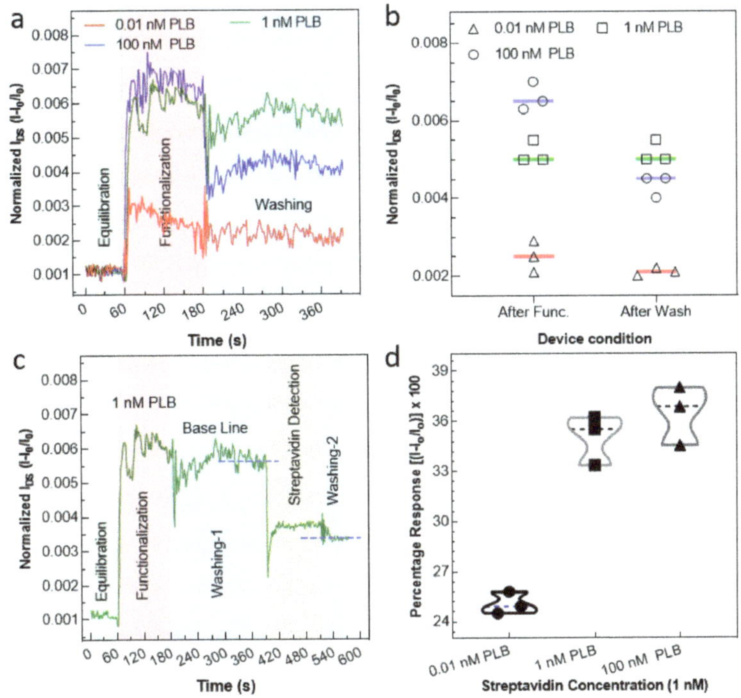

Figure 5. Real-time electrical measurement of the G-FET device. (**a**) The current profile of the device after applying PLB at various concentrations ranging from 0.01 nM to 100 nM. After the current became constant, washing was carried out using the same buffer to get rid of the PLB molecules stacked onto the surface with bonds other than the graphene surface. (**b**) The normalized current for three different devices was measured right after functionalization and after applying the washing step. The experimental investigation revealed that the 1 nM PLB concentration was adequate to functionalize all the graphene channel surfaces, avoiding any surface screening and loss of material due to stacking. (**c**) Finally, the same device was used to detect the streptavidin, since the PLB construct was biotinylated. (**d**) The graph was plotted to observe the percentage response of devices functionalized using various concentrations of PLB. For all the measurements, the concentration of streptavidin was kept constant. The optimum response, i.e., obtaining a max response without wasting material (PLB), was obtained at 1 nM PLB.

4. Conclusions

In this work, we demonstrated a way to efficiently functionalize (non-covalently) a graphene-channel-based field-effect transistor using our self-engineered pyrene-based supporter construct (PLB). The monomolecular self-assembly of our peptide construct was made possible due to $\pi-\pi$ stacking of free electrons with graphene's Bravais lattice, allowing ultra-thin monomolecular self-assembly via the graphene lattice. It is worth mentioning that functionalizing 2D materials, especially graphene, is not new; however, we demonstrated a way to achieve an ultra-thin receptor using our engineered construct, which enabled high gate coupling between the PLB construct and graphene channel. The thickness profile was measured and analyzed via various routes, such as Raman spectra, AFM, and electrical measurements. Owing to our novel PLB construct and efficient way to screen out the desired concentration for efficient functionalization, our device demonstrated excellent results in real time, and a prompt inherent response was observed. For the practical application of the device as a biosensor, streptavidin protein was also detected using the graphene-based FET after exploring the functionalization technique. Our developed strategy can help in evaluating the best way to functionalize any 2D material surface efficiently with good surface density, avoiding any degradation [9] of material and surface screening effects.

Supplementary Materials: The following supporting information can be downloaded at: https://www.mdpi.com/article/10.3390/chemosensors11020083/s1, Figure S1: The Raman spectra peak position analysis of graphene-based substrate before and after functionalization; Figure S2: The Raman spectra analysis of graphene-based substrate before and after functionalization. (a) The analysis of intensity ratio of main, second resonance, and defect peak before and after functionalization. (b) The FWHM of the peak before and after functionalization. All the data were plotted using N = 3 to eliminate possible variation in the marked value; Figure S3: The optical image of graphene on Si/SiO_2 substrate, confirming unbroken graphene layer transfer via wet transfer method; Figure S4: The sketch of our device for electrical measurements, created using CREO 6.0; Figure S5: The comparison of results for streptavidin detection of our device; Figure S6: A comparison of streptavidin and bovine serum albumin (BSA) at various concentrations is plotted with error bars with an accuracy of 95%.

Author Contributions: G.D., S.N., M.S. and Z.M.S. designed the project M.S. and S.N. performed the experimental work. G.D., D.-K.K. and Z.M.S. supervised the project and wrote the final manuscript. N.S., W.A., A.M.A., A.R., M.I. and M.A.A. analyzed the data and reviewed the manuscript. All the authors reviewed and revised the final manuscript. M.S. and S.N. share 1st authorship equally. All authors have read and agreed to the published version of the manuscript.

Funding: This research was funded by the National Research Foundation (2022R1G1A1009887), Korea, and Sejong University, Seoul 05006, Korea.

Data Availability Statement: The data presented in this study are available on request from the corresponding authors.

Acknowledgments: This research was funded by the National Research Foundation (2022R1G1A100-9887), Korea, and Sejong University, Seoul 05006, Korea, and all the authors are thankful for this support. M.A.A. extends their appreciation to the Ministry of Education in KSA for funding this research work through project number KKU-IFP2-DA-1.

Conflicts of Interest: The authors declare no conflict of interest.

References

1. Georgakilas, V.; Tiwari, J.N.; Kemp, K.C.; Perman, J.A.; Bourlinos, A.B.; Kim, K.S.; Zboril, R.J. Noncovalent functionalization of graphene and graphene oxide for energy materials, biosensing, catalytic, and biomedical applications. *Chem. Rev.* **2016**, *116*, 5464–5519. [CrossRef] [PubMed]
2. Su, Q.; Pang, S.; Alijani, V.; Li, C.; Feng, X.; Müllen, K.J. Composites of graphene with large aromatic molecules. *Adv. Mater.* **2009**, *21*, 3191–3195. [CrossRef]
3. Kim, J.; Song, S.H.; Im, H.G.; Yoon, G.; Lee, D.; Choi, C.; Kim, J.; Bae, B.S.; Kang, K.; Jeon, S. Moisture barrier composites made of non-oxidized graphene flakes. *Small* **2015**, *11*, 3124–3129. [CrossRef] [PubMed]
4. Forsyth, R.; Devadoss, A.; Guy, O.J. Graphene field effect transistors for biomedical applications: Current status and future prospects. *Diagnostics* **2017**, *7*, 45. [CrossRef] [PubMed]
5. Zafar, M.S.; Dastgeer, G.; Kalam, A.; Al-Sehemi, A.G.; Imran, M.; Kim, Y.H.; Chae, H. Precise and Prompt Analyte Detection via Ordered Orientation of Receptor in WSe2-Based Field Effect Transistor. *Nanomaterials* **2022**, *12*, 1305. [CrossRef] [PubMed]
6. Kim, J.E.; No, Y.H.; Kim, J.N.; Shin, Y.S.; Kang, W.T.; Kim, Y.R.; Kim, K.N.; Kim, Y.H.; Yu, W.J. Highly sensitive graphene biosensor by monomolecular self-assembly of receptors on graphene surface. *Appl. Phys. Lett.* **2017**, *110*, 203702. [CrossRef]
7. MacLeod, J.; Lipton-Duffin, J.; Cui, D.; De Feyter, S.; Rosei, F. Substrate effects in the supramolecular assembly of 1, 3, 5-benzene tricarboxylic acid on graphite and graphene. *Langmuir* **2015**, *31*, 7016–7024. [CrossRef]
8. Stradi, D.; Garnica, M.; Díaz, C.; Calleja, F.; Barja, S.; Martín, N.; Alcamí, M.; de Parga, A.L.V.; Miranda, R.; Martín, F. Controlling the spatial arrangement of organic magnetic anions adsorbed on epitaxial graphene on Ru (0001). *Nanoscale* **2014**, *6*, 15271–15279. [CrossRef]
9. Seo, G.; Lee, G.; Kim, M.J.; Baek, S.-H.; Choi, M.; Ku, K.B.; Lee, C.-S.; Jun, S.; Park, D.; Kim, H.G.; et al. Rapid Detection of COVID-19 Causative Virus (SARS-CoV-2) in Human Nasopharyngeal Swab Specimens Using Field-Effect Transistor-Based Biosensor. *ACS Nano* **2020**, *14*, 5135–5142. [CrossRef]
10. Fathi-Hafshejani, P.; Azam, N.; Wang, L.; Kuroda, M.A.; Hamilton, M.C.; Hasim, S.; Mahjouri-Samani, M. Two-Dimensional-Material-Based Field-Effect Transistor Biosensor for Detecting COVID-19 Virus (SARS-CoV-2). *ACS Nano* **2021**, *15*, 11461–11469. [CrossRef]

11. Zhang, X.; Liu, T.; Boyle, A.; Bahreman, A.; Bao, L.; Jing, Q.; Xue, H.; Kieltyka, R.; Kros, A.; Schneider, G.F.; et al. Dielectric-Modulated Biosensing with Ultrahigh-Frequency-Operated Graphene Field-Effect Transistors. *Adv. Mater.* **2022**, *34*, 2106666. [CrossRef] [PubMed]
12. Nakatsuka, N.; Yang, K.-A.; Abendroth, J.M.; Cheung, K.M.; Xu, X.; Yang, H.; Zhao, C.; Zhu, B.; Rim, Y.S.; Yang, Y.; et al. Aptamer–field-effect transistors overcome Debye length limitations for small-molecule sensing. *Science* **2018**, *362*, 319–324. [CrossRef] [PubMed]
13. Huang, W.; Diallo, A.K.; Dailey, J.L.; Besar, K.; Katz, H.E. Electrochemical processes and mechanistic aspects of field-effect sensors for biomolecules. *J. Mater. Chem. C Mater.* **2015**, *3*, 6445–6470. [CrossRef] [PubMed]
14. Dastgeer, G.; Nisar, S.; Shahzad, Z.M.; Rasheed, A.; Kim, D.-K.; Jaffery, S.H.A.; Wang, L.; Usman, M.; Eom, J. Low-Power Negative-Differential-Resistance Device for Sensing the Selective Protein via Supporter Molecule Engineering. *Adv. Sci.* **2022**, *10*, 2204779. [CrossRef] [PubMed]
15. Dastgeer, G.; Shahzad, Z.M.; Chae, H.; Kim, Y.H.; Ko, B.M.; Eom, J. Bipolar Junction Transistor Exhibiting Excellent Output Characteristics with a Prompt Response against the Selective Protein. *Adv. Funct. Mater.* **2022**, *32*, 2204781. [CrossRef]
16. Jeong, J.H.; Kang, S.; Kim, N.; Joshi, R.K.; Lee, G.-H. Recent trends in covalent functionalization of 2D materials. *Phys. Chem. Chem. Phys.* **2022**, *24*, 10684–10711. [CrossRef] [PubMed]
17. Amieva, E.J.C.; López-Barroso, J.; Martínez-Hernández, A.L.; Velasco-Santos, C. Graphene-based materials functionalization with natural polymeric biomolecules. *Recent Adv. Graphene Res.* **2016**, *1*, 257–298.
18. Xie, T.; Liu, Y.; Xie, J.; Luo, Y.; Mao, K.; Huang, C.; Li, Y.; Zhen, S. Catalyzed Hairpin Assembly-Assisted DNA Dendrimer Enhanced Fluorescence Anisotropy for MicroRNA Detection. *Chemosensors* **2022**, *10*, 501. [CrossRef]
19. Ameku, W.A.; Negahdary, M.; Lima, I.S.; Santos, B.G.; Oliveira, T.G.; Paixão, T.R.L.C.; Angnes, L. Laser-Scribed Graphene-Based Electrochemical Sensors: A Review. *Chemosensors* **2022**, *10*, 505. [CrossRef]
20. Li, X.; Cai, W.; Colombo, L.; Ruoff, R.S. Evolution of graphene growth on Ni and Cu by carbon isotope labeling. *Nano Lett.* **2009**, *9*, 4268–4272. [CrossRef]
21. Li, X.; Magnuson, C.W.; Venugopal, A.; An, J.; Suk, J.W.; Han, B.; Borysiak, M.; Cai, W.; Velamakanni, A.; Zhu, Y.; et al. Graphene films with large domain size by a two-step chemical vapor deposition process. *Nano Lett.* **2010**, *10*, 4328–4334. [CrossRef] [PubMed]
22. Deokar, G.; Avila, J.; Razado-Colambo, I.; Codron, J.-L.; Boyaval, C.; Galopin, E.; Asensio, M.-C.; Vignaud, D. Towards high quality CVD graphene growth and transfer. *Carbon* **2015**, *89*, 82–92. [CrossRef]
23. Deng, B.; Liu, Z.; Peng, H. Toward mass production of CVD graphene films. *Adv. Mater.* **2019**, *31*, 1800996. [CrossRef] [PubMed]
24. Li, X.; Magnuson, C.W.; Venugopal, A.; Tromp, R.M.; Hannon, J.B.; Vogel, E.M.; Colombo, L.; Ruoff, R.S. Large-area graphene single crystals grown by low-pressure chemical vapor deposition of methane on copper. *J. Am. Chem. Soc.* **2011**, *133*, 2816–2819. [CrossRef]
25. Yan, Z.; Peng, Z.; Tour, J.M. Chemical vapor deposition of graphene single crystals. *Acc. Chem. Res.* **2014**, *47*, 1327–1337. [CrossRef]
26. Gao, L.; Ren, W.; Zhao, J.; Ma, L.-P.; Chen, Z.; Cheng, H.-M. Efficient growth of high-quality graphene films on Cu foils by ambient pressure chemical vapor deposition. *Appl. Phys. Lett.* **2010**, *97*, 183109. [CrossRef]
27. Wang, C.; Chen, W.; Han, C.; Wang, G.; Tang, B.; Tang, C.; Wang, Y.; Zou, W.; Chen, W.; Zhang, X.-A.; et al. Growth of Millimeter-Size Single Crystal Graphene on Cu Foils by Circumfluence Chemical Vapor Deposition. *Sci. Rep.* **2014**, *4*, 4537. [CrossRef]
28. Yang, Q.; Zhang, Z.; Zhu, W.; Wang, G. Growth of Large-Area High-Quality Graphene on Different Types of Copper Foil Preannealed under Positive Pressure H2 Ambience. *ACS Omega* **2019**, *4*, 5165–5171. [CrossRef]
29. Xu, S.; Zhan, J.; Man, B.; Jiang, S.; Yue, W.; Gao, S.; Guo, C.; Liu, H.; Li, Z.; Wang, J.; et al. Real-time reliable determination of binding kinetics of DNA hybridization using a multi-channel graphene biosensor. *Nat. Commun.* **2017**, *8*, 14902. [CrossRef]
30. Yang, Y.; Asiri, A.M.; Tang, Z.; Du, D.; Lin, Y. Graphene based materials for biomedical applications. *Mater. Today* **2013**, *16*, 365–373. [CrossRef]
31. Wang, X.; Hao, Z.; Olsen, T.R.; Zhang, W.; Lin, Q. Measurements of aptamer–protein binding kinetics using graphene field-effect transistors. *Nanoscale* **2019**, *11*, 12573–12581. [CrossRef] [PubMed]

chemosensors

Article

Fast Protein and Metabolites (Nucleotides and Nucleosides) Liquid Chromatography Technique and Chemical Sensor for the Assessment of Fish and Meat Freshness

Artur Kuznetsov [1], Aleksandr Frorip [1], Alar Sünter [1,2], Nensi Kasvand [1], Vadim Korsakov [3], Georgii Konoplev [4], Oksana Stepanova [4], Linda Rusalepp [2], Dea Anton [2], Tõnu Püssa [2], Mati Roasto [2], Liubov Abramova [5], Andrey Kozin [5], Lauri Toom [6], Soeren Hirsch [7] and Nikolay Mukhin [7,*]

1 AS Ldiamon, 50411 Tartu, Estonia
2 Veterinary Biomedicine and Food Hygiene, Estonian University of Life Sciences, 51006 Tartu, Estonia
3 Jeko Disain OÜ, 51014 Tartu, Estonia
4 Department of Photonics, Saint Petersburg Electrotechnical University "LETI", Saint Petersburg 197022, Russia
5 Research Institute of Fisheries and Oceanography, Moscow 105187, Russia
6 Institute of Chemistry, University of Tartu, Ravila 14A, 50411 Tartu, Estonia
7 Department of Engineering, University of Applied Sciences Brandenburg, 14770 Brandenburg an der Havel, Germany
* Correspondence: nikolay.mukhin@ovgu.de

Abstract: Fast protein and metabolite liquid chromatography (FPLMC) was introduced years ago to enable the easy separation of high-molecular compounds such as proteins from small molecules and the identification of the low-molecular substances. In this paper, the method is applied for the rapid evaluation of freshness and monitoring the aging of animal meat and fish. A novel chromatographic sensor was developed with a deep UV LED-based photometric detection unit (255–265 nm), an original flow cuvette and registration scheme; the processing of a chromatogram with the sensor takes approximately 15 min. Strict isochronism between the elution of ATP metabolites, mainly hypoxanthine (Hx) and inosine monophosphate (IMP), and the time of maturation of meat or fish, was discovered. A new freshness index H* = [Hx]/[IMP] was introduced, which is proportional to the instrumental delay time in the FPMLC chromatograms: the H* index < 0.5 indicates the presence of inosine monophosphate (IMP) and the high quality of the meat or fish. Reasonably strong correlations were revealed between data obtained by FPMLC and total volatile basic nitrogen TVB-N (for fish) or volatile fatty acids VFA (for meat) content. Moreover, putative nucleotide salvage and an increase in the concentration of IMP were observed in fish after heat treatment using the FPMLC sensor and NMR technique.

Keywords: fast protein liquid chromatography; UV detection; meat freshness; fish freshness; inosine monophosphate; hypoxanthine; nucleotide salvage

Citation: Kuznetsov, A.; Frorip, A.; Sünter, A.; Kasvand, N.; Korsakov, V.; Konoplev, G.; Stepanova, O.; Rusalepp, L.; Anton, D.; Püssa, T.; et al. Fast Protein and Metabolites (Nucleotides and Nucleosides) Liquid Chromatography Technique and Chemical Sensor for the Assessment of Fish and Meat Freshness. *Chemosensors* **2023**, *11*, 69. https://doi.org/10.3390/chemosensors11010069

Academic Editor: Xiaolong Yang

Received: 30 November 2022
Revised: 5 January 2023
Accepted: 11 January 2023
Published: 14 January 2023

1. Introduction

Fast protein liquid chromatography (FPLC) is a form of liquid chromatography that was developed in 1982 by Pharmacia LKB (Uppsala, Sweden) for proteins as a partial alternative to high-performance liquid chromatography (HPLC) [1]. Initially, FPLC was considered exclusively as a preparative technique for the purification and fractionation of proteins and other biopolymers, but later it was widely demonstrated that the method can be successfully used in analytical tasks as a more affordable and cost-effective method than HPLC. The price of an FPLC column is approximately ten times lower than the price of a corresponding HPLC column, and the total cost of the test can be up to 30 times lower [2,3]. The positive features of FPLC include high loading capacity, biocompatible aqueous buffer systems, high flow rates, and the availability of stationary phases for most common chromatographic modes (e.g., ion exchange, gel filtration, reversed phase, and affinity chromatography) [4]. The use of FPLC as a characterization and analytical technique can be particularly relevant in the fields where the availability of rapid and affordable analytical methods and tools is crucial, e.g., routine test in healthcare or the food industry [2,3,5–9]. In this context, FPLC can be modified and enhanced with the ability to fractionate and detect low and middle molecular weight metabolites: ATP metabolites nucleotides and nucleosides, advanced glycation end products (AGE), uric acid, and other metabolic products originating from various biological samples [5,7–9]. In our opinion, the term fast protein and metabolites liquid chromatography (FPMLC) is more appropriate for such applications and will continue be used in this paper.

The ability of FPMLC to separate and detect ATP metabolites can be particularly important in food science and technology, because relative changes in the concentrations of ATP metabolites: adenosine diphosphate (ADP), adenosine monophosphate (AMP), inosine (Ino), inosine monophosphate (IMP), and hypoxanthine (Hx) during the post-mortem catabolism of adenosine triphosphate (ATP) in muscle tissue have been proven to be a reliable indicator of the freshness of animal or fish meat. The chain of ATP transformations after animal slaughter or harvesting fish is accompanied by a gradual decrease in the molecular weights of the ATP metabolites (the molecular weights in Daltons are in parentheses) [10,11]:

$$\text{ATP}(507) \rightarrow \text{ADP}(427) \rightarrow \text{AMP}(347) \rightarrow \text{IMP}(348) \rightarrow \text{Ino}(268) \rightarrow \text{Hx}(136). \quad (1)$$

The first step of the chain is relatively fast: the biochemical processes of ATP, ADP and AMP transformation to IMP are generally limited to 24 h. As a result, muscle tissue rapidly accumulates IMP, which is partly responsible for the pleasant taste (umami) and high nutritional value of meat and fish products [12]. In the following slower step, IMP is further degrading into Ino and Hx; the process is often associated with loss of freshness. In the final stage of bacterial spoilage, Hx is converted to xanthine, uric acid, and other circle cleavage end products [10,11]. The freshness index K is calculated as the ratio the concentrations of ATP metabolites concentrations (usually presented in μmol/L) [13]:

$$K = \frac{[\text{Ino}] + [\text{Hx}]}{[\text{ATP}] + [\text{ADP}] + [\text{AMP}] + [\text{IMP}] + [\text{Ino}] + [\text{Hx}]} \cdot 100\% \quad (2)$$

This index was first introduced in 1959 in Japan and has been widely used worldwide to evaluate fish freshness [14]. This correlates with fish taste and flavor: very high-grade products have a K-value of less than 10%; high-grade products have K-values lower than 20% or 30%, depending on fish species; fish with a K-value up to 50% are of medium grade; K-values larger than 50–70% are obtained for low-grade samples [15]. Multiple alternative indices based on ATP catabolism (K, K_0, K_I, H, P, G, Fr, Hx, etc.) were also proposed for various fish and shellfish species [10]. The index K_I is of particular interest, because it was specifically designed to assess freshness after more than 24 h post-mortem, when almost all ATP, ADP and AMP have already been catabolized:

$$K_I = \frac{[\text{Ino}] + [\text{Hx}]}{[\text{IMP}] + [\text{Ino}] + [\text{Hx}]} \cdot 100\%. \quad (3)$$

Starting from the 1980s, the K-value and, in some cases, the index K_I have been also applied for pork, beef, rabbit, and poultry [16–19]. Nevertheless, there are still no universal approaches to freshness evaluation based on ATP breakdown, which are equally applicable for meat and fish testing, and leading researchers in these fields work independently [15,19].

The main convenience of the K-value approach compared with universally accepted standard methods, e.g., determination of total volatile basic nitrogen (TVB-N), trimethylamine nitrogen (TMA-N) or volatile fatty acids (VFA), is the ability to detect changes in food freshness during storage at an early stage of quality declination, before any evidence of microbial spoilage appears [20,21]. Despite the obvious advantages of this approach, it is still used quite rarely in routine food quality control because conventional laboratory methods for the determination of ATP metabolites, including HPLC, nuclear magnetic resonance (NMR) spectroscopy, and mass spectrometry, are cumbersome, time consuming, and require extremely expensive analytical instruments operated by highly qualified staff [22,23]. By their nature, these instrumental methods cannot ensure rapid on-site freshness testing, and sensory evaluation by organoleptic methods, which are strictly regulated by national standards (GOST 7631-2008 in Russia and Council Regulation (EC) No 2406/96 in EU), are still dominant in the industry. An alternative approach to the EU standard is the quick Quality Index Method (QIM) [24].

Biosensor and microfluidic technologies have demonstrated great potential in the field of medical testing and food quality control [25–27]. The development of biosensors for the express testing of fish freshness is particularly active, as low testing time and the possibility of on-site analysis play a crucial role here. Electrochemical and enzymatic biosensors are capable of detecting marker substances such as xanthine, hypoxanthine, histamine, and uric acid; gas biosensors can detect volatile amines [19,28–31]. Moreover, multienzyme biosensors were developed for the simultaneous detection of ATP and its post-mortem breakdown products, which can directly measure the freshness indices K and K_I [19,29–36]. In addition to enzymatic biosensors, more conventional assay kits based on enzymatic reactions with microplate spectrophotometric detection are available on the market, but the price per test is quite high [37].

Undoubtedly, biosensors and microfluidic technologies can significantly simplify sampling procedures, detection efficiency, and portability of analytical systems for fish and meat freshness evaluation. At the same time, obvious difficulties in achieving high reproducibility and stability of characteristics during biosensor manufacturing and storage, the complexity of detection systems, and the high cost of disposable biosensors, comparable with the cost of some fish species [28], prevent their wider practical application as a convenient and cost-effective alternative to conventional analytical techniques and organoleptic evaluation. Prospects for the development of microfluidic biosensor chips are also inextricably linked with progress in chip technology [38] and on-wafer techniques for characterization purposes [39,40].

The obvious gap between sophisticated and expensive laboratory analytical instruments and miniaturized biosensing microfluidic devices can be filled with compact, simplified mid-range instruments based on classical methods of analysis, but which can be used on-site, outside the laboratory, do not require complex sample preparation, and are suitable for express testing. The aim of this work is the development and validation of a compact, affordable, non-disposable optical chemical sensor and analytical protocol for the rapid on-site evaluation of meat and fish freshness using FPMLC with UV photometric detection complemented with the ability to determine the relative content of the nutritional nucleotide and nucleoside originating from ATP post-mortem degradation.

2. Materials and Methods

2.1. FPLMC Optical Chemical Sensor

An optical chromatographic sensor based on FPMLC with UV photometric detection at a wavelength of 260 nm was designed and manufactured by Ldiamon AS (Tartu, Estonia) for meat and fish freshness evaluation. Previously, optical sensors based on the same principles were developed by the authors of this work for point-of-care diagnosis and screening of proteinuria, and assessment of protein loss during continuous ambulatory peritoneal dialysis; the instruments were clinically tested on urine and effluent peritoneal dialysate samples, respectively [6,7].

The chromatographic sensor is adapted for low-cost commercially available PD-10 protein desalting columns (Code No. 17-0851-01) from GE Healthcare® Bio-Sciences AB (Uppsala, Sweden), repurposed for the separation and detection of nutritional nucleotides and nucleosides. The PD-10 is a gel-filtration column filled with Sephadex G-25 Medium chromatographic gel originally designed for protein purification; the column is a single-use type according to the instruction for use, but it was demonstrated that it can be successfully regenerated and used multiple times up to several hundred tests [7]. The photometric detection unit employs a narrow-band (255–265 nm) deep UV LED with a quartz collimator and a visible-blind UV photodetector (Figure 1a). The detection wavelength corresponds to the UV absorption maxima of the ATP and its main metabolites: ATP, ADP, and AMP (259 nm); IMP and Ino (248 nm); Hx (249 nm) [41]. The UV absorption spectrum of ATP aqueous solution (concentration 0.3 mM, cuvette thickness 10 mm) and the normalized spectrum of the UV LED emission (maximum emission 262 nm, FWHM 12 nm) measured with the AvaSpec-2048 spectrophotometer from Avantes B.V. (Apeldoorn, The Netherlands) are presented in Figure 1b. The photo of the device is presented in the Figure 2.

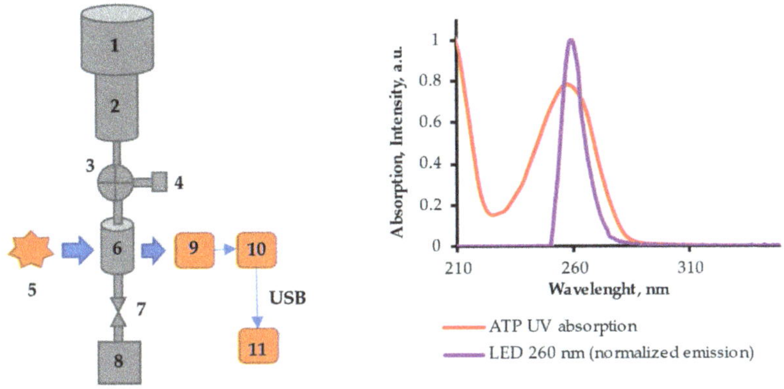

(a) (b)

Figure 1. The general principles of the sensor operation: (**a**) Schematic diagram of the sensor; (**b**) ATP UV absorption and UV LED emission spectra (1–LabMate buffer reservoir (Code No. 18-3216-03) from GE Healthcare® Bio-Sciences AB (Uppsala, Sweden); 2–PD-10 column; 3–three-way valve; 4–service port; 5–UV LED; 6–flow cell; 7–flow rate regulator; 8–drain vessel; 9–photodetector; 10–electronic unit, 11–laptop PC).

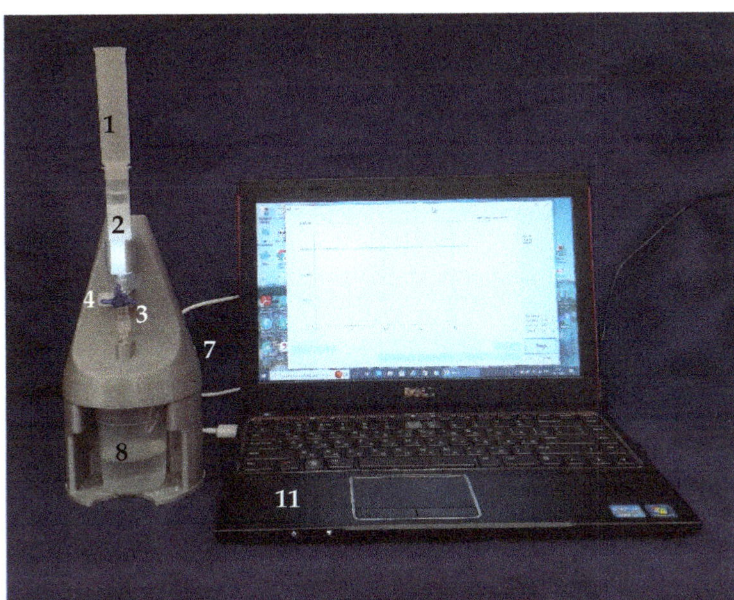

Figure 2. The photo of the sensor.

The sensor is a gravity flow chromatograph instrument: reservoir 1 is filled with TRIS buffer (pH 8.0), which flows via column 2 into flow cell 6 of the photometric detection unit into drain vessel 8. A mechanical regulator 7 is necessary to adjust buffer flow rate: in normal operation, it is about 2 mL/min when the buffer reservoir is full and about 1 mL/min when the reservoir is almost empty. The three-way valve 3 can be set to three positions: (a) the column is locked (standby mode), (b) the column is connected to the flow cell (chromatogram-recording or column-regeneration modes), and (c) the flow cell is connected to the service port 4 for manual cleansing or air bubble removal (a Luer-Lock syringe filled with buffer solution must be connected to the service port for this operation). A more detailed description of the device can be found in Kuznetsov et al., 2022 [7].

2.2. Chemicals

TRIS buffer (pH 8.0) contained 10 mM of tris(hydroxymethyl)aminomethane (TRIS), 150 mM of sodium chloride (NaCl), and 2 mM of ethylenediaminetetraacetic acid disodium salt (EDTA-Na_2); washing solution (pH 13.0) was 200 mM of sodium hydroxide (NaOH). TRIS (Product No. GE17-1321-01), EDTA (Product No. ED), sodium hydroxide (Product No. S5881), and sodium chloride (Product No. S9888) were purchased from Sigma-Aldrich (Darmstadt, Germany).

Standards of bovine serum albumin (BSA), ATP, IMP, Ino and Hx were also from Sigma-Aldrich., TRIS buffer, acetonitrile of ultra lc grade for LC-MS was from Romil PC (Cambridge, UK); ultrapure water was prepared by EASYpure RF, Barnstead type.

2.3. Sample Preparation

As samples of animal products for freshness testing with the FPMLC device, porcine *longissimus dorsi* muscle (striploin) obtained from a slaughterhouse within 24 h post-slaughter and completely fresh or within 24 h after catch farmed fish fillets of Trout (*Oncorhynchus mykiss*), Carp (*Cyprinus carpio*) were used. To analyze the effects, the thermal processing and putative nucleotide salvage defrosted fish fillets of Alaska pollack (*Theragra chalcogramma)*, Hake (*Merluccius)*, Norway haddock (*Sebastes)*, Pangasius (*Pangasianodon hypophthalmus)*, Wolffish (*Anarhichas)* from a local supermarket were used as samples of consumer products. Alaska pollack was frozen on 10 January 2021and the best before day (BBD) was 2 August 2023; Wolffish was frozen on 22 September 2021and the BBD was 22 September 2022; the experiments with Wolffish were carried out late August 2022; the other fish fillets had no data on the packaging.

The muscle tissue was finely minced with a meat grinder or cut into small pieces, approximately, of 2 mm$\times$ 2 mm $\times$ 2 mm, two grams of the ground meat (pH = 5.6) were placed into a 15 mL test tube with 6 mL of TRIS buffer (pH = 8.0). The mixture was shaken for 10–15 min in a rotator Biosan Multi RS60 (BioSan, Riga, Latvia), and used a syringe filter Whatman® GF/B (Product No. Z242195) from Merck KGaA (Darmstadt, Germany). For thermal processing, the samples were put in a microwave oven for 1–2 min at 400 W or steam cooked at +100 °C for 15–60 min; in this case, the fish fillets were cut into slightly larger pieces of 5 mm$\times$ 5 mm $\times$ 2 mm.

When working in the fields, a simplified version of the just-described sample preparation protocol is possible; for instance, when a rotator is not available, the tube with sample pieces can be manually shaken.

2.4. pH Determination

pH values of the samples were determined in homogenates composed of 1 g of sample and 9 mL of distilled water. Readings were taken with Consort C833 digital pH-meter (Consort, Turnhout, Belgium) at room temperature, and the pH meter calibration was regularly checked.

2.5. FPMLC Measurement Protocol and New Freshness Index H*

The FPMLC device was operated in the same mode as previously described in [42,43]. A total of 6 drops (200–300 μL) of the extract prepared according to the procedure described in Section 2.3 were placed directly on the gel surface in the PD-10 column previously washed with the NaOH wash solution (pH 13.0) and regenerated with 25 mL of TRIS buffer (pH 8.0). After the sample was completely absorbed by the gel, another 25 mL of TRIS buffer was added to the reservoir, and chromatograms were measured according to the procedure described in the work [7].

Gel filtration separates the molecules in the extract based on their size: proteins and other macromolecules larger than the pores of the gel matrix (the exclusion limit of the PD-10 column is M_r = 5000 Da) are quickly eluted with the void buffer volume outside the matrix, while smaller molecules, e.g., nucleotides and nucleosides, enter the pores and elute later at different times depending on the molecular weight [44,45]. The photometric detection unit of the sensor continuously measures the UV transmittance $T(t)$ of the eluate at a wavelength of 260 nm and records it at a time interval of 1 s; the optical transmission signal $T(t)$ is then converted into the optical density of the eluate (OD); the recording time of one chromatogram is about 15 min.

FPMLC chromatograms of both fish and mammalian meat samples consist of two main parts: a sharp protein peak and a broad post-protein band (Figure 3), which is formed by the merger of individual peaks of the main nucleotide and nucleoside actors. During meat storage, ATP is broken down by the enzymes into metabolites with a lower molecular weight according to the chain of transformations presented in (1), as a result of which, the retention time of the metabolites increases. The main parameter of FPMLC used in the measurements is the interval between the retention times of the broad band maxima of the metabolites (FPMLC fraction II) and the sharp protein peak (fraction I) on the FPMLC chromatograms (Figure 3), hereafter referred to as the index Time. This parameter directly corresponds to the relative content of different ATP metabolites and increases gradually during meat storage.

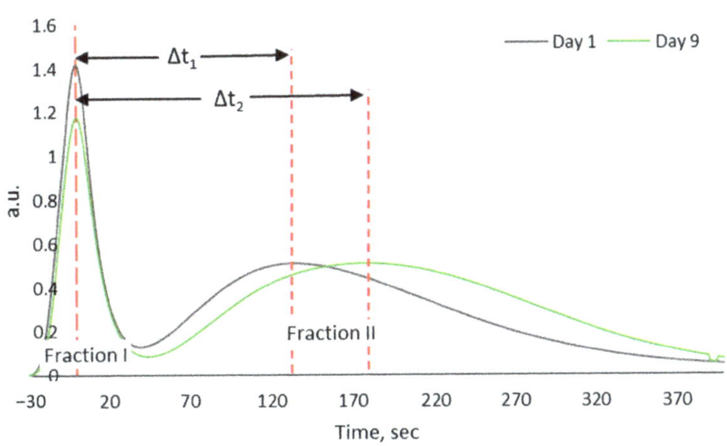

Figure 3. Comparison of two FPMLC chromatograms obtained on days 1 and 9 of a horse meat sample storage.

The first stage of this process, often described as absolute freshness, when ATP, ADP and AMP are still detected in a sample in significant amounts, lasts only several hours post-mortem; thus, fresh meat or fish are extremely rare in the food industry. In the following stages, IMP, Ino, and Hx become dominant and determine the shape of chromatograms and the index Time. In the final stage of massive bacterial contamination and obvious spoilage, Hx is mainly detected, and the maximum retention time for FPMLC fraction II is reached.

To establish the relationship between FPLMC data and the relative content of the ATP metabolites, Equation (3) can be further simplified, and a new freshness index H*, proportional to the instrumental lag time, can be introduced:

$$H* = \frac{[Hx]}{[IMP]}. \quad (4)$$

This assumption is based on the anticorrelation between the concentrations of inosine monophosphate [IMP] and hypoxanthine [Hx]. The anticorrelation phenomenon makes it possible to level out random deviations in the concentrations of Hx and IMP and increase the accuracy of measurements to the value Stdev/Mean ≈ 3%. H* index < 0.5 indicates the presence of IMP and high quality of the meat/fish.

2.6. *Validation Methods*

Results obtained by FPLMC were validated by direct identification and quantification of ATP metabolites by LC-DAD MS or NMR spectroscopy, and by comparison with widely recognized standard meat (VFA) and fish (TVB-N) quality control test methods.

2.6.1. LC-DAD MS

Liquid chromatographic analysis of the broad band of metabolites fraction II from FPMLC was carried out on a 1290 Infinity system (Agilent Technologies, Waldbronn, Germany) coupled to an Agilent 6450 Q-ToF mass spectrometer equipped with a Jetstream ESI source. Samples were subjected to a Zorbax 300SB-C18 column 2.1 × 150 mm; 5 μm (Agilent Technologies), and maintained at 40 °C. A gradient of 0.1% of formic acid in water (A) and 5% of water in acetonitrile (B) was used to separate the compounds as follows: 0.0 min 1% B, 3.0 min 1% B, 3.01 min 99% B, 11.1 min 99% B, 11.01 min 1% B, and a regeneration time of 8 min. The eluent flow rate was 0.3 mL/min and the injection volume was 2.5 μL. The mass-spectrometer was operated in the negative ion mode in the mass-to-charge ratio (m/z) range of 100–1000 Da. UV absorbance was measured at λ = 250 nm. Data acquisition and initial data processing were performed with MassHunter software (Agilent Technologies).

Identification of IMP, Ino and Hx in the fraction II of the FPMLC eluate samples was performed by comparing MS/MS and UV spectra of these substances with respective parameters of analytical standards. IMP, Ino, and Hx were quantified by UV absorption at 250 nm using the external calibration curve method. Methanolic standard solutions were prepared at concentrations of 3.125, 6.25, 12.5, 25, 50 and 100 μM for analytical standards of all three metabolites. Calibration curves were characterized by a high correlation coefficient (R^2 = 1) [46].

2.6.2. NMR Spectroscopy

The water-soluble polar metabolites of fish samples were extracted with 7.5% trichloroacetic acid (TCA) solution, as described in [47,48]. For this purpose, 25 g of fish muscle was added to 50 mL of 7.5% TCA and homogenized with a vertical homogenizer. The homogenate was filtered through a paper filter and the filtrate was neutralized with 9 M solution of KOH up to a pH of 7.8. The solution was filtered through a regular paper filter and stored at minus 40 °C.

NMR measurements were performed using a Bruker Avance III 700 NMR spectrometer equipped with a 5 mm BBO probe. The ^{1}H Larmor frequency was 700.08 MHz. The ^{1}H NMR spectra were measured at 298 K with solvent suppression using the noesypr1d pulse sequence. The acquisition time was 3.67 s; the recycle delay was set to 6.00 s. For every ^{1}H NMR spectra, 1520 scans were collected. NMR solutions were prepared by adding approximately 20% of D_2O containing 0.05% (w/v) sodium 2,2,3,3-tetradeuterio-3-trimethylsilylpropanoate (TSP-d_4) to the aqueous samples. All spectra were referenced to TSP-d_4 (0 ppm). Corrections of phase and baseline were performed with the Bruker Topspin 3.6.2 (Bruker, Rheinstetten, Germany). The free induction decays (FIDs) were multiplied by a line-broadening function of 0.3 Hz prior to Fourier transformation.

2.6.3. TVB-N and VFA Determination

The evaluation of VFA (for pork) and TVB-N (for fish) was carried out the at National Centre for Laboratory Research and Risk Assessment (before 1 January 2023–Estonian Veterinary and Food Laboratory, Tartu, Estonia, https://labris.agri.ee/en, accessed on 2 January 2023), according to the requirements of standard EVS-EN ISO/IEC 1705:2017 and Laboratory of Fish Products Quality Control, Russian Federal Research Institute of Fisheries and Oceanography (Moscow, Russia, http://vniro.ru/en/, accessed on 2 January 2023) independently. The measurements of TVB-N were conducted according to the method EU 2074/2016 (Estonia) or GOST 7636-85 (Russia); VFA content has been measured according to GOST 23392-78-2/1980.

3. Results

3.1. FPLMC Sensor Calibration

As a general rule in gel-filtration chromatography, the elution time of chemical substances with a molecular weight higher than the column exclusion limit depends on the molecular weight. For ATP metabolites, this means that the lowest elution time belongs to ATP, which has the largest molecular weight (M = 507) among the substances in the transformation chain (1), and the highest elution time to Hx, which has the lowest molecular weight (M = 136).

On the contrary, all proteins and other high molecular weight substances, such as free RNA and DNA, are synchronously eluted shortly after the process is started, as larger molecules cannot penetrate the gel pores. Since various water-soluble proteins are always present in animal and fish products, the first protein peak could be used as a reference point (zero on the timescale) from which elution times of nucleotides and nucleosides are counted.

To calibrate the FPMLC device, 100 μL of 1 g/l BSA and 100 μL of ATP, IMP, Ino, or Hx standard solutions were applied to the PD-10 column, and respective chromatograms were recorded (Figure 4) according to the procedure described in [7,42,43]. The mean time intervals between the BSA peak and the peaks of ATP, IMP, Ino, and Hx are presented in the Table 1: as predicted, the elution time monotonously increases in inverse proportion to the decrease in the molecular weight.

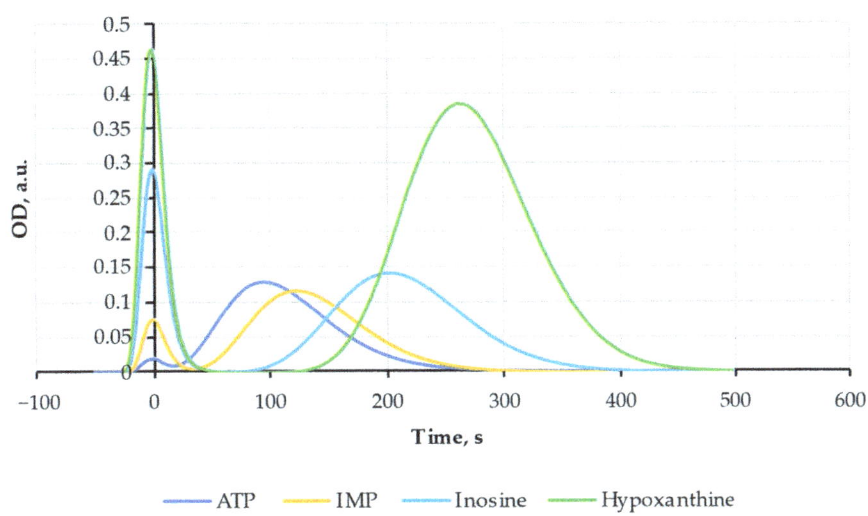

Figure 4. FPLMC sensor calibration with BSA and the nutritional nucleotides and nucleosides standards.

Table 1. Elution times of the main nutritional nucleotides and nucleosides.

Substance	Molecular Weight, Da	Elution Time, s
ATP	507	95
IMP	348	130
Ino	268	200
Hx	136	275

In case of routine measurements, this procedure should be repeated regularly, at least with hypoxanthine, and corresponding to the final point on the timescale; the time between BSA and Hx peaks should be kept constant near 275 ± 5 s by adjusting the buffer flow rate if necessary.

3.2. Verification of Compatibility of New Index H and FPMLC Data*

The next step was to establish the correlation between the independently obtained new freshness index H* = [Hx]/[IMP] (LC-DAD MS technique) and the index Time (FPMLC technique). This experiment can be also considered as a verification procedure of the FPMLC method by the LC-DAD MS technique [46]. Figure 5 shows the interdependences of the indices Time and H* measured during the tests of pork, both for raw samples stored aerobically (Figure 5a) and for the same samples after cooking (Figure 5b). The largest values exceeding the threshold H* >> 1 were excluded from the datasets for the diagrams in Figure 5.

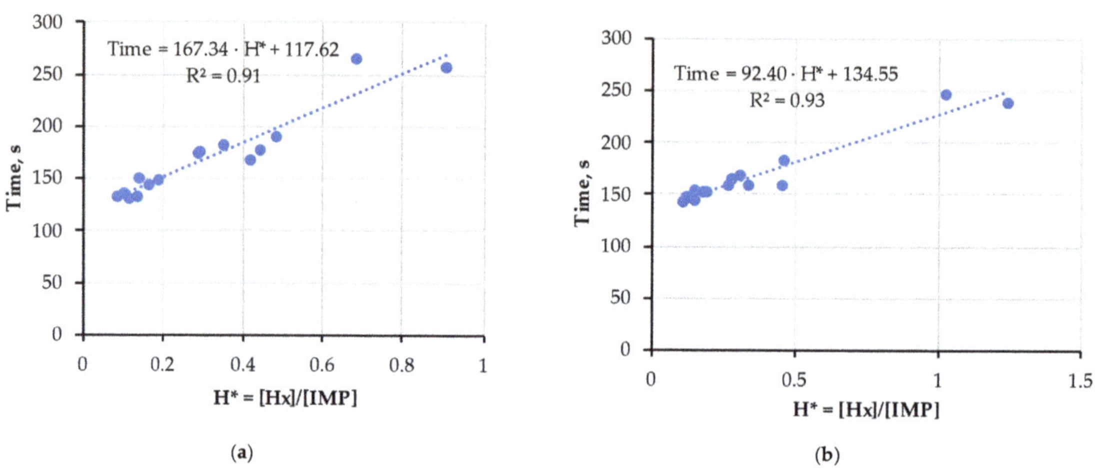

Figure 5. Correlation between the index H* and the index Time for raw (**a**) and cooked (**b**) minced pork stored or previously stored aerobically.

The correlations in Figure 5a,b are similar, but still have important peculiarities. In case of H* ≈ 0, the meat is absolutely fresh and the value of Time is correspondingly the smallest. For raw meat (Figure 5a), this value derived from the correlation formula is 117.6 s. The Special Time measurement within four hours after slaughter gave Time = 114 ± 5.7 s for another pork in pieces. The good agreement between the extrapolated and directly measured values strengthens confidence in the validity and perfect compatibility of the H* and Time indices.

3.3. Results of Validation of FPMLC Data by VFA and TVB-N Methods

The test results obtained with the FPMLC device were verified by the TVB-N method with farmed trout and carp samples during storage from the day 0 to day 14 in aerobic conditions at +4 °C. Figure 6 shows the dynamics of the average index Time and TVB-N values obtained with farmed trout caught in spring and autumn seasons. It can be observed that both curves are close in shape and correlate with the coefficient r = 0.85. The onset of the strong TVB-N increase observed on day 11 and later on correlates well with the published data [37,49–51]. Moreover, the value TVB-N at day 7 in Figure 6 (the very onset of bacterial contamination according to the FAO scheme [52]) of 16 ± 2.83 mg/100g agrees very well with the average value that can be deduced from the data in [37,49–51], i.e., 16.25 ± 4.2 mg/100g. This means that quite typical trout fish have been used.

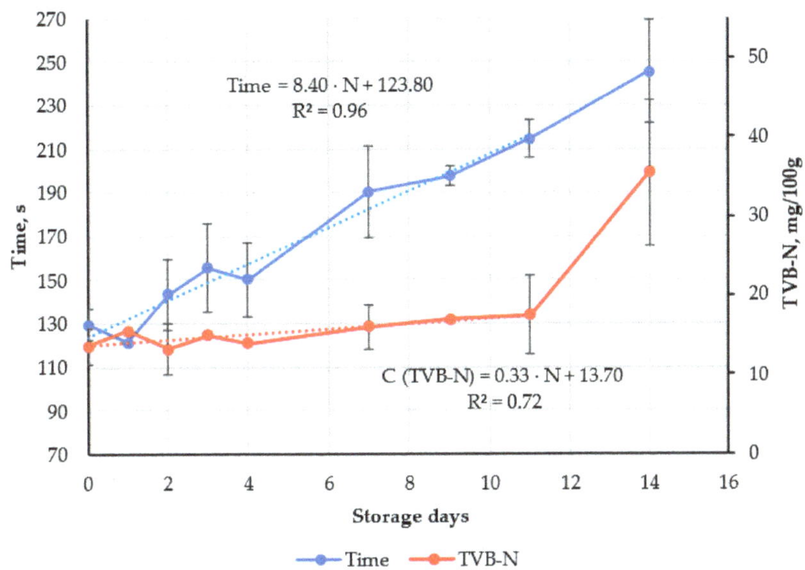

Figure 6. Averaged index Time obtained with FPMLC device and TVB-N content values for farmed trout during storage.

A similar comparison of Time and TVB-N values was conducted for carp and a correlation r = 0.8 was obtained. This, it can be concluded that the FPMLC device can provide relevant data for freshness, at least, of the well-known trout and carp fish.

The compatibility of the freshness test results with the FPMLC device with widely accepted meat quality control method based on VFA content was evaluated for porcine *longissimus dorsi* muscle (loin), obtained from a slaughterhouse the day after slaughter. The fillet sample was minced by means of a manual grinder with a sieve with 3 mm orifices. Minced meat was stored aerobically (in a plastic box) at +4 °C for 14 days. Measurements with the FPMLC device and by the VFA method were performed twice in each measurement during the 14-day period synchronously, with an accuracy not worse than 2 h. The relative standard deviations (Stdev/Mean) of the Time index and TVB-N content were in the interval of 0–5% usually around 2.5%. Bacterial contamination (CFU/g) was also estimated, but microbiological tests were synchronized more freely (within the same day's work shift) with the FPLMC and VFA measurements, due to the fixed working schedule of the microbiological laboratory.

The evolution of the Time index and VFA content during the storage period is presented in the Figure 7: the data demonstrate that VFA content increased slowly during the first 9 days, then both the Time index and VFA jump sharply, accompanied by an increase in bacterial contamination to 1.6×10^8 CFU/g from the initial level 7.2×10^3 CFU/g and pH 7.0 from the initial value of pH 5.6.

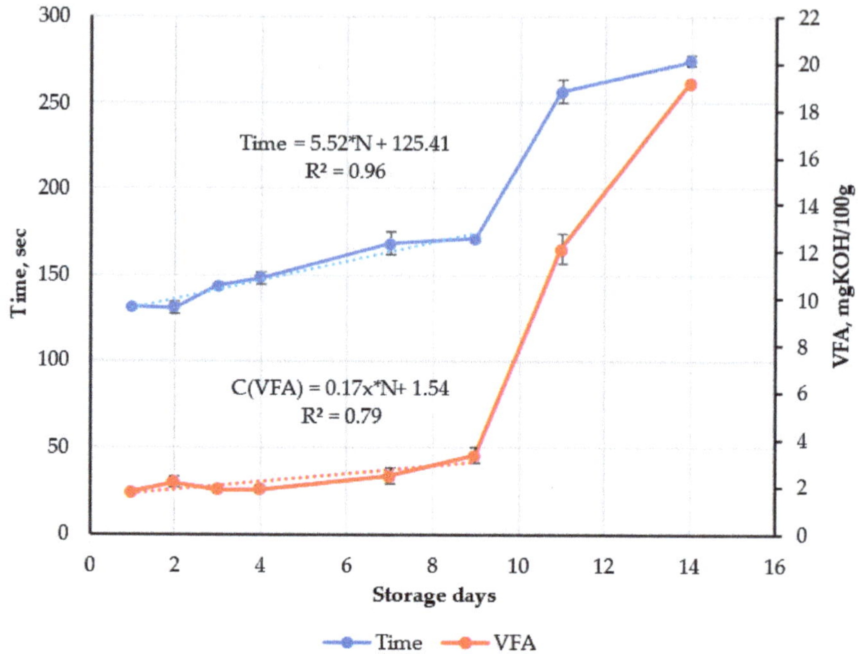

Figure 7. Temporal change of the index Time obtained with FPMLC device and VFA content during storage of minced pork at +4 °C under aerobic conditions.

Figure 7 shows the strictly parallel trajectories of the Time and VFA curves during the first 9 days of storage, which is confirmed by the high correlation coefficient r = 0.96 between these two parameters. The results obtained by the FPLMS method are quite compatible with the data of microbiological safety, which prove that the product is safe for consumption until the 9th day of storage at +4 °C. After the 11th day, the minced meat had an unpleasant off-smell of spoilage and a bad appearance, which made it unacceptable for culinary use.

A more detailed comparison of the curves at the initial stage of storage highlights a significant fact: when the meat is still in the autolysis phase (bacterial contamination on the 7th day was 2×10^6 CFU/g, which is below the threshold level 10^7 [53]) the trajectories of the curves are parallel, but the irregularities of the Time index are much less obvious than for VFA data, and the Time index correlates much better with the duration of storage (R^2 = 0.96) than VFA content (R^2 = 0.79)

It can be argued that for fresh pork, the Time is more preferable as a freshness index than VFA content, and testing meat with the FPMLC device is much faster, cheaper, and simpler than the complicated determination of VFA. We expect that after further research, this conclusion may become more general and be extended to the meat of other animals.

3.4. *Effects of Heat Treatment and Putative Nucleotide Salvage*

There is a number of evidences for a peculiar effect resulting in an increase in AMP and IMP content in animal meat after heat treatment (Nucleotide salvage) [54–58]. We had not found any published research data on a similar effect in fish meat and, hence, conducted a series of experiments with defrosted fish fillets (Alaska pollack, Hake, Norway haddock, Pangasius, Wolffish) from a local supermarket (Tartu, Estonia). All samples had a Time index about 200 s, corresponding to the Phase 3 of FAO UN Quality Score (see also Discussion) [52]. For heat treatment, the samples were placed in a microwave oven for 1–2 min at 400W or steam cooked at +100 °C for 15–60 min; fish fillets were cut into pieces of 5 × 5 × 2 mm.

For all fish samples, we observed the shift of the second metabolite peak (FPMLC fraction II) in the chromatograms towards lower retention times and a consequent decrease in the Time index, i.e., nucleotide relative content after thermal processing changed to become more to that of fresh fish. The shifts of the Time index ΔTime after 15–60 min of steam cooking at 100 °C are presented in Table 2.

Table 2. Decrease in the Time index values for various fish species after 15–60 min of steam cooking at 100 °C.

Fish Specie	ΔTime, s	Note
Alaska pollack	−21	-
Pangasius	−17	-
Norway haddock	−18	Two peaks at 130 and 260 s emerged after treatment
Wolffish	−37	Unstable shifts of the broad band ±10 s
Hake	−30	-

The chromatograms of Alaska pollack before (raw) and after cooking (steam) at 100 °C for 60 min are shown in Figure 8, and the difference between the two chromatograms (raw-steam) is also presented. Two extrema (besides protein) are clearly observed for the difference curve (raw-steam): one maximum at 135 s, which probably corresponds to an increase in the IMP concentration after thermal treatment, and another minimum at 282 s, which could be related to a decrease in Hx concentration.

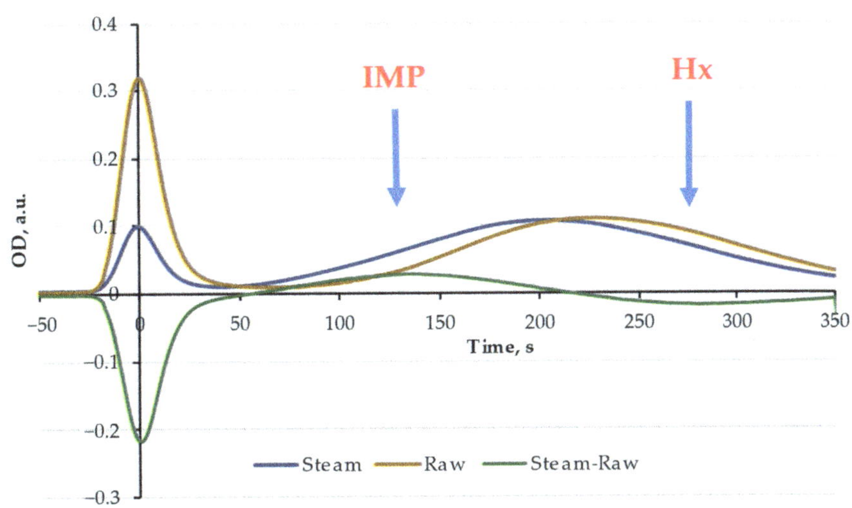

Figure 8. Chromatograms of Alaska Pollock samples before (Raw) and after (Steam) cooking at 100 °C for 60 min, and difference of the two chromatograms (Steam–Raw).

To verify this assumption, NMR spectra of two samples of raw and steam cooked Alaska Pollock samples, previously used for FPMLC testing, were measured. The measurements were carried out according to 2.6.2. The preliminary assignment of the main signals has been conducted in agreement with the data of [59–61], for strict identification spiking tricks with IMP and Inosine have been performed.

In fact, only Inosine and Hypoxanthine lines were detected in raw pollock meat (Figure 9a), but the presence of IMP was also observed in the cooked meat (Figure 9b). The singlet at 8.58 and the doublet near it at 6.15 ppm belong to IMP, as observed in the spectrum with added IMP. Moreover, after the addition of IMP, another line of IMP was revealed: this line overlaps very closely with the Inosine line at 8.24 ppm and remains overshadowed in the unspiked liquid. Thus, the formation of IMP (Figure 9c) in the course of the heating of this pollock fish has been registered. Such a manifestation of the possible nucleotide salvage or nucleotide synthesis in cooked fish (pollock) was observed, to the best of our knowledge, for the first time. Further experiments are under way to confirm this hypothesis. The FPMLC sensor is proved to be a very useful instrument in this work for the rapid pre-selection of samples for more precise but expensive and cumbersome evaluation methods as NMR, HPLC, MS, etc.

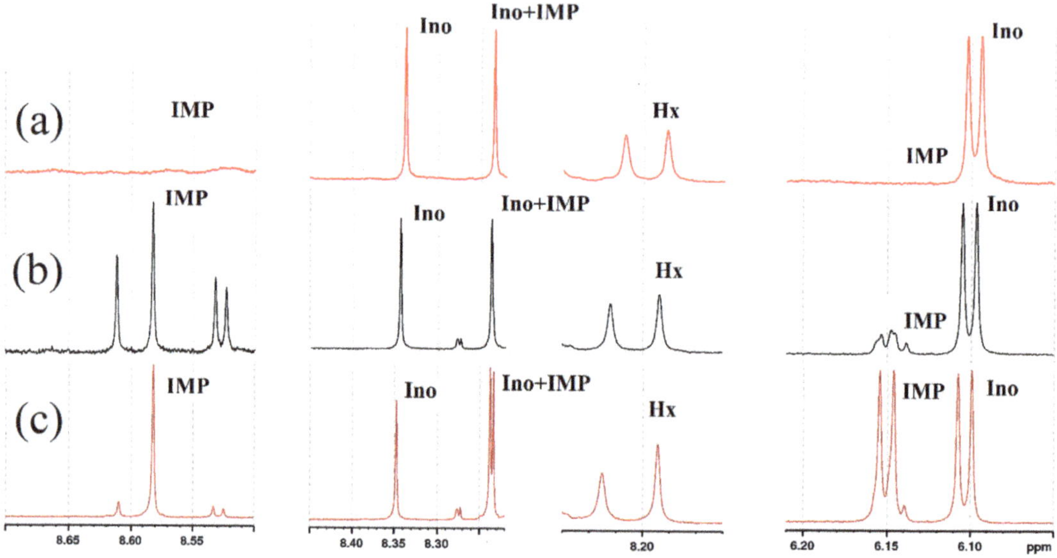

Figure 9. The low-field region of ^{1}H NMR spectra of Alaska Pollock samples: (**a**) before thermal treatment, (**b**) after steam cooking at 100 °C for 60 min, and (**c**) after steam cooking with added IMP and Ino.

4. Discussion

The results obtained for in-house-validation by TVB-N and VFA standard methods show that the FPMLC sensor can be used equally successfully to evaluate the freshness of both fish and animal meat. This conclusion is not surprising, since in both cases, freshness in this study is determined by the formulas (2) and (3) based on the main metabolites of ATP as the variables. These metabolites are absolutely the same in fish and meat, which reflects the common features of the ATP post-mortem catabolism.

The classic freshness indices K, K_I and some others have a certain drawback. Their limit value is 1 (or 100%) and approaches this limit slowly and asymptotically as the sample deteriorates. As a result, the difference between the index values for a product still acceptable for consumption and an irreversibly spoiled product can be as little as 10% or less. This situation means a low sensitivity and too late warning of imminent danger, for example, in the form of intense bacterial contamination.

The index H* = [Hx]/[IMP] proposed in this work takes into account only Inosine monophosphate IMP and Hypoxanthine Hx, but these metabolites are the most important determinants during the loss of freshness. There are no known examples of IMP content increasing with the aging of fish or meat; outside the first day, there is always a decrease in IMP. This means that H* can take values even as H* >> 1. A similar situation also occurred in this study when a rather early critical bacterial contamination (1.6×10^{8} CFU/g) took place on day 9 of the experiment with minced pork (Figure 7). This resulted in dramatic growth in the Time index from 172 to 257 sec in the interval of 9–11 days of storage, accompanied by a sharp increase in H* index from 0.44 to 0.91, reflecting the rapid deterioration of the freshness status.

Our proposal for the H* index builds on previous recommendations to improve the freshness index situation. In particular, in [62], the K* index was proposed, which differs from the K index in that it does not take into account inosine at all due to its change during the loss of freshness with an inconsistent trend—at first a slow rise in concentration followed by a slow decline. The authors of the work [62] came to the conclusion that the K* index can better predict the shelf life of refrigerated striped bass than the K or K_I indices.

Since the early 1960s, the use of hypoxanthine [Hx] as an independent index of fish spoilage has been proposed, as it is the longest-lived metabolite with a constant upward trend over time [63,64], and it was also noted that hypoxanthine concentration [Hx] increased during storage almost linearly [65]. However, the growth rate during shelf-life was too low to be useful as an index of freshness. In this regard, the H* index does not suffer from the noted drawback, since at [IMP] $\rightarrow 0$, which always happens during prolonged storage, the ratio [Hx]/[IMP] $\rightarrow \infty$.

It can also be noted that the H* index has a strict linear relationship with Time readings of the FPMLC sensor in the most important and initial freshness region (Figure 5) and, thus, an important rule in instrumentation, i.e., linear response to a stimulus, is fulfilled. Of significant interest is the comparison of the technical parameters and capabilities of the device with the organoleptic characteristics of foods. A good pattern to achieve this goal is the Phases diagram of Quality score of the FAO of UN [52] (Figure 10). This diagram was elaborated on during the organoleptic examination of cod fishes cooked in steam. To the best of our knowledge, this is the most valuable scheme of the subject under consideration.

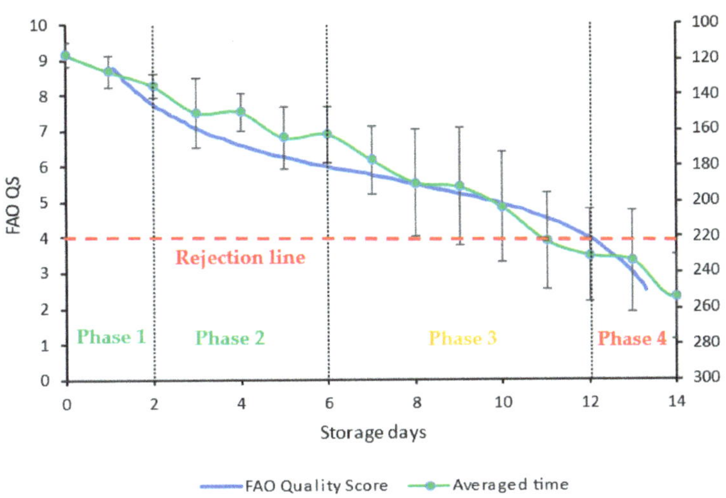

Figure 10. Comparison of UN FAO Quality Score curve and mean values of Time obtained with a number of fish and meat samples stored aerobically at +2–4 °C. Phase 1: The fish is very fresh and has a sweet, algal, and delicate taste. The taste can be very slightly metallic; Phase 2: There is a loss of the characteristic odor and taste. The flesh becomes neutral but has no off-flavors. The texture is still pleasant; Phase 3: There is sign of spoilage and a range of volatile, unpleasant-smelling substances is produced, depending on the fish species and type of spoilage (aerobic or anaerobic); Phase 4: The fish can be characterized as spoiled and putrid [52].

We have carried out a number of Time index measurements with fish and meat in exactly the same mode as the other experiments in this study but precisely knowing the day 0 or even the hour 0 (exact day and time of animal slaughter or fish harvest). For this purpose, the work has been conducted in a slaughterhouse with absolutely fresh pork and beef to obtain the corresponding values of Time for the Phase 1. Farmed trout and carp were used as fish samples. The obtained Time data is placed at the FAO diagram in a randomized mode, i.e., without indicating whether it was fish or animal meat (carp data are given only for the days 0 and 1). It can be observed that the series of Time points are fairly anticorrelated with the FAO curve, with a very high coefficient of r = −0.97.

The technical data obtained with the FPMLC sensor, compiled with the Phase diagram of UN FAO [52] and the experiences of ordinary food consumers, are presented in Table 3. The data therein cannot be used as an instruction or guidelines; they are only for orientation in the ocean of information about foods in the form of animal meat and fish as well. We hope that Table 3 together with the data in 3.4 encourage the researchers to look for phenomena occurring during the thermal treating of raw products in relation to the phenomenon of nucleotides salvage, which can enhance the usefulness of foods that stand near the end of shelf life. This, in turn could help consumers and retailers to reduce food waste and bring it down below the current level of 30%. The approach of this study and the FPMLC method and device can be of help in this work.

Table 3. Correlation between the consumer properties of meat and fish, Time and H* indices, and bacterial contamination CFU/g.

CFU/g Unsterile Pork, Trout Initial pH 5.5–5.8	Time, s	Index K	Index H*	Characteristic and Accordance to FAO Phases	Objects	Suits for	Some Other Characteristics
– [1]	<130	<0.3	H* < 0.1	Absolutely fresh Phase 1	All rapidly frozen fish or meat	Many purposes. Aging can be needed	Quickly (in hours) ending condition in fridge at +4 °C
$3 \times 10^3 \div 10^4$	130÷160	0.3 < K < 0.5	0.1 < H* < 0.2	Fresh Phase 2	High quality meat and fish	Haute cuisine	Rich in endogenous IMP; no need for artificial E630
$\geq 10^4 \div 10^5$	160÷190	0.5 < K < 0.75	0.2 < H* < 0.45	Rather fresh Status "Use before" The first half of Phase 3	Stored or thawed meat and fish	Ordinary consumption	Often coincides with end of bacterial lag phase
$\geq 10^5 \div 10^7$ [2]	190 ÷210	K ≈ 0.8	H* ≈ 0.5	Start or middle point of exponential bacterial growth Ending of Phase 3	Can be all kinds of samples	Standard shelf-life end	May be in quickly changing condition
$\geq 10^7 \div 10^8$	210÷240	0.8 < K < 0.95	0.8 < H*≈ 0.9	Heavy contamination but may not be spoiled irreversibly Overlapping of Phases 3 and 4	Often reduced-price foods or aged meat	Intense temperature treatment needed	Bacterial stationary phase: oscillating contamination and crash
$>10^8$	>240	K ≈ 1	1 ≈ H* >> 1	Phase 4	Spoiled	Rejection	Smell, bad appearance

[1] Not counted. [2] Contamination level 10^7 is often called "Borderline" (see, e.g., [53]).

In addition to freshness control, the developed device and analytical technique are well in line with the modern attitude towards the nutritional nucleotides, which is also gaining interest as immunity boosters [66–68]. A perspective that in the near future there will be introduced a nutritive index of nucleotides [68] and a "green transition" to a healthy umamiization of foods achieved [12] makes the matter highly interesting. For this purpose, however, simple and reliable instruments for assessing the quality of foods and just the degree of their umamization should be available.

5. Concluding Remarks

It was demonstrated that sensing technology based on fast protein and metabolites liquid chromatography (FPLMC) provides rapid on-site assessment of animal meat and fish freshness with low-cost, affordable, multiple use, easily operated optical chemical sensors. In was demonstrated that direct readings of FPMLC sensor (the index Time, difference between the retention times of proteins and nutritional nucleotide and nucleoside pools, mainly hypoxanthine and inosine monophosphate), and the new freshness index H* = [Hx]/[IMP] strongly correlated with TVB-N (for fish) and VFA (for meat) content, which are considered reliable indicators of product spoilage. Moreover, it was demonstrated that the indices Time and H* more accurately reflect freshness deterioration at the early stages of product storage than conventional TVB-N and VFA, which are slightly sluggish and strongly react only at the late stages, when the sample is approaching close to spoilage, failing to detect the changes in muscle tissue during autolysis (Figures 6 and 7). The developed approach is free of this drawback: the averaged data obtained with the FPMLC sensor are in good concordance with the Phases diagram of Quality score of the FAO of UN based on organoleptic examination and consumer properties of fish. In addition to the routine control of meat and fish quality, the relative content of dietary nucleotides and nucleosides determined with the FPMLC sensor, especially inosine monophosphate, is extremely important for product taste (umami taste) and nutritional value. In this context, a quite peculiar effect of nucleotide salvage has been observed in fish muscle tissue and validated by NMR spectroscopy.

It should be emphasized, that most, if not all, portable low-cost instruments for the evaluation of meat and fish freshness are of indicator-type disposable sensors envisaged mainly for distinguishing between the low and high bacterial contamination, while more sophisticated tools, e.g., multienzyme biosensors, electronic noses and tongues, or near-infrared (NIR) spectroscopic analytical systems are expensive, single-use (biosensors), sensitive to contaminated laboratory environment (electronic nose), or dependent on complex object-specific chemometrics algorithms (NIR). It is very difficult, if not impossible, to monitor with simple and affordable sensors the stages of autolysis, discriminate between the Phase 1 and Phase 2 (Figure 10) and predict the residual shelf life. This problem can be easily solved with the FPMLC sensor due to its high precision (the relative standard deviation of the main instrumental parameter Time is less than 5%).

In a comprehensive review [22], as many as 41 freshness evaluation methods, including the FPLC approach presented in the current article (based on a brief previous description in a conference paper [42]), were discussed. The authors points to the obvious fact that this method still needs universal acceptance and more research to be conducted, but at the same time, mention seven positive features compared to other techniques:

- Quickest determination of fish freshness available until now,
- Can be used in labs as well as in retail chains,
- Environment friendly,
- Low cost and reliable,
- Less qualified staff can also work,
- Non-destructive and less time consuming, and
- Quickly determine shelf life of almost all varieties of fish.

The main point is that the FPMLC technique and sensor are oriented towards the future, towards the development and introduction into practice of the nutritive nucleotide index (NNI) [12,66]. Indeed, the Time values can be easily recalibrated into the values of NNI for raw or ready-to-eat foods. In any case, the FPMLC method invented and developed here would be in a more strict and direct functional connection with nucleotides' presence in subjects than the methods based on the correlation between another parameters as it is, e.g., in the classical methods of TVB-N, TMA-N, VFA, etc. Moreover, the FPMLC sensors are potentially capable of the simultaneous qualitative determination of protein and nucleotides content provided a dual-wavelength UV detection unit is installed, and the development of such devices is underway.

Author Contributions: Conceptualization, A.K. (Artur Kuznetsov) and G.K.; methodology, A.K. (Artur Kuznetsov), T.P. and L.A.; funding acquisition, A.F., M.R., and S.H.; software, V.K.; validation, A.S., A.K. (Andrey Kozin) and L.T.; supervision, A.K. (Artur Kuznetsov) and M.R.; formal analysis, N.M.; investigation, A.S., N.K., L.R., D.A. and L.T.; resources, L.A. and A.F.; data curation, A.K. (Andrey Kozin) and M.R.; writing—original draft preparation, A.K. (Artur Kuznetsov), G.K. and T.P.; visualization, O.S. and A.K. (Andrey Kozin); writing—review and editing, N.M., S.H., O.S., T.P. and A.F.; project administration, A.F. and M.R. All authors have read and agreed to the published version of the manuscript.

Funding: This research was partially financially supported by the Estonian Research Council grant (PRG 1441) of the Estonian University of Life Sciences (A.S., L.R., D.A., T.P., M.R.) and by DFG (Deutsche Forschungsgemeinschaft), project number 315440263, project title: "Verhalten von Proteinen in mikrofluidischen Kanälen (Behavior of proteins in microfluidic channels)" (N.M., S.H.).

Institutional Review Board Statement: Not applicable.

Informed Consent Statement: Not applicable.

Data Availability Statement: Data available on request due to Ldiamon AS commercial secret.

Acknowledgments: The authors express their gratitude to the staff of National Centre for Laboratory Research and Risk Assessment (before 1 January 2023–Estonian Veterinary and Food Laboratory) for help and valuable comments and to Roman Korsakov for assistance in software development.

Conflicts of Interest: The authors declare no conflict of interest.

References

1. Richey, J. FPLC: A comprehensive separation technique for biopolymers. *Am. Lab.* **1982**, *14*, 104–129.
2. Tangvarasittichai, S.; Tangvarasittichai, O.; Jermnim, N. Comparison of fast protein liquid chromatography (FPLC) with HPLC, electrophoresis & microcolumn chromatography techniques for the diagnosis of beta-thalassaemia. *Indian J. Med. Res.* **2009**, *129*, 242–248. [PubMed]
3. Verbeke, K.; Verbruggen, A. Usefulness of fast protein liquid chromatography as an alternative to high performance liquid chromatography of 99mTc-labelled human serum albumin preparations. *J. Pharm. Biomed. Anal.* **1996**, *14*, 1209–1213. [CrossRef]
4. Madadlou, A.; O'Sullivan, S.; Sheehan, D. Fast Protein Liquid Chromatography. *Methods Mol. Biol.* **2017**, *1485*, 365–373. [CrossRef]
5. Deori, S.; Hurri, E.; Karkehabadi, S.; Morrell, J.M. Fast protein liquid chromatography profiles of seminal plasma proteins in young bulls: A biomarker of sperm maturity? *Livest. Sci.* **2021**, *250*, 104600. [CrossRef]
6. Sünter, A.; Frorip, A.; Korsakov, V.; Kurruk, R.; Kuznetsov, A.; Ots-Rosenberg, M. Optical method for screening and a new proteinuria focus group. *J. Biomed. Photon. Eng.* **2015**, *1*, 236–247. [CrossRef]
7. Kuznetsov, A.; Frorip, A.; Sünter, A.; Korsakov, V.; Konoplev, G.; Stepanova, O.; Roschina, N.; Ovsyannikov, N.; Lialin, D.; Gerasimchuk, R.; et al. Optical Chemical Sensor Based on Fast-Protein Liquid Chromatography for Regular Peritoneal Protein Loss Assessment in End-Stage Renal Disease Patients on Continuous Ambulatory Peritoneal Dialysis. *Chemosensors* **2022**, *10*, 232. [CrossRef]
8. Noonin, C.; Kapincharanon, C.; Sueksakit, K.; Kanlaya, R.; Thongboonkerd, V. Application of tandem fast protein liquid chromatography to purify intact native monomeric/aggregated Tamm-Horsfall protein from human urine and systematic comparisons with diatomaceous earth adsorption and salt precipitation: Yield, purity and time-consumption. *Anal. Methods* **2021**, *13*, 3359–3367. [CrossRef]
9. Tsay, F.-R.; Haidar Ahmad, I.A.; Henderson, D.; Schiavone, N.; Liu, Z.; Makarov, A.A.; Mangion, I.; Regalado, E.L. Generic anion-exchange chromatography method for analytical and preparative separation of nucleotides in the development and manufacture of drug substances. *J. Chromatogr. A* **2019**, *1587*, 129–135. [CrossRef]
10. Hong, H.; Regenstein, J.M.; Luo, Y. The importance of ATP-related compounds for the freshness and flavor of post-mortem fish and shellfish muscle: A review. *Crit. Rev. Food Sci. Nutr.* **2017**, *57*, 1787–1798. [CrossRef]
11. Howgate, P. A review of the kinetics of degradation of inosine monophosphate in some species of fish during chilled storage. *Int. J. Food Sci. Technol.* **2006**, *41*, 341–353. [CrossRef]
12. Mouritsen, O.G. Umamification of food facilitates the green transition. *Soil Ecol. Lett.* **2022**, *5*, 9. [CrossRef]
13. Zhang, Z.; Sun, Y.; Sang, S.; Jia, L.; Ou, C. Emerging Approach for Fish Freshness Evaluation: Principle, Application and Challenges. *Foods* **2022**, *11*, 1897. [CrossRef] [PubMed]

14. Saito, T.; Arai, K.-I.; Matsuyoshi, M. A New Method for Estimating the Freshness of Fish. *Nippon Suisan Gakkaishi* **1959**, *24*, 749–750. [CrossRef]
15. García, M.R.; Ferez-Rubio, J.A.; Vilas, C. Assessment and Prediction of Fish Freshness Using Mathematical Modelling: A Review. *Foods* **2022**, *11*, 2312. [CrossRef] [PubMed]
16. Fujita, T.; Hori, Y.; Otani, T.; Kunita, Y.; Sawa, S.; Sakai, S.; Tanaka, Y.; Takagahara, I.; Nakatani, Y. Applicability of the K0 Value as an Index of Freshness for Porcine and Chicken Muscles. *Agric. Biol. Chem.* **1988**, *52*, 107–112. [CrossRef]
17. Batlle, N.; Aristoy, M.-C.; Toldrá, F. ATP Metabolites During Aging of Exudative and Nonexudative Pork Meats. *J. Food Sci.* **2001**, *66*, 68–71. [CrossRef]
18. Nakatani, Y.; Fujita, T.; Sawa, S.; Otani, T.; Hori, Y.; Takagahara, I. Changes in ATP-Related Compounds of Beef and Rabbit Muscles and a New Index of Freshness of Muscle. *Agric. Biol. Chem.* **1986**, *50*, 1751–1856. [CrossRef]
19. Nanda, P.K.; Bhattacharya, D.; Das, J.K.; Bandyopadhyay, S.; Ekhlas, D.; Lorenzo, J.M.; Dandapat, P.; Alessandroni, L.; Das, A.K.; Gagaoua, M. Emerging Role of Biosensors and Chemical Indicators to Monitor the Quality and Safety of Meat and Meat Products. *Chemosensors* **2022**, *10*, 322. [CrossRef]
20. Bekhit, A.E.-D.A.; Holman, B.W.B.; Giteru, S.G.; Hopkins, D.L. Total volatile basic nitrogen (TVB-N) and its role in meat spoilage: A review. *Trends Food Sci. Technol.* **2021**, *109*, 280–302. [CrossRef]
21. Bleicher, J.; Ebner, E.E.; Bak, K.H. Formation and Analysis of Volatile and Odor Compounds in Meat-A Review. *Molecules* **2022**, *27*, 6703. [CrossRef]
22. Prabhakar, P.K.; Vatsa, S.; Srivastav, P.P.; Pathak, S.S. A comprehensive review on freshness of fish and assessment: Analytical methods and recent innovations. *Food Res. Int.* **2020**, *133*, 109157. [CrossRef]
23. Franceschelli, L.; Berardinelli, A.; Dabbou, S.; Ragni, L.; Tartagni, M. Sensing Technology for Fish Freshness and Safety: A Review. *Sensors* **2021**, *21*, 1373. [CrossRef] [PubMed]
24. Sveinsdottir, K.; Martinsdottir, E.; Hyldig, G.; Jorgensen, B.; Kristbergsson, K. Application of Quality Index Method (QIM) Scheme in Shelf-life Study of Farmed Atlantic Salmon (*Salmo salar*). *J. Food Sci.* **2002**, *67*, 1570–1579. [CrossRef]
25. Bakhchova, L.; Jantaree, P.; Gupta, A.; Isermann, B.; Steinmann, U.; Naumann, M. On-a-Chip-Based Sensitive Detection of Drug-Induced Apoptosis in Polarized Gastric Epithelial Cells. *ACS Biomater. Sci. Eng.* **2021**, *7*, 5474–5483. [CrossRef] [PubMed]
26. Wang, J.; Bei, J.; Guo, X.; Ding, Y.; Chen, T.; Lu, B.; Wang, Y.; Du, Y.; Yao, Y. Ultrasensitive photoelectrochemical immunosensor for carcinoembryonic antigen detection based on pillar5arene-functionalized Au nanoparticles and hollow PANI hybrid BiOBr heterojunction. *Biosens. Bioelectron.* **2022**, *208*, 114220. [CrossRef]
27. Wang, J.; Zhou, L.; Bei, J.; Zhao, Q.; Li, X.; He, J.; Cai, Y.; Chen, T.; Du, Y.; Yao, Y. An enhanced photo-electrochemical sensor constructed from pillar 5arene functionalized Au NPs for ultrasensitive detection of caffeic acid. *Talanta* **2022**, *243*, 123322. [CrossRef]
28. Xiong, X.; Tan, Y.; Mubango, E.; Shi, C.; Regenstein, J.M.; Yang, Q.; Hong, H.; Luo, Y. Rapid freshness and survival monitoring biosensors of fish: Progress, challenge, and future perspective. *Trends Food Sci. Technol.* **2022**, *129*, 61–73. [CrossRef]
29. Wu, L.; Pu, H.; Sun, D.-W. Novel techniques for evaluating freshness quality attributes of fish: A review of recent developments. *Trends Food Sci. Technol.* **2019**, *83*, 259–273. [CrossRef]
30. Nimbkar, S.; Auddy, M.; Manoj, I.; Shanmugasundaram, S. Novel Techniques for Quality Evaluation of Fish: A Review. *Food Rev. Int.* **2021**, *1996*, 1–24. [CrossRef]
31. Mustafa, F.; Othman, A.; Andreescu, S. Cerium oxide-based hypoxanthine biosensor for Fish spoilage monitoring. *Sens. Actuators B Chem.* **2021**, *332*, 129435. [CrossRef]
32. Wang, G.; Liu, J.; Yue, F.; Shen, Z.; Xu, D.; Fang, H.; Chen, W.; Wang, Z.; Li, P.; Guo, Y.; et al. Dual enzyme electrochemiluminescence sensor based on in situ synthesis of ZIF-67@AgNPs for the detection of IMP in fresh meat. *LWT* **2022**, *165*, 113658. [CrossRef]
33. Ghosh, S.; Sarker, D.; Misra, T.N. Development of an amperometric enzyme electrode biosensor for fish freshness detection. *Sens. Actuators B Chem.* **1998**, *53*, 58–62. [CrossRef]
34. Gil, L.; Barat, J.M.; Garcia-Breijo, E.; Ibañez, J.; Martínez-Máñez, R.; Soto, J.; Llobet, E.; Brezmes, J.; Aristoy, M.-C.; Toldrá, F. Fish freshness analysis using metallic potentiometric electrodes. *Sens. Actuators B Chem.* **2008**, *131*, 362–370. [CrossRef]
35. Hu, S.; Liu, C.-C. Amperometric sensor for fish freshness based on immobilized multi-enzyme modified electrode. *Electroanalysis* **1997**, *9*, 1229–1233. [CrossRef]
36. Itoh, D.; Sassa, F.; Nishi, T.; Kani, Y.; Murata, M.; Suzuki, H. Droplet-based microfluidic sensing system for rapid fish freshness determination. *Sens. Actuators B Chem.* **2012**, *171–172*, 619–626. [CrossRef]
37. Postmortem ATP Catabolism in Fish Muscle. ATP Breakdown in Post Mortem Muscle. Available online: https://www.novocib.com/ATP_breakdown_salmon.html (accessed on 26 November 2022).
38. Konoplev, G.; Agafonova, D.; Bakhchova, L.; Mukhin, N.; Kurachkina, M.; Schmidt, M.-P.; Verlov, N.; Sidorov, A.; Oseev, A.; Stepanova, O.; et al. Label-Free Physical Techniques and Methodologies for Proteins Detection in Microfluidic Biosensor Structures. *Biomedicines* **2022**, *10*, 207. [CrossRef]
39. Wojnowski, M.; Issakov, V.; Knoblinger, G.; Pressel, K.; Sommer, G.; Weigel, R. High-Q embedded inductors in fan-out eWLB for 6 GHz CMOS VCO. In Proceedings of the 2011 IEEE 61st Electronic Components and Technology Conference (ECTC), Lake Buena Vista, FL, USA, 31 May–3 June 2011; pp. 1363–1370. [CrossRef]

40. Issakov, V.; Wojnowski, M.; Thiede, A.; Weigel, R. Considerations on the de-embedding of differential devices using two-port techniques. *Int. J. Microw. Wirel. Technol.* **2010**, *2*, 349–357. [CrossRef]
41. Zur Nedden, S.; Eason, R.; Doney, A.S.; Frenguelli, B.G. An ion-pair reversed-phase HPLC method for determination of fresh tissue adenine nucleotides avoiding freeze-thaw degradation of ATP. *Anal. Biochem.* **2009**, *388*, 108–114. [CrossRef]
42. Kuznetsov, A.; Frorip, A.; Korsakov, R.; Korsakov, V.; Püssa, T.; Sünter, A.; Kasvand, N. New Freshness Index, Method and Device to Determine Freshness Status in Fish and Meat According to FAO Recommendations. Available online: https://www.researchgate.net/publication/321213587_New_freshness_index_method_and_device_to_determine_freshness_status_in_fish_and_meat_according_to_FAO_recommendations#fullTextFileContent (accessed on 11 January 2020).
43. Püssa, T.; Frorip, A.; Kuznetsov, A.; Sünter, A.; Anton, D.; Raudsepp, P. New Freshness Index, Method and Device to Determine Freshness Status of Meat and Fish. Available online: https://www.researchgate.net/publication/326059014_NEW_FRESHNESS_INDEX_METHOD_AND_DEVICE_TO_DETERMINE_FRESHNESS_STATUS_OF_MEAT_AND_FISH (accessed on 11 January 2021).
44. Walls, D.; Loughran, S.T. *Protein Chromatography: Methods and Protocols*, 2011th ed.; Humana Press: Totowa, NJ, USA, 2011; p. 681. ISBN 9781607619123.
45. Determann, H. *Gel Chromatography, Gel Filtration, Gel Permeation, Molecular Sieves: A Laboratory Hand Book*, 2nd ed.; Springer: New York, NY, USA, 1969; p. 208, ISBN 978-3642950841.
46. Kasvand, N.; Frorip, A.; Kuznetsov, A.; Püssa, T.; Rusalepp, L.; Sünter, A. A new approach for evaluation of meat freshness. In Proceedings of the 5th International Conference on Optical Characterization of Materials (OCM 2021), Karlsruhe, Germany, 17–18 March 2021; Beyerer, J., Längle, T., Eds.; KIT Scientific Publishing: Karlsruhe, Germany, 2021; pp. 21–32. [CrossRef]
47. Ciampa, A.; Picone, G.; Laghi, L.; Nikzad, H.; Capozzi, F. Changes in the amino acid composition of Bogue (Boops boops) fish during storage at different temperatures by ^{1}H-NMR spectroscopy. *Nutrients* **2012**, *4*, 542–553. [CrossRef]
48. Abramova, L.S.; Kozin, A.V.; Shashkov, A.S. NMR—Spectroscopy technique for salmon fish species quality assessment. *Food Syst.* **2019**, *2*, 4–9. [CrossRef]
49. Rodrigues, B.L.; Alvares, T.d.S.; Sampaio, G.S.L.; Cabral, C.C.; Araujo, J.V.A.; Franco, R.M.; Mano, S.B.; Conte Junior, C.A. Influence of vacuum and modified atmosphere packaging in combination with UV-C radiation on the shelf life of rainbow trout (*Oncorhynchus mykiss*) fillets. *Food Control* **2016**, *60*, 596–605. [CrossRef]
50. Shen, S.; Jiang, Y.; Liu, X.; Luo, Y.; Gao, L. Quality assessment of rainbow trout (*Oncorhynchus mykiss*) fillets during super chilling and chilled storage. *J. Food Sci. Technol.* **2015**, *52*, 5204–5211. [CrossRef]
51. Ježek, F.; Buchtová, H. The effect of vacuum packaging on physicochemical changes in rainbow trout (*Oncorhynchus mykiss*) during cold storage. *Acta Vet. Brno* **2014**, *83*, S51–S58. [CrossRef]
52. Quality and Quality Changes in Fresh Fish. Postmortem Changes in Fish. Available online: https://www.fao.org/3/v7180e/V7180e06.htm#5.1%20Sensory%20changes (accessed on 26 November 2022).
53. Guidelines for the Interpretation of Results of Microbiological Testing of Ready-to-Eat Foods Placed on the Market (Revision 4). Available online: https://www.fsai.ie/food_businesses/micro_criteria/guideline_micro_criteria.html (accessed on 15 November 2022).
54. Arya, S.S.; Parihar, D.B.; Vijayaraghavan, P.K. Changes in free nucleotides, nucleosides and bases during preparation of pre-cooked dehdyrated minced meats. *Nahrung* **1979**, *23*, 495–499. [CrossRef]
55. Rotola-Pukkila, M.K.; Pihlajaviita, S.T.; Kaimainen, M.T.; Hopia, A.I. Concentration of Umami Compounds in Pork Meat and Cooking Juice with Different Cooking Times and Temperatures. *J. Food Sci.* **2015**, *80*, C2711–C2716. [CrossRef] [PubMed]
56. Madruga, M.S.; Elmore, J.S.; Oruna-Concha, M.J.; Balagiannis, D.; Mottram, D.S. Determination of some water-soluble aroma precursors in goat meat and their enrolment on flavour profile of goat meat. *Food Chem.* **2010**, *123*, 513–520. [CrossRef]
57. Sasaki, K.; Motoyama, M.; Mitsumoto, M. Changes in the amounts of water-soluble umami-related substances in porcine longissimus and biceps femoris muscles during moist heat cooking. *Meat Sci.* **2007**, *77*, 167–172. [CrossRef]
58. Cambero, M.I.; Pereira-Lima, C.I.; Ordoez, J.A.; Garca de Fernando, G.D. Beef broth flavour: Relation of components with the flavour developed at different cooking temperatures. *J. Sci. Food Agric.* **2000**, *80*, 1519–1528. [CrossRef]
59. Belton, P.S.; Hills, B.P.; Webb, G.A. *Advances in Magnetic Resonance in Food Science*; Elsevier: Amsterdam, The Netherlands, 1999; ISBN 9781855737730.
60. Lou, X.; Zhai, D.; Yang, H. Changes of metabolite profiles of fish models inoculated with Shewanella baltica during spoilage. *Food Control* **2021**, *123*, 107697. [CrossRef]
61. Shumilina, E.; Ciampa, A.; Capozzi, F.; Rustad, T.; Dikiy, A. NMR approach for monitoring post-mortem changes in Atlantic salmon fillets stored at 0 and 4 °C. *Food Chem.* **2015**, *184*, 12–22. [CrossRef]
62. Handumrongkul, C.; Silva, J.L. Aerobic Counts, Color and Adenine Nucleotide Changes in CO_2 Packed Refrigerated Striped Bass Strips. *J. Food Sci.* **1994**, *59*, 67–69. [CrossRef]
63. Jones, N.R.; Murray, J.; Livingston, E.I.; Murray, C.K. Rapid estimations of hypoxanthine concentrations as indices of the freshness of chill-stored fish. *J. Sci. Food Agric.* **1964**, *15*, 763–774. [CrossRef]
64. Spinelli, J.; Eklund, M.; Miyauchi, D. Measurement of Hypoxanthine in Fish as a Method of Assessing Freshness. *J. Food Sci.* **1964**, *29*, 710–714. [CrossRef]
65. Kyrana, V.R.; Lougovois, V.P.; Valsamis, D.S. Assessment of shelf-life of maricultured gilthead sea bream (*Sparus aurata*) stored in ice. *Int. J. Food Sci. Technol.* **1997**, *32*, 339–347. [CrossRef]
66. Ding, T.; Song, G.; Liu, X.; Xu, M.; Li, Y. Nucleotides as optimal candidates for essential nutrients in living organisms: A review. *J. Funct. Foods* **2021**, *82*, 104498. [CrossRef]

67. Carver, J.D. Dietary nucleotides: Effects on the immune and gastrointestinal systems. *Acta Paediatr.* **1999**, *88*, 83–88. [CrossRef]
68. Hess, J.R.; Greenberg, N.A. The role of nucleotides in the immune and gastrointestinal systems: Potential clinical applications. *Nutr. Clin. Pract.* **2012**, *27*, 281–294. [CrossRef]

chemosensors

Article

Impedimetric Biosensor Coated with Zinc Oxide Nanorods Synthesized by a Modification of the Hydrothermal Method for Antibody Detection

Nikita Sitkov [1,2,*], Andrey Ryabko [3,*], Alexey Kolobov [2,4], Alexsandr Maximov [1], Vyacheslav Moshnikov [1], Stanislav Pshenichnyuk [5], Alexei Komolov [6], Andrey Aleshin [3] and Tatiana Zimina [1,2]

[1] Department of Micro and Nanoelectronics, Saint Petersburg Electrotechnical University "LETI", Saint Petersburg 197022, Russia; aimaximov@mail.ru (A.M.); vamoshnikov@mail.ru (V.M.); tmzimina@gmail.com (T.Z.)

[2] Centre for Digital Telecommunication Technologies, Saint Petersburg Electrotechnical University "LETI", Saint Petersburg 197022, Russia; alexey.kolobov.spb@gmail.com

[3] Laboratory of Nonequilibrium Processes in Semiconductors, Ioffe Institute, 26 Politekhnicheskaya, Saint Petersburg 194021, Russia; aleshin.transport@mail.ioffe.ru

[4] Laboratory of Peptide Chemistry, Institute of Human Hygiene, Occupational Pathology and Ecology, Saint-Petersburg 188663, Russia

[5] Laboratory of Physics of Atomic Collisions, Institute of Molecule and Crystal Physics, Ufa Federal Research Centre, Russian Academy of Sciences, Prospekt Oktyabrya 151, Ufa 450075, Russia; sapsh@anrb.ru

[6] Solid State Electronics Department, Saint Petersburg State University, Saint Petersburg 199034, Russia; a.komolov@spbu.ru

* Correspondence: sitkov93@yandex.ru (N.S.); a.a.ryabko93@yandex.ru (A.R.)

Abstract: Impedimetric biosensors are used for detecting a wide range of analytes. The detection principle is a perspective for the development of new types of analytical devices for biomolecular diagnosis of diseases. Of particular interest are biosensors with very high sensitivities, capable of detecting trace amounts of biomarkers or drugs in biological fluids. Impedimetric biosensors possess a potential for increased sensitivity, since their electrodes can be modified with nanostructured materials, in particular zinc oxide. In this work, a miniature biosensor with an array of zinc oxide nanorods synthesized by the hydrothermal method has been created. Protein A was immobilized on the resulting structure, which was previously tested for binding to omalizumab by capillary electrophoresis. Using impedance spectroscopy, it was possible to detect the binding of omalizumab at concentrations down to 5 pg/mL. The resulting structures are suitable for creating reusable biosensor systems, since ZnO-coated electrodes are easily cleaned by photocatalytic decomposition of the bound molecules. The biosensor is promising for use in Point-of-Care systems designed for fast, multimodal detection of molecular markers of a wide range of diseases.

Keywords: impedimetric biosensor; zinc oxide; express detection; nanorods; antibodies; label-free detection; hydrothermal synthesis

Citation: Sitkov, N.; Ryabko, A.; Kolobov, A.; Maximov, A.; Moshnikov, V.; Pshenichnyuk, S.; Komolov, A.; Aleshin, A.; Zimina, T. Impedimetric Biosensor Coated with Zinc Oxide Nanorods Synthesized by a Modification of the Hydrothermal Method for Antibody Detection. *Chemosensors* **2023**, *11*, 66. https://doi.org/10.3390/chemosensors11010066

Academic Editor: Xiaolong Yang

Received: 10 December 2022
Revised: 28 December 2022
Accepted: 11 January 2023
Published: 13 January 2023

1. Introduction

The use of micro- and nano-technologies in biology and medicine makes it possible to develop new miniature devices for expressing the control of biological parameters, including multiparametric molecular biomarker monitoring. The main areas of development are biosensors [1,2], lab-on-a-chip (LoC) [3,4] devices and, based on these, portable diagnostic devices in the Point-of-Care Testing (PocT) [5] class. A biosensor is an analytical device that converts a chemical or physical stimulus derived by an interaction with a biological component into a measurable signal. The following main components form the structure of a biosensor: a biorecognition element (antibody, protein, aptamer, nucleic acid, microorganism, etc.), which has a specific complementarity to the analyzed object; a transducer (optical, electrochemical, magnetic, etc.), which converts the biointeraction into a recorded and measured signal; and a data processing and analysis system that visualizes and converts experimental data into a form convenient for the operator. Such devices should provide high sensitivity and speed of analysis, use a small sample volume and be of low cost [6].

Antibodies are one of the most popular objects for fast diagnostic devices due to their high specificity and high binding strength to the target. Antibodies are either monoclonal or polyclonal. Monoclonal antibodies are produced by B cells that are clones of the same parent cell and are monovalent when they recognize the same antigen epitope [7]. This type of antibody is produced ex vivo in tissue culture. To create polyclonal antibodies, an antigen is injected into an animal that has an immune response, after which they are collected directly from the serum and purified. They can be used as markers for diagnosing chronic [8], neurodegenerative [9,10], oncological [11] and infectious [12] diseases, etc. Traditional methods for the detection of protein biomarkers, such as immunoassays (ELISA, immunochemiluminescent assays), despite their high sensitivity and selectivity, are expensive, time-consuming and multi-stage procedures performed in specialized diagnostic laboratories [13]. Therefore, there is a need to create highly sensitive, fast diagnostic biosensor systems capable of detecting the level of various antibodies at extremely low concentrations.

Optical and electrochemical detection systems are widely used to create miniature devices for protein detection [14,15]. Optical detection methods, which are considered very sensitive and specific, are based on a change in the phase, amplitude, polarization or frequency of incoming light in response to biorecognition processes. Often, these methods use specialized labels, which make it possible to increase sensitivity, but complicate and increase the cost of the manufacturing technology. In addition, the instrumental implementation of individual methods can be large and expensive and require specially trained personnel to conduct tests.

Electrochemical biosensors are one of the most widespread classes of biosensor devices, which use electrodes with recognition elements immobilized on their surface capable of selectively binding to target molecules. Detection of the binding of the target to the recognizer on the electrode is carried out by registering changes in current and/or voltage. According to the measurement principle, electrochemical biosensors are divided into potentiometric, amperometric and impedimetric detection systems that convert a chemical reaction parameter into a measurable electrical signal. Due to low operating voltages, fabrication costs and ease of miniaturization, electrochemical biosensors have great prospects for various biomedical applications, especially for fast, multimodal biosensor systems.

Biosensor systems based on impedance detection are currently one of the most popular solutions for the detection of binding of antibody biomarkers. Unlike other electrochemical methods, impedimetry works on low-amplitude electrical signal disturbances; therefore, this method is considered non-destructive [16]. Since the signal measurement principle of impedance spectroscopy is based on modifications of the electrode surface, it is largely sensitive to the organization of the near-electrode monolayer and can be used for measurements at low concentrations or in studying enzymatic processes [17]. The low cost, speed and convenience of analysis, the possibility of eliminating the influence of third-party substances on the test sample, as well as the ease of miniaturization of such sensors make them promising for integration into LoC. The work [18] is an example of an impedimetric biosensor for the qualitative detection of antibodies of COVID-19 in serum by using gold electrodes deposited on a SiO_2 substrate modified with the SARS-Cov-2 spike protein. The resulting device made it possible to register the specific interaction between the spike protein and antibodies in the studied samples. Soma et al. [19] reported the development of an impedimetric biosensor for the detection of the norovirus with a detection limit of 60 μg/l, where a gold electrode was modified with polyaniline and streptococcus to improve the electron transfer process and the conjugation of a biotinylated monoclonal antibody, which improves the electrochemical response and provides enhanced active centers for targeted analytes.

Modern biosensors are an example of the convergence of various scientific and technical areas. The use of a variety of nanomaterials (magnetic nanoparticles, nanorods, carbon nanotubes, graphene, quantum dots, etc.) is one of the main means of increasing the sensitivity and selectivity of biosensors [20–22]. Zinc oxide nanostructures are promising in electrochemical biosensors, as they serve in immobilizing a bioselective element and distributing it over the entire electrode area [23–30]. In [23], an example of using zinc oxide nanorods as a layer on the surface of a biosensor is presented, which makes it possible to detect the HRP antigen and nonspecific antigen in the buffer. This result was confirmed by capacitive measurements. The authors found the maximum response for their structure was in the range of 5–6 kHz, which is promising for the creation of a single-frequency measuring system for the analysis of multiple samples. Shanmugam et al. [25] used zinc oxide nanostructures to create a biosensor with multiplex detection of a panel of cardio-biomarkers. The nanostructures were hydrothermally grown, functionalized with specific antibodies and thus prepared to detect cardiac Troponin I and Troponin T with a detection limit of 1 pg/mL.

Zinc oxide is a direct-gap semiconductor with a band gap $Eg \approx 3.3$ eV, n-type conductivity, piezoelectric properties and a high electron mobility [31]. However, interest in zinc oxide is due not only to its electrophysical properties, but also due to the possibility of synthesizing ZnO in the form of nanorods using the hydrothermal method at low (<100 °C) temperatures. This eliminates the use of autoclaves and makes the synthesis of ZnO nanorods scalable and inexpensive, which is extremely important for the creation of disposable biochips. Low-temperature hydrothermal synthesis may be preferable in comparison to physical methods of synthesis, not only due to the possibility of scaling; an advantage of the method is the ability to introduce impurities (dopants) into the synthesis process. The introduction of salts of other metals into the growth solution can modify the surface of nanoparticles with new adsorption centers to improve further binding.

ZnO nanoparticles in the form of nanorods are efficient when treating the substrate surface, since the surface area for immobilization of the bioselective element is considerably higher. This feature makes ZnO nanorods attractive for use in gas sensors [32,33]. ZnO nanorods have also been successfully used as a photocatalyst, where ZnO with high concentrations of OH groups on its surface demonstrated a high photocatalytic activity [34,35]. We assume that such a surface should facilitate immobilization. It should be noted that the possibility of photocatalytic decomposition of molecular units conjugated on the surface of ZnO nanorods can probably ensure the reusability of electrodes coated with ZnO nanorods. The process of biorecognition will be followed by the illumination of the electrodes in the ultraviolet (<380 nm) region and photocatalytic decomposition on the surface of ZnO. After the photocatalysis, the stages of purification and immobilization can be repeated. Therefore, we assume that the use of photocatalytically active materials is especially promising for implementation of reusable sensor chips.

The synthesis methods of ZnO nanorods can be divided into two main approaches: physical gas-phase synthesis methods (including the vapor–liquid–solid method [36], pulsed laser deposition [37], chemical vapor deposition [38,39], molecular beam epitaxy [40] and others) and chemical methods (including such methods as the electrochemical method [41], electrospinning process [42] and others). Nevertheless, the most flexible chemical method is hydrothermal [43,44]. This method, convenient for laboratory application as well as for scaling up, has generated additional interest in the wide study and application of zinc oxide in the form of nanorods. During hydrothermal synthesis, ZnO nanorods are formed in solution with precursors, and if a substrate with a seed layer is placed in the solution, nanorods will grow on the substrate. Therefore, to create biosensors, ZnO nanoparticles are applied directly from a suspension [24,45,46] or formed on a substrate with preformed seed layers [23,25,28]. The advantage of application from a suspension is the possibility of obtaining a given dispersion of particles by separating the particles according to size by methods such as centrifugation or filtration. However, this approach seems unpromising for integration with microfluidic systems due to the poor adhesion of particles on substrates with electrodes. ZnO nanorods grown on a substrate seems to be more promising for this purpose. However, during the hydrothermal synthesis of nanorods on a substrate, they grow not only on the substrate, but also in the bulk of the solution, forming agglomerates which can be fixed on the sample surface. This worsens the reproducibility of the technology, since the process is random to a certain extent. In addition, the seed layer is applied by centrifugation [23,28], which, unlike the hydrothermal method, is not so convenient for scaling. Additionally, the use of methods for depositing a ZnO seed layer in a vacuum, such as the Magnetron sputtering technique [25], also somewhat reduces the advantages of the hydrothermal method, which does not require vacuum systems.

In this study, the hydrothermal synthesis of ZnO nanorods on the substrate surface was carried out with the suppression of nucleation in the bulk of the solution, making it more reproducible. Additionally, the ZnO seed layer was deposited by ultrasonic spray pyrolysis, which does not require a vacuum and is easily scaled up, similar to the hydrothermal method. We believe that cost effectiveness and scalability play an important role in terms of future prospects for creating commercial sensor platforms. Therefore, the purpose of this work was to study the possibility of using this technique for the synthesis of coatings from nanorods to develop an impedimetric biosensor on interdigitated electrodes.

2. Materials and Methods

2.1. In Vitro Study of Complex Formation between Protein A and Omalizumab

In this study, we decided to demonstrate the use of zinc oxide modified electrodes for the detection of omalizumab using surface-bound protein A.

Omalizumab (MW 149 kDa) is a humanized anti-IgE monoclonal IgG1 antibody used in the treatment of bronchial asthma [47]. It was produced in soluble form in the modified CHO cell line, then purified from the supernatant with Protein A affinity chromatography and with SEC to remove aggregates. The resulting purity of the omalizumab specimen used in this work was 96.8%, as determined with RP-HPLC.

Protein A (M_W 50 kDa in S-S dimer form) is a surface protein of *S. aureus*, able to bind IgG [48] and modified with His-tag and several amino acid substitutions. It was produced in a soluble form in the modified *E. coli* strain, then purified from the cell lysate with metal-chelate affinity chromatography on Ni-NTA resin and then with cation-exchange chromatography. The resulting purity of the protein A specimen used in this work was 97.6% as determined with RP-HPLC.

Both omalizumab and protein A specimens were kindly provided by S.V. Rodin.

To show their ability for complementary interaction, we conducted studies using Kapel-105M (Lumex Ltd., Saint Petersburg, Russia) capillary electrophoresis, with acapillary inner diameter of 75 μm, a total length of 60 cm, an effective length of 50 cm, a temperature of +30 °C and a spectrophotometric detector with detection wavelength at 200 nm. To dilute the stock solutions, a 10 mM Tris-HCl buffer with a pH of 7.2 was used.

Electropherograms of stock solutions showed a single electrophoretic peak at 4.8 min for omalizumab (Figure 1) and 9.9 min for protein A (Figure 2).

The areas of both peaks were directly proportional to the concentration of the initial solutions.

2.2. Formation of an Electrode Coating from ZnO Nanorods

The biosensor was based on a ceramic substrate with NiCr/Ni/Au interdigitated electrodes (Sensor Platform, Tesla Blatna, a.s.). The width of the electrodes and the distance between the electrodes deposited on the substrate were 25 μm. Coating of the ZnO nanorods was performed by a two-stage procedure, consisting of ultrasonic spray pyrolysis of a zinc oxide seed layer and the low-temperature hydrothermal synthesis of nanorods with suppression of nucleation in the bulk solution. This technique allows a selective growth of nanorods to be realized only on the substrate surface. In the first stage, an aqueous solution of 0.05 M zinc acetate (purity ≥ 99%, Vekton, Saint Petersburg, Russia) was used to deposit a ZnO seed film by ultrasonic spray pyrolysis for 5 min. The substrate heating temperature was maintained at +380 °C. A control sample of the ZnO seed layer was formed on the polished side of the silicon substrate. The morphology of the seed layer was studied using atomic force microscopy (Atomic Force Microscope Probe nanolaboratory INTEGRA- TERMA, NT-MDT, Saint Petersburg, Russia) (Figure 3).

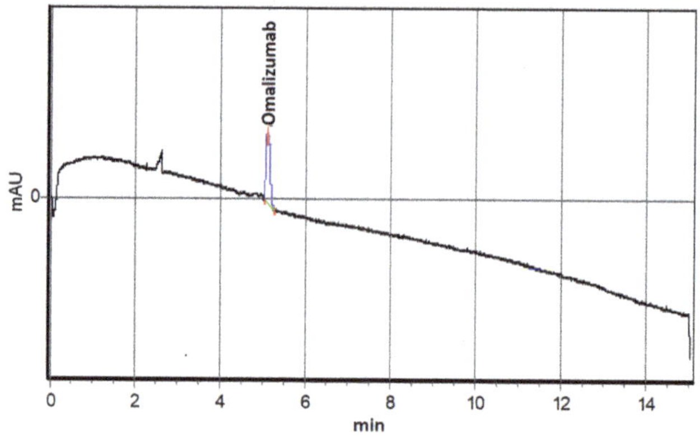

Figure 1. Electropherogram of an 0.6 μM Omalizumab solution.

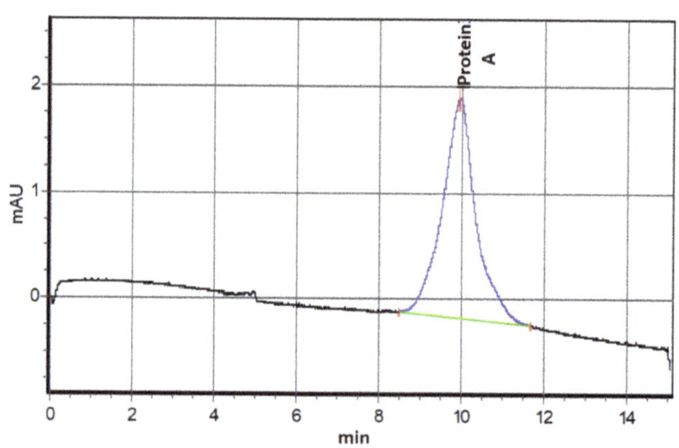

Figure 2. Capillary electrophoresis of 1.28 μM Protein A solution.

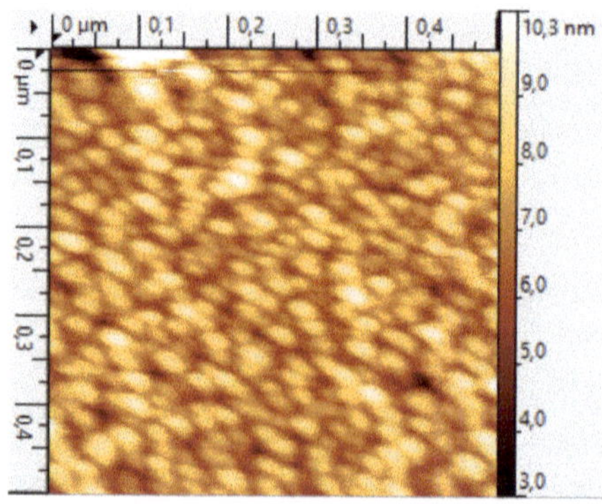

Figure 3. AFM image of the topography of a seed layer obtained by ultrasonic spray pyrolysis.

The roughness of the ZnO seed layer was ~10 nm, and the size of the crystallite width was 25–30 nm.

Low-temperature hydrothermal synthesis of ZnO nanorods from a seed layer was carried out using an aqueous solution of zinc nitrate (purity ≥ 98%, Vekton, Saint Petersburg Russia) and hexamethylenetetramine (purity ≥ 98%, Vekton, Saint Petersburg, Russia) with an equimolar concentration of 25 mM, to which ammonia water (purity ≥ 98%, Vekton, Saint Petersburg, Russia) and polyethyleneimine (branched, Mw ~ 800, Sigma-Aldrich, St. Louis, MO, USA) were added to ensure ZnO nucleation in the bulk of the solution. The features of the synthesis of the ZnO nanorods have been described in more detail in [49]. After synthesis, the ZnO nanorods were annealed at a temperature of +500 °C for 5 min. The results of the scanning electron microscopy of a control sample on a silicon substrate are shown in Figure 4.

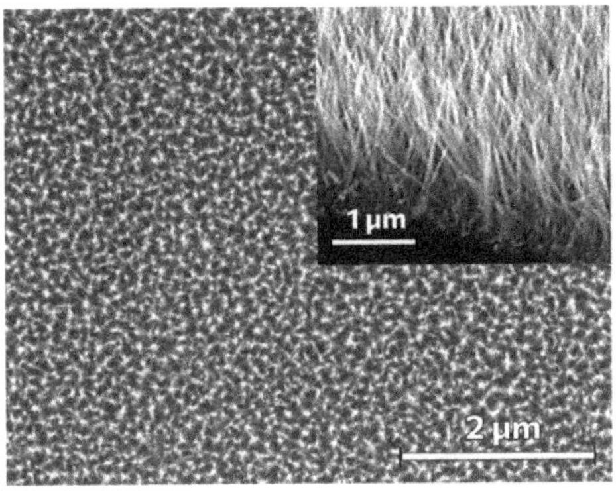

Figure 4. SEM micrograph of a coating of ZnO nanorods formed by a low-temperature hydrothermal method with suppression of nucleation in the bulk solution and the use of a seed layer formed by ultrasonic spray pyrolysis. The inset shows the cross-section of the sample.

As can be seen from the SEM micrograph (Figure 4), a uniform coating of ZnO nanorods of ~1 μm in length and with a diameter which does not exceed 50 nm was formed.

The surface of the ZnO nanorods was studied using X-ray photoelectron spectroscopy (XPS). The XPS study was carried out under ultrahigh vacuum conditions (~10^{-7} Pa) on an Escalab 250Xi complex photoelectron spectrometer (Thermo Fisher Scientific Inc., Waltham, MA, USA) at the excitation photon energy, AlKα, of 1486 eV. The XPS spectra were processed using the CasaXPS Version 2.3.24 software.

2.3. *Immobilization of Ligands on ZnO Nanorods*

In the development and fabrication of a biosensor, an important step is the integration of a recognizing element into its structure that is capable of selectively capturing an analyte. The recognizing element must be immobilized on the surface of the electrode. For that, the electrodes surface must be functionalized with some reactive chemical groups, usually NH_2- or SH- groups. The easiest way to functionalize the ZnO surface with NH_2- groups is by treatment with aminosilanes, which can be achieved in many different ways with different silanes [50,51]. In this study, we used the method of (3-aminopropyl)trimethoxysilane (APTMS) treatment in water solutions. This method could be easily scaled-up for mass production of biosensor chips. The NH_2- groups on surface could be then used for direct conjugation of various organic compounds, including antibodies, onto the substrate surface, or as an intermediate layer in multi-step conjugations.

In addition to chips with electrodes for the purpose of technological control, we also carried out silanization of glasses coated with a layer of zinc oxide nanorods using the same technology. ZnO was annealed to glass slides by heating at +180 °C for 10 min. Then, glasses were treated with 2% APTMS (Sigma-Aldrich 281778, >97%, St. Louis, MO, USA) in isopropanol for 60 min, washed in isopropanol and dried in air at room temperature.

After silanization, electrodes and glass slides were functionalized with SH-groups by treatment with *m*-maleimidobenzoyl-*N*-hydroxysuccinimide ester (MBS) (Thermo Fisher Scientific 22311, Waltham, MA, USA) solution. The presence of SH- groups allows conjugation of ligands with more constant orientation as SH- groups are much less abundant in proteins that NH_2 groups. To prepare a solution of MBS, 5 mg of MBS was dissolved in 5 mL of dimethyl sulfoxide (DMSO), and then the mixture was dissolved in 20 mL of PBS (phosphate-buffered saline, 20 mM monosodium phosphate, 0.9% NaCl, pH 7.4, adjusted with NaOH). The substrate was treated with this solution for 70 min, washed with PBS and deionized water, and then dried in air at room temperature.

In the last step, protein A was conjugated to the chips and the glass slides treated with MBS through formation of S-S bonds. First, the SH- groups of protein A were reduced with dithiothreitol (DTT) (Sigma-Aldrich 111474, >99%, St. Louis, MO, USA). An amount of 5 mg of DTT was added to 100 µL of 10 mg/mL protein A solution in PBS, this solution was incubated at +40 °C for 1 h, then dissolved with PBS to 2 mg/mL and dialyzed against PBS in a dialysis bag with MWCO of 14 kDa (Sigma D-92777). Second, the reduced protein A was conjugated to chips and glasses treated with MBS in two steps: short-time conjugation with concentrated ligand for a short time and then long-time conjugation with diluted ligand for a long time. Initially, only some small areas on the chips which will be in further contact with analytes were treated with protein A solution after dialysis for 1 h. Next, 10 mL of PBS was added to the Petri dish with chips, and the samples were left on a shaker for 8 h. This was done to ensure the conjugation of protein A to the surface. After immobilization of protein A, the chips and slides were washed in buffer and water, dried in air and stored at +4 °C.

2.4. Impedance Spectroscopy

Impedance spectroscopy was implemented on the biochips for detection of the binding of the target substance, omalizumab, and was performed in the frequency range from 1 Hz to 500 kHz, with a voltage amplitude of 100 mV and without applying an additional DC bias (impedance meter Z500P, Elins, Chernogolovka, Russia). We decided to use this amplitude upon analysis of the experiments of scientific groups that detected cardiac troponin [24], as well as colleagues who worked on a similar device [52]. In the future, we plan to use the developed chip for multiparametric detection of protein biomarker arrays for diagnosis of chronic diseases. The measurement procedure (Figure 5) consisted of immersing the immobilized sensor in a 1.5 mL solution of omalizumab in PBS buffer (pH 7.4) with a concentration of 25 mM, where the sensor was kept for 5 min. Then, the sensor was further rinsed in 1.5 mL of deionized water to remove residual buffer solution and non-crosslinked omalizumab, and finally it was immersed in deionized water for impedance spectrum measurements. The sensor chip contacts were connected to the impedance meter via miniature clamps and shielded wires to minimize noise.

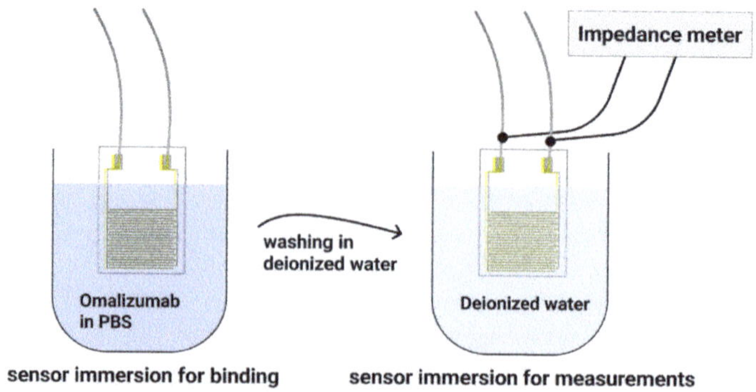

Figure 5. Schematic presentation of the measurement procedure stages.

3. Results and Discussion

The zinc oxide nanorods were obtained using hydrothermal synthesis on substrates with comb electrodes. The SEM micrograph of the nanorods on the surface of a ceramic substrate with an electrode is shown in Figure 6.

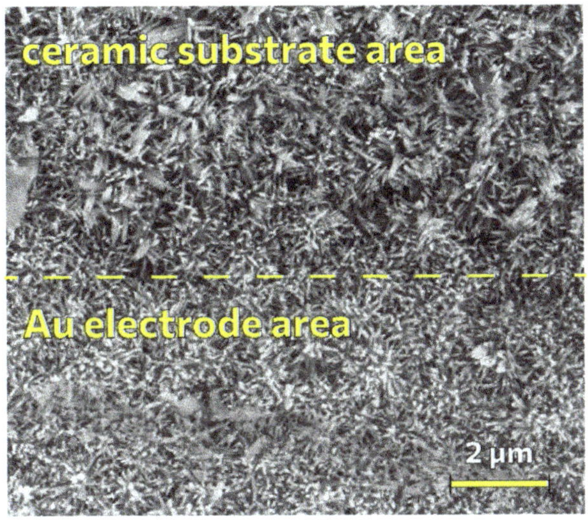

Figure 6. SEM micrograph of ZnO nanorods on the surface of a ceramic substrate with interdigitated electrodes.

It can be seen from Figure 6 that the morphology of the coating of ZnO nanorods on the surface of the ceramic substrate with interdigitated electrodes differs from the coating of nanorods on a silicon substrate, this is due to the morphology of the initial ceramic substrate, which consists of larger grains. The grain edges determine the general directivity for several nanorods at once.

It should be noted that the distance between the ZnO nanorods is large enough to effectively immobilize the nanorods, with protein A binding to omalizumab. The characteristic dimensions of such biological objects are in units of nanometers [53], while the distance between nanorods is tens of nanometers (Figures 4 and 6). Nevertheless, the nanorod diameter of ~50 nm is comparable with the total thickness of bounded biorecognition elements (>10 nm), since they are crosslinked from all sides of the nanorods. The resulting system can be considered as a composite material consisting of a zinc oxide phase and a phase of biorecognition elements, obviously differing in dielectric constant, ε, and polarization mechanisms. Moreover, since the crosslinking processes occur in the buffer solution, a double electric layer should form on the biosensor surface, which also affects the polarization processes. All these factors indicate that the geometric parameters of the nanostructured ZnO layer can have a significant effect on the features of the impedance spectra and the sensitivity of the sensor. The optimization of these parameters is a goal of our further research.

The results of X-ray photoelectron spectroscopy of the surface of ZnO nanorods are shown in Figure 7.

It can be seen from the survey XPS spectrum that, in addition to the core levels of zinc and oxygen, the core level of carbon C1s is observed, which is probably associated with the adsorption of carbon-containing compounds from the air. A detailed interpretation of the overview spectrum was made according to [54]. In addition to the main peaks of the Zn2p, O1s and C1s core levels, less intense peaks of photoelectrons of the Zn2s, Zn3p and Zn3d core levels of zinc, as well as Auger electrons O KLL and Zn LMM, were noted. The ratio of zinc atoms to oxygen was ~1.13. In this case, obviously, part of the oxygen is contained in the surface-adsorbed groups, which makes the ratio of zinc to oxygen atoms even higher than 1.13. The detailed spectrum of the core oxygen level, O1s, shows the main oxygen peak in the zinc oxide crystal lattice (~530.9 eV) and a distinct second peak, which, according to the literature [55–58], is associated with oxygen in the adsorbed OH groups on the ZnO surface. Since carbon is observed on the surface of ZnO, part of the O1s spectrum must be due to oxygen in adsorbed compounds with carbon. The deconvolution of the O1s spectrum was carried out according to [55,56].

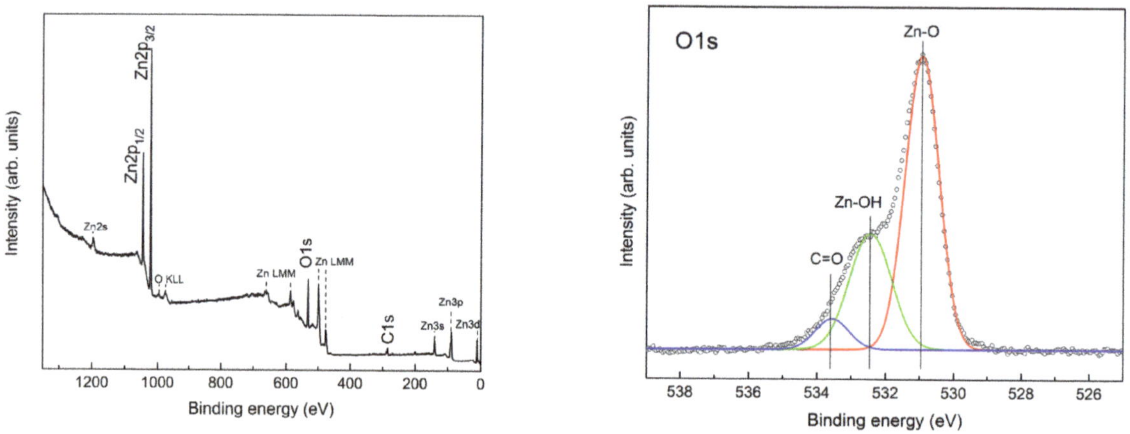

(a) **(b)**

Figure 7. XPS spectra of the surface of ZnO nanorods: (**a**) survey spectrum of the ZnO surface; (**b**) spectrum of the core oxygen level, O1s.

We assume that the high concentration of oxygen vacancies and the predominance of zinc atoms in the ZnO surface layer promote the adsorption of -OH groups. In turn, the obtained ratio of oxygen atoms to zinc and oxygen can be associated not only with the processes of assembly of ZnO nanocrystals during hydrothermal synthesis, but to a greater extent to the process of annealing the samples at 500 °C after synthesis. A high concentration of OH groups should contribute to the successful silanization process in our experiment.

As noted above, a high concentration of OH groups on the surface of ZnO is also accompanied by a high photocatalytic activity. This may be due to the formation of additional defective levels within the band gap, which are more energetically favorable for the formation of reactive oxygen species and OH radicals. As can be seen from the XPS data, the composition of the near-surface region of the synthesized ZnO is indeed significantly deviated from the stoichiometry. The ZnO nanorods used in this work can potentially have good photocatalytic activity. At the same time, standard glasses and some optically transparent polymeric materials that are used to create microfluidic systems will be transparent for the absorption edge of ZnO (~380 nm). Thus, photocatalytic decomposition on ZnO or similar metal oxide photocatalysts can be easily implemented for microfluidic systems.

Calibration relationships for omalizumab and protein A were obtained using the following solutions:

- Omalizumab: 0.2; 0.4; and 0.6 μM (Figure 8a);
- Protein A: 0.64; 1.28; 3.2; and 6.4 μM (Figure 8b).

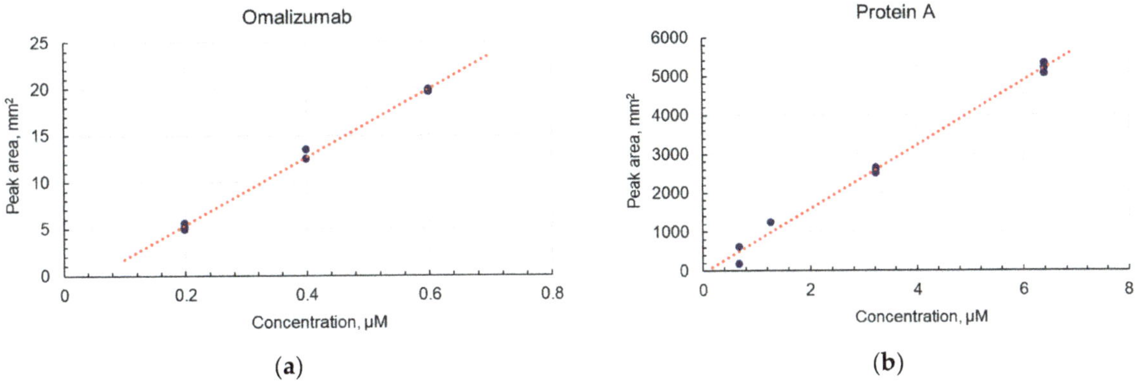

Figure 8. Concentration (μM) dependence of the electrophoretic peak area (mm^2): (**a**) omalizumab, y = 36.434x − 1.9043 (R^2 = 0.9982); (**b**) protein A, y = 821.88x − 53.334 (R^2 = 0.9974).

The complex formation of omalizumab and protein A in different molar ratios was studied with capillary electrophoresis. An example electropherogram of a mixture with a ratio of protein A/omalizumab of 1.06 is shown in Figure 9. All electropherograms of mixtures, regardless of the ratio of components, contained three zones, two of which were previously assigned to the initial components, while the third intermediate zone represents the resulting protein–antibody complex.

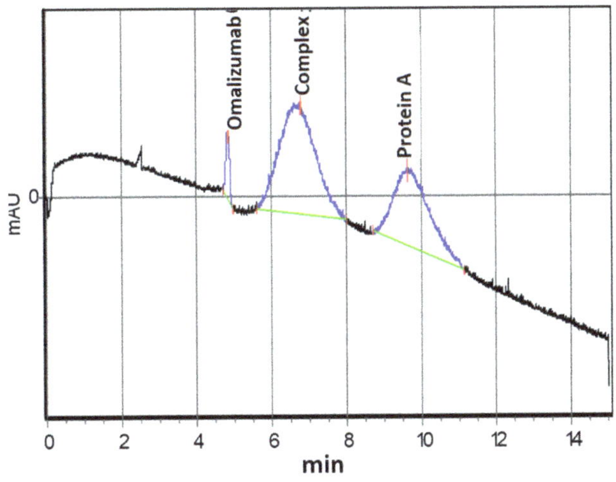

Figure 9. Electropherogram of a solution containing omalizumab (0.6 μM) and protein A (0.64 μM).

Using the capillary electrophoresis data, the average value of the effective dissociation constant of the 1:1 complex was determined, at Kd = (5.1 ± 2.1) × 10^{-7} (pKd = 6.3). Thus, the formation of a complex between omalizumab and protein A was demonstrated, which proved its further application for quality control of biosensor substrates and the sensitivity of the formed nanostructured biosensor element.

We tested the effectiveness of protein A immobilization on zinc oxide nanorods by detecting the binding of antibodies labeled with horseradish peroxidase. For this, glass substrates were washed in BBST pH 8.2 (20 mM borate buffer with 0.9% NaCl and 0.1% Tween-20) and treated with 0.2 μg/mL horseradish peroxidase antibody conjugate in BBST with 1 mg/mL BSA (Sigma-Aldrich, >98%) for 1 h. Then, the substrates were washed in BBST and developed in 3,3′,5,5′-tetramethylbenzidine (TMB). As a negative control, a glass with nanorods of zinc oxide subjected only to silanization without treatment with MBS and protein A was used, and one glass with conjugated protein A was taken as the sample (Figure 10).

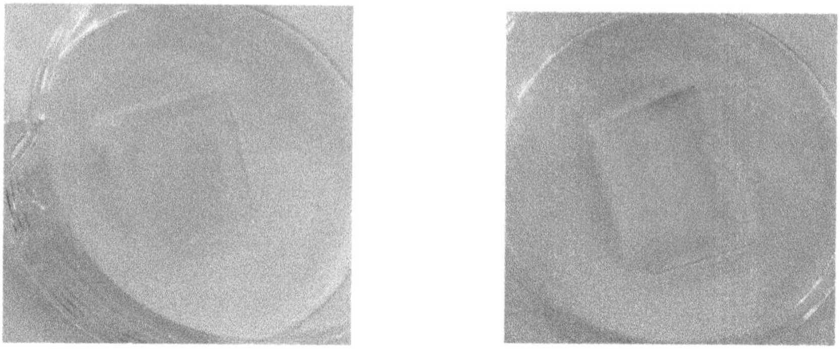

Figure 10. Detection of binding of horseradish peroxidase-conjugated antibodies to the surface of ZnO-coated glasses with bound protein A using TMB treatment: (**a**) glass with ZnO nanorods without MBS treatment, only silanized: negative control; (**b**) glass coated with zinc oxide nanorods silanized with MBS treatment.

A drop of TMB on the glass with protein A quickly began to turn blue and then yellow, which meant that TMB reacted with horseradish peroxidase. Horseradish peroxidase is associated with antibodies; therefore, this observation indicates that the antibodies have bound, meaning that protein A has been adsorbed onto the glass. Slight traces of staining are visible on the glass without treatment with protein A, which may indicate a small proportion of physical adsorption of the antibody. The obtained data indicate that protein A is adsorbed on the ZnO–silane–MBS structure in the operable state and orientation.

As a result of the study of the frequency range for measuring the impedance spectra of various concentrations of omalizumab, we determined the frequency range where the concentration dependence of the impedance is clearly visible (Figure 11).

The characteristic frequency range of the supplied alternating signal for the created biochip was approximately from 7 to 500 kHz in the total investigated range of 1 Hz–500 kHz. As can be seen from Figure 11, an increase in the concentration of the target antibody in the sample leads to a change in the impedance spectrum in the Nyquist coordinates of the biochip after it is immersed in the analyzed solution. With an increase in the concentration of the target antibody, omalizumab, a shift in the circular arc in the impedance spectrum is observed.

The impedance spectrum in the Nyquist coordinates for such systems of semiconductor nanorods has a characteristic semicircle shape and can be described by an equivalent RC scheme [52,59]. For composite organo–inorganic structures that are detected by electrochemical nanobiosensors, part of the impedance spectrum usually also has a distinct semicircle region [60–62]. However, as we assumed from the design of the experiment, despite the stage of washing and measurements in deionized water, the binding processes are also accompanied by the transfer of ions in the buffer solution (surrounding the proteins in the buffer). The impedance spectra indicate (Figure 11) an outline typical for an electrochemical system [60], where there is a circular arc region corresponding to the resistance of the electrolyte and ZnO nanorods and a characteristic "tail" of the corresponding electric double layer (EDL) diffusion region.

A decrease in the arc of the circle is characteristic of an increase in the concentration of electrolyte ions [52]. In our case, a decrease in the circular arc is observed with an increase in the concentration of the target protein, and, apparently, is accompanied by an increase in the concentration of bound buffer ions on the protein surface.

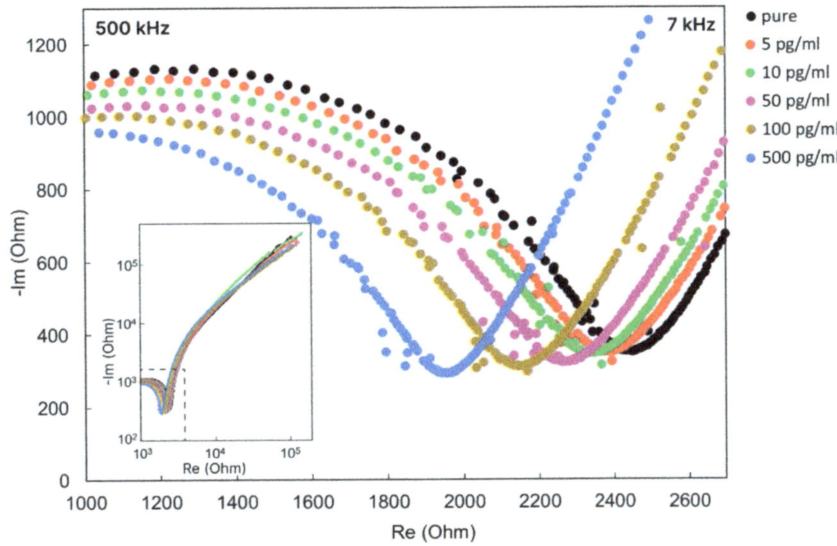

Figure 11. Nyquist plot for the biochip after immersion in omalizumab solutions of various concentrations.

Thus, the binding of omalizumab to protein A redistributes the charge in the near-electrode electric double layer (EDL). Tracking the EDL changes indicates biomolecular binding events [25]. The changes that occur with the formation of the near-electrode EDL at the boundary between the biosensor surface and the surrounding liquid can be successfully detected using electrical impedance spectroscopy. At the same time, the identification of the frequency range in which the maximum sensitivity and response is observed will make it possible to implement a simplified measurement scheme in the future, in the form of a portable device with a disposable line of sensitive electrodes and immobilized biorecognizing elements.

Based on the results of the impedance measurements, we determined the frequency range from 7 to 50 kHz, which allowed us to observe a linear change in the impedance in response to a change in the analyte concentration. Figure 12 shows the change in impedance of biochips with bound omalizumab versus the logarithm of concentration at a frequency of 12 kHz. The dependence for omalizumab impedance is linear for the studied concentration range and satisfies the linear equation: $y = 220.11x - 134.87$, with $R^2 = 0.9421$. The concentration limit of detection (LOD) for omalizumab was 5 pg/mL, which was calculated by a standard procedure using standard deviation and slope values.

In addition, for the same range of concentrations, we demonstrated a similar dependence for BSA solutions. The selection of BSA was motivated by the presence of this protein in blood samples in large concentrations; thus, it can be used to demonstrate the absence of non-specific binding. Compared to omalizumab solutions, the developed chips in BSA solutions showed almost the same impedance response over the studied concentration range, which indicates the absence of nonspecific binding between protein A and BSA. Therefore, we have demonstrated the operability of the developed biosensor structure and have determined the operating frequency range and good sensitivity. Further research will be carried out to create multisensor impedimetric microsystems for detecting various protein markers of diseases.

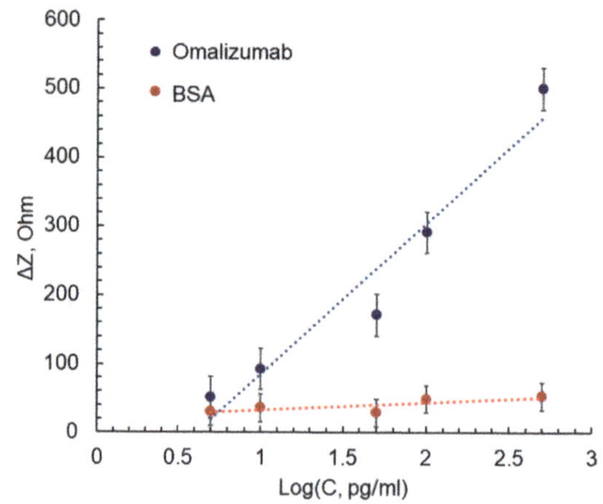

Figure 12. Dependences of the impedance change at a frequency of 10 kHz on the logarithm of the concentration of omalizumab (blue line, y = 220.11x − 134.87, R^2 = 0.9421) and BSA (red line, y = 11.096x + 21.432, R^2 = 0.6464).

4. Conclusions

Micro- and nano-technologies make it possible to implement the automatic control of the stages of analytical procedures using integrated functional modules in microsystems on the microscale. They enable flexible modeling of the topology and design of such systems for performing a specific analysis, and furthermore enable the creation of multichannel systems for parallel analysis of multiple samples.

Impedimetric sensors can be successfully combined with microfluidic systems [63,64], which promises the prospect of their use as sensor nodes in miniature analytical systems, such as in lab-on-a-chip systems. Unlike other electrochemical sensors, impedimetric sensors are less demanding in regard to the quality of the electrodes. They can be chemically stable in a variety of media and offer the ability to work in opaque samples, unlike most optical biosensors.

Modern biosensor systems should provide high sensitivity and selectivity, high detection rate, small sample volume, be small in size and have low cost. Ensuring this set of characteristics is possible due to the heterogeneous integration of functional modules into a single system, the architectonics of which determines a set of design and technological solutions implemented at the micro- and nano-levels, including those using biological media. The physical and technical principles of detection underlying the created biosensor system are often determined by the nature and properties of the analyzed biological component, in particular, protein structures identified as markers of diseases and physiological conditions. The development of registration methods without the use of specialized labels, which complicate and increase the cost of the design and technology of forming biosensor systems, is an object of interest for many researchers [65–67].

In this study, we demonstrated the operability of a label-free impedimetric biosensor for detecting antibodies using electrodes coated with zinc oxide nanorods. We performed a preliminary study of the binding of protein A and omalizumab using capillary electrophoresis. Zinc oxide nanorods were deposited by spray pyrolysis on comb electrodes, after which they were silanized and conjugated with protein A on the surface. Using impedance spectroscopy, it was possible to detect the binding of omalizumab at concentrations down to 5 pg/mL. We assume that the coating of electrodes with ZnO nanorods in the future will not only serve as a matrix for immobilization and increase the effective area of electrodes, but will also ensure the reusability of the biochip due to the cleaning of electrodes coated with ZnO by photocatalytic decomposition of the bound molecules.

Further development of this work should go towards the creation of highly sensitive multiparametric biosensor systems for the diagnosis of protein markers of diseases. To this end, we plan to create detection sites using peptide aptamers with spatial complementarity to target proteins. We have a set of tools and methods for an in silico search for such structures [68,69] and are currently testing a number of samples for protein markers of cardiovascular and inflammatory diseases. This is particularly important in relation to the prospect of using such biosensors in Point-of-Care systems.

Author Contributions: Conceptualization, N.S., A.R., A.K. (Alexey Kolobov), V.M. and T.Z.; methodology, N.S., A.R., A.K. (Alexey Kolobov), A.A., A.M., S.P., A.K. (Alexei Komolov), V.M. and T.Z.; software, S.P. and A.K. (Alexei Komolov); validation, N.S., A.R. and A.K. (Alexey Kolobov); formal analysis, A.A., A.M., S.P., A.K. (Alexei Komolov), V.M. and T.Z.; investigation, N.S., A.R., A.K. (Alexey Kolobov) and A.K. (Alexei Komolov); resources, N.S., A.R., A.K. (Alexei Komolov), A.K. (Alexey Kolobov), A.A., V.M. and T.Z.; data curation, A.A., A.M., A.K. (Alexei Komolov), S.P., V.M. and T.Z.; writing—original draft preparation, N.S. and A.R.; writing—review and editing, N.S., A.R. and T.Z.; visualization, N.S., A.R., A.K. (Alexey Kolobov), A.M. and A.K.; supervision, V.M. and T.Z.; project administration, N.S., A.R., V.M. and T.Z.; funding acquisition, N.S., A.R., A.K. (Alexey Kolobov), V.M., T.Z., A.K. (Alexei Komolov) and S.P. All authors have read and agreed to the published version of the manuscript.

Funding: N.S., A.K. (Alexey Kolobov) and T.Z. acknowledge support of the Russian Science Foundation, (Project No. 21-79-20219) for the material and technical support of research. A.K. (Alexei Komolov). and S.P. acknowledge the support of the Russian Science Foundation (Project No. 19-13-00021) for XPS measurements and spectral analysis. The measurements were partly taken using the equipment of the Research Park of St. Petersburg State University "Physical methods of surface investigation (Project No. 93021679)".

Institutional Review Board Statement: Not applicable.

Informed Consent Statement: Not applicable.

Data Availability Statement: Not applicable.

Acknowledgments: The authors acknowledge with thanks Y.A. Skorik (Almazov National Medical Research Centre, Akkuratova 2, 197341 St. Petersburg, Russia) for their help with the electrophoretic study of omalizumab and protein A interactions.

Conflicts of Interest: The authors declare no conflict of interest.

References

1. Samuel, V.; Rao, K.J. A Review on Label Free Biosensors. *Biosens. Bioelectron. X* **2022**, *11*, 100216. [CrossRef]
2. Sohrabi, H.; Bolandi, N.; Hemmati, A.; Eyvazi, S.; Ghasemzadeh, S.; Baradaran, B.; Oroojalian, F.; Reza Majidi, M.; de la Guardia, M.; Mokhtarzadeh, A. State-of-the-Art Cancer Biomarker Detection by Portable (Bio) Sensing Technology: A Critical Review. *Microchem. J.* **2022**, *177*, 107248. [CrossRef]
3. Alhalaili, B.; Popescu, I.N.; Rusanescu, C.O.; Vidu, R. Microfluidic Devices and Microfluidics-Integrated Electrochemical and Optical (Bio)Sensors for Pollution Analysis: A Review. *Sustainability* **2022**, *14*, 12844. [CrossRef]
4. Dkhar, D.S.; Kumari, R.; Malode, S.J.; Shetti, N.P.; Chandra, P. Integrated Lab-On-a-Chip Devices: Fabrication Methodologies, Transduction System for Sensing Purposes. *J. Pharm. Biomed. Anal.* **2023**, *223*, 115120. [CrossRef] [PubMed]
5. Jaisankar, A.; Krishnan, S.; Rangasamy, L. Recent Developments of Aptamer-Based Lateral Flow Assays for Point-of-Care (POC) Diagnostics. *Anal. Biochem.* **2022**, *655*, 114874. [CrossRef] [PubMed]
6. Sitkov, N.; Zimina, T.; Kolobov, A.; Sevostyanov, E.; Trushlyakova, V.; Luchinin, V.; Krasichkov, A.; Markelov, O.; Galagudza, M.; Kaplun, D. Study of the Fabrication Technology of Hybrid Microfluidic Biochips for Label-Free Detection of Proteins. *Micromachines* **2022**, *13*, 20. [CrossRef]
7. Calidonio, J.M.; Hamad-Schifferli, K. Biophysical and Biochemical Insights in the Design of Immunoassays. *Biochim. Biophys. Acta (BBA)—Gen. Subj.* **2023**, *1867*, 130266. [CrossRef]
8. Warren, A.D.; Kwong, G.A.; Wood, D.K.; Lin, K.Y.; Bhatia, S.N. Point-of-Care Diagnostics for Noncommunicable Diseases Using Synthetic Urinary Biomarkers and Paper Microfluidics. *Proc. Natl. Acad. Sci. USA* **2014**, *111*, 3671–3676. [CrossRef]
9. Serin, M.; Kara, P. Biosensing Strategies (Approaches) for Diagnosis and Monitoring of Multiple Sclerosis. *Talanta* **2023**, *252*, 123794. [CrossRef]

10. Avgerinos, K.I.; Ferrucci, L.; Kapogiannis, D. Effects of Monoclonal Antibodies against Amyloid-β on Clinical and Biomarker Outcomes and Adverse Event Risks: A Systematic Review and Meta-Analysis of Phase III RCTs in Alzheimer's Disease. *Ageing Res. Rev.* **2021**, *68*, 101339. [CrossRef]
11. Banerjee, S.; Drapkin, R.; Richardson, D.L.; Birrer, M. Targeting NaPi2b in Ovarian Cancer. *Cancer Treat. Rev.* **2023**, *112*, 102489. [CrossRef] [PubMed]
12. Roberts, A.; Gandhi, S. A Brief Review on Novel Biomarkers Identified and Advanced Biosensing Technologies Developed for Rapid Diagnosis of Japanese Encephalitis Virus. *Proc. Indian Natl. Sci. Acad.* **2022**, *88*, 617–662. [CrossRef]
13. Chen, C.; Lehr, J. Label-free Selective Detection of Protein Markers in the Picomolar Range via a Convenient Voltammetric Sensing Strategy. *Electroanalysis* **2021**, *33*, 563–567. [CrossRef]
14. Qureshi, A.; Gurbuz, Y.; Niazi, J.H. Biosensors for Cardiac Biomarkers Detection: A Review. *Sens. Actuators B Chem.* **2012**, *171–172*, 62–76. [CrossRef]
15. Kimura, H.; Asano, R. Strategies to Simplify Operation Procedures for Applying Labeled Antibody-Based Immunosensors to Point-of-Care Testing. *Anal. Biochem.* **2022**, *654*, 114806. [CrossRef]
16. Menon, S.; Mathew, M.R.; Sam, S.; Keerthi, K.; Kumar, K.G. Recent Advances and Challenges in Electrochemical Biosensors for Emerging and Re-Emerging Infectious Diseases. *J. Electroanal. Chem.* **2020**, *878*, 114596. [CrossRef]
17. Alshanski, I.; Sukhran, Y.; Mervinetsky, E.; Unverzagt, C.; Yitzchaik, S.; Hurevich, M. Electrochemical Biosensing Platform Based on Complex Biantennary N-Glycan for Detecting Enzymatic Sialylation Processes. *Biosens. Bioelectron.* **2021**, *172*, 112762. [CrossRef]
18. Zeng, J.; Duarte, P.A.; Ma, Y.; Savchenko, O.; Shoute, L.; Khaniani, Y.; Babiuk, S.; Zhuo, R.; Abdelrasoul, G.N.; Charlton, C.; et al. An Impedimetric Biosensor for COVID-19 Serology Test and Modification of Sensor Performance via Dielectrophoresis Force. *Biosens. Bioelectron.* **2022**, *213*, 114476. [CrossRef]
19. Soma, F.N.; Khoris, I.M.; Chowdhury, A.D.; Boonyakida, J.; Park, E.Y. Impedimetric Biosensor of Norovirus with Low Variance Using Simple Bioconjugation on Conductive Polymer-Au Nanocomposite. *SSRN Electron. J.* **2022**. [CrossRef]
20. Solaimuthu, A.; Vijayan, A.N.; Murali, P.; Korrapati, P.S. Nano-Biosensors and Their Relevance in Tissue Engineering. *Curr. Opin. Biomed. Eng.* **2020**, *13*, 84–93. [CrossRef]
21. Zimina, T.M.; Sitkov, N.O.; Gareev, K.G.; Fedorov, V.; Grouzdev, D.; Koziaeva, V.; Gao, H.; Combs, S.E.; Shevtsov, M. Biosensors and Drug Delivery in Oncotheranostics Using Inorganic Synthetic and Biogenic Magnetic Nanoparticles. *Biosensors* **2022**, *12*, 789. [CrossRef] [PubMed]
22. Eivazzadeh-Keihan, R.; Bahojb Noruzi, E.; Chidar, E.; Jafari, M.; Davoodi, F.; Kashtiaray, A.; Ghafori Gorab, M.; Masoud Hashemi, S.; Javanshir, S.; Ahangari Cohan, R.; et al. Applications of Carbon-Based Conductive Nanomaterials in Biosensors. *Chem. Eng. J.* **2022**, *442*, 136183. [CrossRef]
23. Sanguino, P.; Monteiro, T.; Bhattacharyya, S.R.; Dias, C.J.; Igreja, R.; Franco, R. ZnO Nanorods as Immobilization Layers for Interdigitated Capacitive Immunosensors. *Sens. Actuators B Chem.* **2014**, *204*, 211–217. [CrossRef]
24. Yang, W.-C.; Liao, S.-Y.; Phan, T.L.; Van Hieu, N.; Chu, P.-Y.; Yi, C.-C.; Wu, H.-J.; Chang, K.-M.; Ching, C.T.-S. An Immunosensor for the Detection of ULBP2 Biomarker. *Micromachines* **2020**, *11*, 568. [CrossRef]
25. Shanmugam, N.R.; Muthukumar, S.; Selvam, A.P.; Prasad, S. Electrochemical Nanostructured ZnO Biosensor for Ultrasensitive Detection of Cardiac Troponin-T. *Nanomedicine* **2016**, *11*, 1345–1358. [CrossRef]
26. Cao, L.; Kiely, J.; Piano, M.; Luxton, R. Facile and Inexpensive Fabrication of Zinc Oxide Based Bio-Surfaces for C-Reactive Protein Detection. *Sci. Rep.* **2018**, *8*, 12687. [CrossRef]
27. Dong, S.; Zhang, D.; Cui, H.; Huang, T. ZnO/Porous Carbon Composite from a Mixed-Ligand MOF for Ultrasensitive Electrochemical Immunosensing of C-Reactive Protein. *Sens. Actuators B Chem.* **2019**, *284*, 354–361. [CrossRef]
28. Gasparotto, G.; Costa, J.P.C.; Costa, P.I.; Zaghete, M.A.; Mazon, T. Electrochemical Immunosensor Based on ZnO Nanorods-Au Nanoparticles Nanohybrids for Ovarian Cancer Antigen CA-125 Detection. *Mater. Sci. Eng. C* **2017**, *76*, 1240–1247. [CrossRef]
29. Jaiswal, N.; Pandey, C.M.; Solanki, S.; Tiwari, I.; Malhotra, B.D. An Impedimetric Biosensor Based on Electrophoretically Assembled ZnO Nanorods and Carboxylated Graphene Nanoflakes on an Indium Tin Oxide Electrode for Detection of the DNA of Escherichia Coli O157:H7. *Microchim. Acta* **2019**, *187*, 1. [CrossRef]
30. Khosravi-Nejad, F.; Teimouri, M.; Jafari Marandi, S.; Shariati, M. The Highly Sensitive Impedimetric Biosensor in Label Free Approach for Hepatitis B Virus DNA Detection Based on Tellurium Doped ZnO Nanowires. *Appl. Phys. A* **2019**, *125*, 616. [CrossRef]
31. Özgür, Ü.; Alivov, Y.I.; Liu, C.; Teke, A.; Reshchikov, M.A.; Doğan, S.; Avrutin, V.; Cho, S.-J.; Morkoç, H. A Comprehensive Review of ZnO Materials and Devices. *J. Appl. Phys.* **2005**, *98*, 041301. [CrossRef]
32. Bobkov, A.; Varezhnikov, A.; Plugin, I.; Fedorov, F.S.; Trouillet, V.; Geckle, U.; Sommer, M.; Goffman, V.; Moshnikov, V.; Sysoev, V. The Multisensor Array Based on Grown-On-Chip Zinc Oxide Nanorod Network for Selective Discrimination of Alcohol Vapors at Sub-Ppm Range. *Sensors* **2019**, *19*, 4265. [CrossRef] [PubMed]
33. Ryabko, A.A.; Nalimova, S.S.; Maximov, A.I.; Moshnikov, V.A. Investigation of the Gas Sensitivity of Nanostructured Layers Based on Zinc Oxide Nanorods under Ultraviolet Irradiation. In Proceedings of the 2021 IEEE Conference of Russian Young Researchers in Electrical and Electronic Engineering (ElConRus), Moscow, Russia, 26–29 January 2021. [CrossRef]
34. De Almeida, J.C.; Corrêa, M.T.; Koga, R.H.; Del Duque, D.M.S.; Lopes, O.F.; da Silva, G.T.S.T.; Ribeiro, C.; de Mendonça, V.R. Crystallization Time in ZnO: The Role of Surface OH Groups in Its Photoactivity. *New J. Chem.* **2020**, *44*, 18216–18224. [CrossRef]

35. Al-Sabahi, J.; Bora, T.; Al-Abri, M.; Dutta, J. Controlled Defects of Zinc Oxide Nanorods for Efficient Visible Light Photocatalytic Degradation of Phenol. *Materials* **2016**, *9*, 238. [CrossRef]
36. Redkin, A.N.; Ryzhova, M.V.; Yakimov, E.E.; Gruzintsev, A.N. Aligned Arrays of Zinc Oxide Nanorods on Silicon Substrates. *Semiconductors* **2013**, *47*, 252–258. [CrossRef]
37. Kawakami, M.; Hartanto, A.B.; Nakata, Y.; Okada, T. Synthesis of ZnO Nanorods by Nanoparticle Assisted Pulsed-Laser Deposition. *Jpn. J. Appl. Phys.* **2003**, *42 Pt 2*, L33–L35. [CrossRef]
38. Wu, J.-J.; Liu, S.-C. Low-Temperature Growth of Well-Aligned ZnO Nanorods by Chemical Vapor Deposition. *Adv. Mater.* **2002**, *14*, 215–218. [CrossRef]
39. Kim, S.-W.; Fujita, S.; Fujita, S. ZnO Nanowires with High Aspect Ratios Grown by Metalorganic Chemical Vapor Deposition Using Gold Nanoparticles. *Appl. Phys. Lett.* **2005**, *86*, 153119. [CrossRef]
40. Tien, L.C.; Norton, D.P.; Pearton, S.J.; Wang, H.-T.; Ren, F. Nucleation Control for ZnO Nanorods Grown by Catalyst-Driven Molecular Beam Epitaxy. *Appl. Surf. Sci.* **2007**, *253*, 4620–4625. [CrossRef]
41. Li, G.-R.; Dawa, C.-R.; Bu, Q.; Zhen, F.; Lu, X.-H.; Ke, Z.-H.; Hong, H.-E.; Yao, C.-Z.; Liu, P.; Tong, Y.-X. Electrochemical Synthesis of Orientation-Ordered ZnO Nanorod Bundles. *Electrochem. Commun.* **2007**, *9*, 863–868. [CrossRef]
42. Di Mauro, A.; Zimbone, M.; Fragalà, M.E.; Impellizzeri, G. Synthesis of ZnO Nanofibers by the Electrospinning Process. *Mater. Sci. Semicond. Processing* **2016**, *42*, 98–101. [CrossRef]
43. Joo, J.; Chow, B.Y.; Prakash, M.; Boyden, E.S.; Jacobson, J.M. Face-Selective Electrostatic Control of Hydrothermal Zinc Oxide Nanowire Synthesis. *Nat. Mater.* **2011**, *10*, 596–601. [CrossRef]
44. Xu, S.; Wang, Z.L. One-Dimensional ZnO Nanostructures: Solution Growth and Functional Properties. *Nano Res.* **2011**, *4*, 1013–1098. [CrossRef]
45. Zhang, Y.; Zhang, Z.; Rong, S.; Yu, H.; Gao, H.; Sha, Q.; Ding, P.; Pan, H.; Chang, D. A Sandwich-Type ECL Immunosensor Based on Signal Amplification Using a ZnO Nanorods-L-Cysteine-Luminol Nano-composite for Ultrasensitive Detection of Prostate Specific Antigen. *Anal. Chim. Acta* **2020**, *1109*, 98–106. [CrossRef] [PubMed]
46. Eveness, J.; Cao, L.; Kiely, J.; Luxton, R. Equivalent Circuit Model of a Non-Faradaic Impedimetric ZnO Nano-Crystal Biosensor. *J. Electroanal. Chem.* **2022**, *906*, 116003. [CrossRef]
47. McCarthy, M.W.; Aguilar-Zapata, D.; Petraitis, V.; Walsh, T.J. Diagnosis, Classification, and Therapeutic Interventions for Sinopulmonary Aspergillosis. *Expert Rev. Respir. Med.* **2017**, *11*, 229–238. [CrossRef] [PubMed]
48. Hober, S.; Nord, K.; Linhult, M. Protein A Chromatography for Antibody Purification. *J. Chromatogr. B* **2007**, *848*, 40–47. [CrossRef]
49. Ryabko, A.A.; Maximov, A.I.; Verbitskii, V.N.; Levitskii, V.S.; Moshnikov, V.A.; Terukov, E.I. Two-Stage Synthesis of Structured Microsystems Based on Zinc-Oxide Nanorods by Ultrasonic Spray Pyrolysis and the Low-Temperature Hydrothermal Method. *Semiconductors* **2020**, *54*, 1496–1502. [CrossRef]
50. Khan, S.; Rasheed, M.A.; Rafiq, M.A.; Shah, G.B.; Rehman, W.; Jamil, A.; Khan, Y. Silanization of ZnO Nanofibers by Tetraethoxysilane. *J. Appl. Polym. Sci.* **2017**, *134*, 45378. [CrossRef]
51. García Núñez, C.; Sachsenhauser, M.; Blashcke, B.; García Marín, A.; Garrido, J.A.; Pau, J.L. Effects of Hydroxylation and Silanization on the Surface Properties of ZnO Nanowires. *ACS Appl. Mater. Interfaces* **2015**, *7*, 5331–5337. [CrossRef]
52. Kondratev, V.M.; Morozov, I.A.; Vyacheslavova, E.A.; Kirilenko, D.A.; Kuznetsov, A.; Kadinskaya, S.A.; Nalimova, S.S.; Moshnikov, V.A.; Gudovskikh, A.S.; Bolshakov, A.D. Silicon Nanowire-Based Room-Temperature Multi-Environment Ammonia Detection. *ACS Appl. Nano Mater.* **2022**, *5*, 9940–9949. [CrossRef]
53. Erickson, H.P. Size and Shape of Protein Molecules at the Nanometer Level Determined by Sedimentation, Gel Filtration, and Electron Microscopy. *Biol. Proced. Online* **2009**, *11*, 32–51. [CrossRef] [PubMed]
54. Iaiche, S.; Djelloul, A. ZnO/ZnAl2O4Nanocomposite Films Studied by X-ray Diffraction, FTIR, and X-Ray Photoelectron Spectroscopy. *J. Spectrosc.* **2015**, *2015*, 836859. [CrossRef]
55. Kwoka, M.; Kulis-Kapuscinska, A.; Zappa, D.; Comini, E.; Szuber, J. Novel Insight on the Local Surface Properties of ZnO Nanowires. *Nanotechnology* **2020**, *31*, 465705. [CrossRef] [PubMed]
56. Al-Gaashani, R.; Radiman, S.; Daud, A.R.; Tabet, N.; Al-Douri, Y. XPS and Optical Studies of Different Morphologies of ZnO Nanostructures Prepared by Microwave Methods. *Ceram. Int.* **2013**, *39*, 2283–2292. [CrossRef]
57. Grånäs, E.; Busch, M.; Arndt, B.; Creutzburg, M.; Semione, G.D.L.; Gustafson, J.; Schaefer, A.; Vonk, V.; Grönbeck, H.; Stierle, A. Role of Hydroxylation for the Atomic Structure of a Non-Polar Vicinal Zinc Oxide. *Commun. Chem.* **2021**, *4*, 7. [CrossRef]
58. Heinhold, R.; Allen, M.W. Polarity-Dependent Photoemission of in Situ Cleaved Zinc Oxide Single Crystals. *J. Mater. Res.* **2012**, *27*, 2214–2219. [CrossRef]
59. Schipani, F.; Miller, D.R.; Ponce, M.A.; Aldao, C.M.; Akbar, S.A.; Morris, P.A.; Xu, J.C. Conduction Mechanisms in SnO2 Single-Nanowire Gas Sensors: An Impedance Spectroscopy Study. *Sens. Actuators B Chem.* **2017**, *241*, 99–108. [CrossRef]
60. Mei, B.-A.; Munteshari, O.; Lau, J.; Dunn, B.; Pilon, L. Physical Interpretations of Nyquist Plots for EDLC Electrodes and Devices. *J. Phys. Chem. C* **2017**, *122*, 194–206. [CrossRef]
61. Tran, D.T.; Vermeeren, V.; Grieten, L.; Wenmackers, S.; Wagner, P.; Pollet, J.; Janssen, K.P.F.; Michiels, L.; Lammertyn, J. Nanocrystalline Diamond Impedimetric Aptasensor for the Label-Free Detection of Human IgE. *Biosens. Bioelectron.* **2011**, *26*, 2987–2993. [CrossRef]
62. Narang, J.; Malhotra, N.; Singh, G.; Pundir, C.S. Electrochemical Impediometric Detection of Anti-HIV Drug Taking Gold Nano-rods as a Sensing Interface. *Biosens. Bioelectron.* **2015**, *66*, 332–337. [CrossRef]

63. Schmidt-Speicher, L.M.; Länge, K. Microfluidic Integration for Electrochemical Biosensor Applications. *Curr. Opin. Electrochem.* **2021**, *29*, 100755. [CrossRef]
64. Nah, J.S.; Barman, S.C.; Zahed, M.A.; Sharifuzzaman, M.; Yoon, H.; Park, C.; Yoon, S.; Zhang, S.; Park, J.Y. A Wearable Microfluidics-Integrated Impedimetric Immunosensor Based on Ti3C2T MXene Incorporated Laser-Burned Graphene for Noninvasive Sweat Cortisol Detection. *Sens. Actuators B Chem.* **2021**, *329*, 129206. [CrossRef]
65. Upasham, S.; Banga, I.K.; Jagannath, B.; Paul, A.; Lin, K.-C.; Muthukumar, S.; Prasad, S. Electrochemical Impedimetric Biosensors, Featuring the Use of Room Temperature Ionic Liquids (RTILs): Special Focus on Non-Faradaic Sensing. *Biosens. Bioelectron.* **2021**, *177*, 112940. [CrossRef] [PubMed]
66. Antiochia, R. Electrochemical Biosensors for SARS-CoV-2 Detection: Voltametric or Impedimetric Transduction? *Bioelectrochemistry* **2022**, *147*, 108190. [CrossRef]
67. Magar, H.S.; Hassan, R.Y.A.; Mulchandani, A. Electrochemical Impedance Spectroscopy (EIS): Principles, Construction, and Biosensing Applications. *Sensors* **2021**, *21*, 6578. [CrossRef]
68. Sitkov, N.; Zimina, T.; Kolobov, A.; Karasev, V.; Romanov, A.; Luchinin, V.; Kaplun, D. Toward Development of a Label-Free Detection Technique for Microfluidic Fluorometric Peptide-Based Biosensor Systems. *Micromachines* **2021**, *12*, 691. [CrossRef]
69. Karasev, V. Data on the Application of the Molecular Vector Machine Model: A Database of Protein Pentafragments and Computer Software for Predicting and Designing Secondary Protein Structures. *Data Brief* **2020**, *28*, 104815. [CrossRef]

chemosensors

MDPI

Communication

Comparison of Colorimetric and Fluorometric Chemosensors for Protein Concentration Determination and Approaches for Estimation of Their Limits of Detection

Anastasiya A. Mamaeva [1], Vladimir I. Martynov [1], Sergey M. Deyev [1] and Alexey A. Pakhomov [1,2,*]

[1] M.M. Shemyakin and Yu.A. Ovchinnikov Institute of Bioorganic Chemistry, Russian Academy of Sciences, 117997 Moscow, Russia
[2] A.N. Nesmeyanov Institute of Organoelement Compounds, Russian Academy of Sciences, 119991 Moscow, Russia
* Correspondence: alpah@mail.ru

Abstract: Here, we present a direct comparison of different dyes and assays for the determination of protein concentrations. We compared the classical Bradford assay with two modern assays based on the fluorogenic dyes QuDye and ProteOrange and showed that the Bradford reagent achieved excellent results in the determination of protein concentrations as compared with more modern rivals. We also showed that standard approaches for determining the limit of detection (LoD) and limit of quantification (LoQ) may not work correctly with the tested dyes. We proposed a new approach that extends the standard algorithm for LoD and LoQ determination. This approach works well with both classical colorimetric and fluorogenic dyes, as well as with nontrivial fluorescent probes.

Keywords: Bradford reagent; ProteOrange; QuDye; fluorescent dye; protein assay; LoD; LoQ

Citation: Mamaeva, A.A.; Martynov, V.I.; Deyev, S.M.; Pakhomov, A.A. Comparison of Colorimetric and Fluorometric Chemosensors for Protein Concentration Determination and Approaches for Estimation of Their Limits of Detection. *Chemosensors* **2022**, *10*, 542. https://doi.org/10.3390/chemosensors10120542

Academic Editor: Xiaolong Yang

Received: 29 November 2022
Accepted: 15 December 2022
Published: 17 December 2022

1. Introduction

In biochemical studies involving the study of proteins, determining their concentration is one of the most frequent tasks encountered [1,2]. The classic method for determining protein concentrations is the Bradford assay, which utilizes the Coomassie® Brilliant Blue G-250 dye [3,4]. In acidic media, the absorption maximum of the dye shifts from 465 nm to 595 nm when binding to a protein. Such behavior of Coomassie G-250 is explained by the protonation of acidic amino acid residues in an acidic medium, resulting in the protein becoming more hydrophobic and positively charged, thus increasing the binding to the hydrophobic part of the dye; on the other hand, the positive charges of arginine and lysine residues stabilize the anionic form of the dye [5]. Recently, it has become increasingly popular to use fluorogenic dyes to determine protein concentrations [1,6,7]. In this case, the fluorescence intensity increases when the dye binds to protein, which results from the chemical modification of the dye [8] or from a decrease in the vibrational freedom of the dye in the protein-bound state and a consequent increase in the quantum yield of fluorescence [9,10]. In this paper, using two commercially available dyes as examples, we compared the effectiveness of new fluorogenic techniques with the classical colorimetric assay for the determination of protein concentrations.

Fluorescent methods are now widely used for the determination of various types of analytes [11–16]. The most important characteristics related to the sensitivity of fluorescent sensors are the limit of detection (LoD) and the limit of quantification (LoQ), which are the concentrations where the response is most probably (95%) above the noise level (LoD) or can be confidently quantified (LoQ). They are defined as the signal gain over the background noise by factors of 3.3 and 10, respectively [17,18]. The LoD and LoQ are usually defined as $LoD = 3.3\ S/b$, $LoQ = 10\ S/b$, where S is the standard deviation (SD) of the response and b is the slope of the calibration curve. In the present work, we showed that this traditional approach may produce incorrect results. To properly estimate the LoD and LoQ, we modified the standard algorithm by accounting for the standard deviations of the response at low concentrations and by adding to the equation a term that takes into account deviations of the experimental data from the linear dependence in the low concentration range.

2. Materials and Methods

Bovine serum albumin (BSA) was purchased from Sigma-Aldrich (St. Louis, MO, USA). Solutions with the defined protein concentrations were obtained by a series of sequential dilutions of 10 mg/mL BSA in phosphate-buffered saline (PBS). The protein concentration was verified by the absorption of the sample at 280 nm (ε280 nm = 43,824 $M^{-1}cm^{-1}$, MW = 66,400 Da). Bradford reagent was used as part of the protein assay kit (Bio-Rad, Hercules, CA, USA); fluorogenic dyes ProteOrange and QuDye were purchased from Lumiprobe (Moscow, Russia). Assays were conducted according to manufacturers instructions. Samples of proteins were diluted with working solutions of the dyes according to Table 1. All samples were prepared in triplicate (for each concentration, three independent dilutions with working solution were produced to take into account possible errors in pipetting and measuring). Measurements were performed in 96-well plates using a plate reader (Tecan Infinite® M1000 Pro, Switzerland).

Table 1. Composition of the assay reaction mixtures.

Dye	Volume of Sample Added, µL	Volume of Working Solution, µL
Coomassie G-250	10	200
ProteOrange	2.5	97.5
QuDye	5	95

For the spectrophotometric measurements of the samples with Bradford reagent, the sample volume was ~200 µL according to the manufacturer's protocol. At this volume, the optical density was close to the optimum 0.3–0.9 [19]. For fluorescent dyes, the final sample volume was 100 µL, instead of the manufacturer's recommended 200 µL. This allowed us to use half of the protein at the same concentration to increase the sensitivity of the method. The plate reader we used had an excitation/detection spot of only 2 mm, so it was not reasonable to take a larger volume. Before measurements, the Z- position of the focusing optics was optimized; settings at which the dispersion of the detected parameters was minimal were used.

For the Bradford assay, the background absorbance of the samples at 800 nm was subtracted from the absorbance at the absorbance maximum of 595 nm. To measure the fluorescence signal of fluorogenic probes, samples were excited at 480 nm, emission was detected at 590 nm, and the slit was set to 10 nm. Absorption and fluorescence emission spectra were measured using the same plate reader. Linear regression was performed using Origin 2021 software (OriginLab).

3. Results and Discussion

We used BSA as a reference protein and prepared a series of dilutions with concentrations ranging from ~10 mg/mL to 0.07 µg/mL (5 orders of magnitude). Then, we measured the response of various dyes to the protein and generated calibration curves (Figure 1). We also measured the absorption and fluorescence emission spectra of the samples for the visual comparison of the dyes properties. For a better representation of the regions of low concentrations, the calibration curves were also plotted in logarithmic scale (Figure 2).

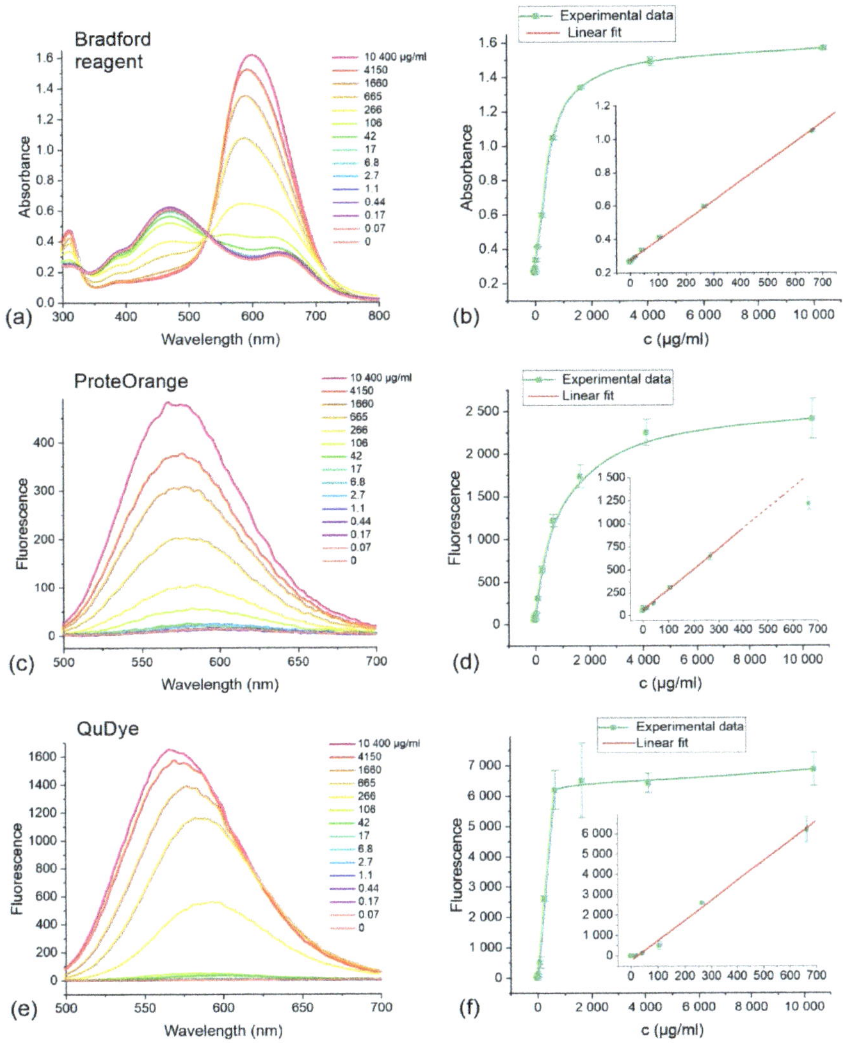

Figure 1. Response spectra of dye to the presence of protein (**a**,**c**,**e**) and calibration curves (**b**,**d**,**f**) for Bradford reagent (**a**,**b**), ProteOrange (**c**,**d**), and QuDye (**e**,**f**). Concentrations of the stock solutions are presented. For each concentration, the experiments were performed in triplicate. The error bars are the SD.

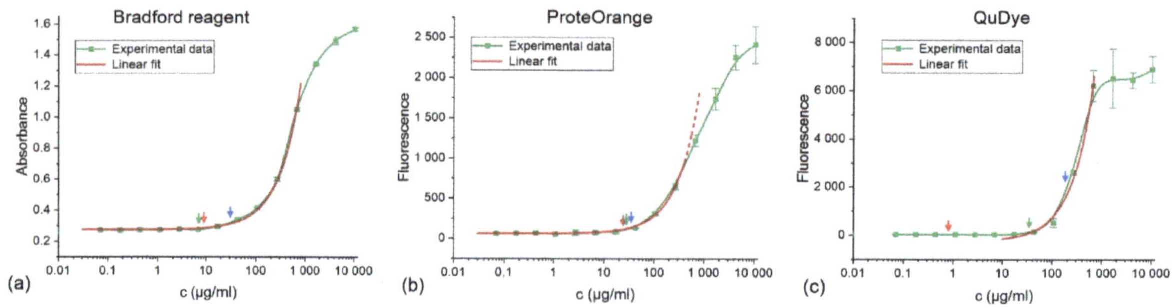

Figure 2. Calibration curves in logarithmic scale for Bradford reagent (**a**), ProteOrange (**b**), and QuDye (**c**). The arrows indicate the LoDs calculated using Equations (1) (in blue), (2) (in red), and (3) (in green).

Bradford's reagent (Coomassie G-250) showed a linear dependence of the growth of light absorption with increasing protein concentration in the range of 0 to ~670 μg/mL. ProteOrange showed linear fluorescence growth in response to the protein in the range of up to 350 μg/mL. It was also clearly visible that the standard deviations increased with increasing concentrations (Figure 1d). In the case of QuDye, a linear response to the protein was observed at concentrations of 40 to 670 μg/mL. The high threshold of the response may indicate the complex character of dye–protein interactions; this was also confirmed by the hypsochromic shift of the emission maximum at higher concentrations (Figure 1e).

To determine the LoD, a calibration line is usually built using linear regression in the range of the linear response of the dye to the analyte; then, the LoD is determined using the following equation:

$$LoD = 3.3\,\frac{S}{b} \qquad (1)$$

where S is the standard deviation of the response (intercept with intensity axis) and b is the slope of the calibration curve.

According to Equation (1) and the results of the linear regression for the Bradford reagent, ProteOrange, and QuDye, the LoDs were 30.5, 34.6, and 182 μg/mL, respectively (Table 2). However, one can see that, at these values, the dye response could reach significant values (Figure 2, blue arrows). In the case of QuDye, it was about one-third of the dynamic range, while the deviations of the signal in the region of "noise" were quite small. In other words, the calculated LoDs seemed to be overestimated.

Table 2. Calculated parameters from the linear regression and blank responses as well as the LoDs calculated by using these parameters and Equations (1)–(3).

Dye	B [1]	SE_{slope} [1]	I_0 [1]	S [1]	I_{blank} [2]	S_{blank} [2]	LoD (1)	LoD (2)	LoD (3)	m (ng)
Coomassie G-250	0.00117	8.6×10^{-6}	0.277	0.0108	0.275	0.00328	30.5	9.24	7.20	72
ProteOrange	2.22	0.046	59.6	23.27	67.1	16.4	34.6	24.3	27.7	69.3
QuDye	9.9	0.43	−305	546	25.5	2.43	182	0.81	34	170

[1] Parameters derived from linear regression; [2] Parameters derived from the response at low concentrations; b—slope of the calibration curve; SE_{slope}—standard error of the slope; I_0—intercept of the calibration curve; S—standard deviation of the intercept; I_{blank}—average response at low concentrations (background noise at concentrations <7 μg/mL); S_{blank}—standard deviation of the response at low concentrations; LoD—limit of detection in μg/mL as calculated using Equations (1), (2), or (3); m (ng)—amount of the protein in the sample corresponding to the LoD (3)

It was noted above that the standard deviation of the response increased with increasing concentration, especially for fluorogenic dyes, while the SD of the response in the low-concentration region (near the actual LoD) was much lower. Therefore, to estimate the LoD, it was more appropriate to use the SD of the response at low concentrations, or the concentrations at which the response did not exceed the background level (S_{blank}). For the tested dyes, it was the area of concentrations below 7 μg/mL. In this case, the detection limit was determined as:

$$LoD = 3.3\,\frac{S_{blank}}{b} \tag{2}$$

According to Equation (2), the LoDs for Bradford reagent, ProteOrange, and QuDye were 9.24, 24.3, and 0.81 μg/mL, respectively. The use of S_{blank} expectedly resulted in lower values of the determined LoDs. However, in the case of QuDye, the LoD was in the region of background noise. This could be explained, on one hand, by the low SD in the noise region and, on the other hand, by the significant discrepancy between the intercept of the regression line and the actual response of the "blank" samples (Table 2).

When considering the basis for the LoD definition, Equation (1) arises from the assumption of Gaussian distribution for the response to an analyte, and that the probability of occurrence of the response values in the background noise did not exceed 5%. At this assumption value of the response at the concentration of the LoD defined as the mean value of the background signal (I_{blank}) increased by 3.3 (rounding of 3.29 [20]) standard deviations of the response in the background (S_{blank}) [17,18,21]. That is,

$$I_{LoD} = I_{blank} + 3.3S_{blank}$$

If we take the equation of linear regression of the response from the concentration

$$I = I_0 + b[C]$$

the concentration at the limit of detection is expressed as

$$[C]_{LoD} = \frac{I_{LoD} - I_0}{b} = \frac{I_{blank} + 3.3S_{blank} - I_0}{b} = \frac{3.3S_{blank}}{b} + \frac{I_{blank} - I_0}{b}$$

thus, the LoD in the general case is expressed by the equation:

$$LoD = 3.3\frac{S_{blank}}{b} + \frac{I_{blank} - I_0}{b} \tag{3}$$

where S_{blank} is the standard deviation of the signal at concentrations below the LoD, i.e., at concentrations at which the response becomes less than the noise; I_{blank} is the mean value of the response at these concentrations; and I_0 and b are the parameters of linear regression of the calibration line (intercept and slope, respectively). For clarity, these parameters are shown in Figure 3 using ProteOrane as an example.

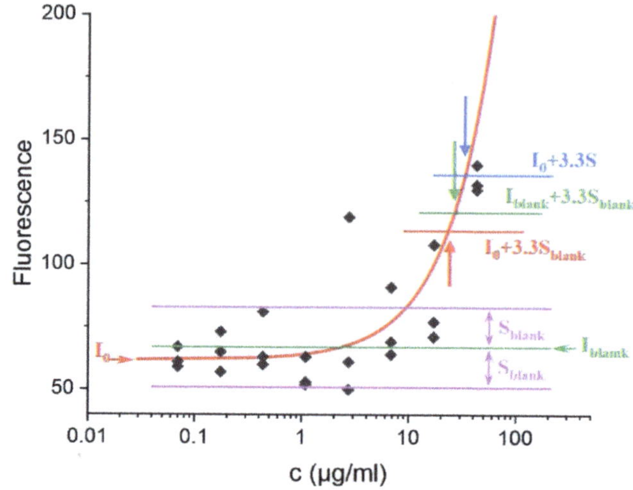

Figure 3. The enlarged region of low concentrations of the ProteOrange calibration curve. Experimental points are shown as black diamonds. The linear regression is shown by the red line. The mean value of I_{blank} and its standard deviation S_{blank} were calculated from the points at concentrations below 7 μg/mL. The fluorescence signal corresponding to the LoD determined by Equation (1) is indicated by the blue line and arrow, by Equation (2) by the red line and arrow, and by Equation (3) by the green line and arrow.

In determining LoD, when the response is measured over the entire dynamic range of the dye, but the impact of nearly zero concentrations is neglected, the I_{blank} values can be assumed to be equal to I_0 and Equation (3) degenerates into Equation (2). If the standard deviations are assumed to be equal in the entire range of measurements, Equation (2) is transformed into Equation (1). Such assumptions are reasonable when the deviations of the response are comparable over the whole range of measurements and the linear regression does not deviate significantly from the experimental data in the region of "blank" samples. In other cases, oversimplifications can result in artifacts that can lead to incorrect LoD determinations. As shown, the deviations can reach orders of magnitude, as in the case of QuDye, for which the I_0 value was even below zero.

The limits of protein detection by Bradford reagent, ProteOrange, and QuDye, as calculated by Equation (3), were 7.20, 27.7, and 34 μg/mL, respectively. When multiplying these values by the volume of sample tested (Table 1), 72, 69.3, and 170 ng of the protein in the sample could be detected by the corresponding methods, respectively. Thus, for determining the protein concentration in solution, the classic Bradford reagent was not worse than newer dyes in terms of the LoD. It also had a large range of linearity of the response in the area of high concentrations. For the measurements, it was necessary to use spectrophotometric equipment, rather than the more expensive spectrofluorimetric equipment, as in the case of fluorescent dyes.

Summing up the results of the LoD determination of the tested dyes, for the proper determination of the LoD of an analyte by dye, linear regression in the region of considerable signals is not enough. It is also necessary to determine the mean value of the signal and its standard deviation at near-zero concentrations. This allows for Equation (3) to be used for a more accurate determination of LoD. Additionally, it should be noted that the results of calculations should be critically interpreted, and the determined parameters should be checked visually using plots of the response dependence on the concentration in the logarithmic scale.

An important characteristic of chemosensors is also the limit of quantification (LoQ). The same linear regression equation as that used for the LoD is usually used to determine the LoQ, but the multiplication factor of the standard deviation of the background is usually 10 instead of 3.3 [18,22]. That is, the LoD is the concentration at which the signal exceeds the noise at this point by 10 times [20]. Analogous to LoD,

$$LoQ = 10\frac{S_{blank}}{b} + \frac{I_{blank} - I_0}{b}$$

However, it is worth noting that, in the case of significant differences in the standard deviations of the response value in the area of the obtained LoQ from S_{blank}, more complex calculations may be required.

It is also worth noting that most programs used to calculate linear regressions (such as Origin or GraphPad Prism), in the results of the calculations display not the standard deviations of the determined parameters, but the standard errors (SE), which are linked to each other by the equation

$$SE = \frac{SD}{\sqrt{N}}$$

where N is the number of points used in the parameter calculation.

It is correct to use SE in specifying the error of a parameter in the form "mean ± SE", but when calculating the LoD and LoQ, the obtained standard errors must be multiplied by $\sqrt{N}$ to obtain the standard deviations.

4. Conclusions

We have shown that the classical Bradford method allows the determination of protein concentrations and is indeed not inferior to contemporary methods. In addition, we showed that the use of standard approaches, including the use of linear regressions to determine the LoD and LoQ, may not work correctly for the tested dyes. We have refined the commonly used algorithm for analyzing experimental data so that it takes into account possible artifacts. The described protocol can be used not only for protein concentration assays using colorimetric and fluorogenic dyes, but also for other systems with deviations in the experimental data from standard linear regressions.

Author Contributions: Conceptualization, methodology, validation, formal analysis, investigation, A.A.M. and A.A.P.; resources and data curation, A.A.M., V.I.M., S.M.D. and A.A.P.; writing—original draft preparation, visualization, A.A.P.; writing—review and editing, A.A.M., V.I.M., S.M.D. and A.A.P.; supervision, project administration, and funding acquisition, V.I.M., S.M.D. and A.A.P. All authors have read and agreed to the published version of the manuscript.

Funding: This research was supported by the Russian Science Foundation (grant No. 19-73-20194).

Institutional Review Board Statement: Not applicable.

Informed Consent Statement: Not applicable.

Data Availability Statement: Not applicable.

Conflicts of Interest: The authors declare no conflict of interest. The funders had no role in the design of the study; in the collection, analyses, or interpretation of data; in the writing of the manuscript; or in the decision to publish the results.

Abbreviations

BSA—Bovine serum albumin; LoD—limit of detection; LoQ—limit of quantification; SD—standard deviation; SE—standard error.

References

1. Olson, B.J.S.C. Assays for Determination of Protein Concentration. *Curr. Protoc. Pharmacol.* **2016**, *73*. [CrossRef] [PubMed]
2. Kielkopf, C.L.; Bauer, W.; Urbatsch, I.L. Methods for Measuring the Concentrations of Proteins. *Cold Spring Harb. Protoc.* **2020**, *2020*, pdb.top102277. [CrossRef] [PubMed]

3. Bradford, M.M. A Rapid and Sensitive Method for the Quantitation of Microgram Quantities of Protein Utilizing the Principle of Protein-Dye Binding. *Anal. Biochem.* **1976**, *72*, 248–254. [CrossRef] [PubMed]
4. Kielkopf, C.L.; Bauer, W.; Urbatsch, I.L. Bradford Assay for Determining Protein Concentration. *Cold Spring Harb. Protoc.* **2020**, *2020*, pdb.prot102269. [CrossRef] [PubMed]
5. Noble, J.E. Chapter Two—Quantification of Protein Concentration Using UV Absorbance and Coomassie Dyes. In *Methods in Enzymology*; Lorsch, J., Ed.; Laboratory Methods in Enzymology: Protein Part A; Academic Press: Cambridge, MA, USA, 2014; Volume 536, pp. 17–26. [CrossRef]
6. Wiśniewski, J.R.; Gaugaz, F.Z. Fast and Sensitive Total Protein and Peptide Assays for Proteomic Analysis. *Anal. Chem.* **2015**, *87*, 4110–4116. [CrossRef] [PubMed]
7. Noble, J.E.; Knight, A.E.; Reason, A.J.; Di Matola, A.; Bailey, M.J.A. A Comparison of Protein Quantitation Assays for Biopharmaceutical Applications. *Mol. Biotechnol.* **2007**, *37*, 99–111. [CrossRef] [PubMed]
8. You, W.W.; Haugland, R.P.; Ryan, D.K.; Haugland, R.P. 3-(4-Carboxybenzoyl)Quinoline-2-Carboxaldehyde, a Reagent with Broad Dynamic Range for the Assay of Proteins and Lipoproteins in Solution. *Anal. Biochem.* **1997**, *244*, 277–282. [CrossRef]
9. Steinberg, T.H.; Jones, L.J.; Haugland, R.P.; Singer, V.L. SYPRO Orange and SYPRO Red Protein Gel Stains: One-Step Fluorescent Staining of Denaturing Gels for Detection of Nanogram Levels of Protein. *Anal. Biochem.* **1996**, *239*, 223–237. [CrossRef]
10. Mora, A.K.; Nath, S. SYPRO Orange—A New Gold Standard Amyloid Probe. *J. Mater. Chem. B* **2020**, *8*, 7894–7898. [CrossRef]
11. Wu, D.; Sedgwick, A.C.; Gunnlaugsson, T.; Akkaya, E.U.; Yoon, J.; James, T.D. Fluorescent Chemosensors: The Past, Present and Future. *Chem. Soc. Rev.* **2017**, *46*, 7105–7123. [CrossRef]
12. Kwon, N.; Hu, Y.; Yoon, J. Fluorescent Chemosensors for Various Analytes Including Reactive Oxygen Species, Biothiol, Metal Ions, and Toxic Gases. *ACS Omega* **2018**, *3*, 13731–13751. [CrossRef] [PubMed]
13. Dongare, P.R.; Gore, A.H. Recent Advances in Colorimetric and Fluorescent Chemosensors for Ionic Species: Design, Principle and Optical Signalling Mechanism. *ChemistrySelect* **2021**, *6*, 5657–5669. [CrossRef]
14. Martynov, V.I.; Pakhomov, A.A. BODIPY Derivatives as Fluorescent Reporters of Molecular Activities in Living Cells. *Russ. Chem. Rev.* **2021**, *90*, 1213–1262. [CrossRef]
15. Li, Y.; Yao, S.; Fang, H.; He, W.; Chen, Y.; Guo, Z. Rational Design of Ratiometric Fluorescent Probe for Zn^{2+} Imaging under Oxidative Stress in Cells. *Chemosensors* **2022**, *10*, 477. [CrossRef]
16. Nootem, J.; Sattayanon, C.; Daengngern, R.; Kamkaew, A.; Wattanathana, W.; Wannapaiboon, S.; Rashatasakhon, P.; Chansaenpak, K. BODIPY-Pyridylhydrazone Probe for Fluorescence Turn-On Detection of Fe^{3+} and Its Bioimaging Application. *Chemosensors* **2021**, *9*, 165. [CrossRef]
17. Desimoni, E.; Brunetti, B. About Estimating the Limit of Detection by the Signal to Noise Approach. *Pharm. Anal. Acta* **2015**, *6*, 4. [CrossRef]
18. Shrivastava, A.; Gupta, V. Methods for the Determination of Limit of Detection and Limit of Quantitation of the Analytical Methods. *Chron. Young Sci.* **2011**, *2*, 21. [CrossRef]
19. Cole, R. Optimum Optical Density in Spectrophotometry. *J. Opt. Soc. Am.* **1951**, *41*, 38. [CrossRef]
20. Currie, L.A. Nomenclature in Evaluation of Analytical Methods Including Detection and Quantification Capabilities (IUPAC Recommendations 1995). *Pure Appl. Chem.* **1995**, *67*, 1699–1723. [CrossRef]
21. Armbruster, D.A.; Pry, T. Limit of Blank, Limit of Detection and Limit of Quantitation. *Clin. Biochem. Rev.* **2008**, *29* (Suppl. 1), S49–S52.
22. Belter, M.; Sajnóg, A.; Barałkiewicz, D. Over a Century of Detection and Quantification Capabilities in Analytical Chemistry—Historical Overview and Trends. *Talanta* **2014**, *129*, 606–616. [CrossRef] [PubMed]

chemosensors

Article

Multi-Endpoint Toxicity Tests and Effect-Targeting Risk Assessment of Surface Water and Pollution Sources in a Typical Rural Area in the Yellow River Basin, China

Fangxu Li [1], Jisui Tan [2], Qian Yang [1], Miao He [2], Ruozhen Yu [3,*], Chun Liu [1,*] and Xiaohong Zhou [2]

1 School of Environmental Science and Engineering, Hebei University of Science and Technology, Shijiazhuang 050000, China
2 State Key Joint Laboratory of ESPC, School of Environment, Tsinghua University, Beijing 10084, China
3 Chinese Research Academy of Environmental Sciences, Beijing 100012, China
* Correspondence: yuruozhen@tsinghua.org.cn (R.Y.); liuchun@hebust.edu.cn (C.L.)

Abstract: Multi-endpoint toxicity tests were used to evaluate the acute toxicity, estrogenic activity, neurotoxicity, genotoxicity, and ecological risks of surface water and sewage from possible pollution sources in rural areas of the Yellow River (China). Toxicity testing results showed that the luminescence inhibition rates of acute toxicity ranged from not detected (ND) to 38%, the 17β-estradiol equivalent (E2-EQ) values of estrogenic activity ranged from 4.8 to 131.0 ng·L^{-1}, neurotoxicity was not detected, and the protein effect level index (PELI) values of genotoxicity ranged from 1 to 6.06. Neither acute toxicity nor genotoxicity were detected in the tributaries of the Yellow River (River 2) flowing through the investigated rural area. The distribution of high estrogenic activity sites was relatively scattered, but mainly located in the tributaries of River 2. Industrial, domestic, and livestock and poultry breeding sewage were all possible sources of toxicity, and the contribution of livestock and poultry to environmental estrogens in the surface water was significant. Furthermore, the potential effect-targeting risks of toxic substances in the surface water for aquatic organisms were assessed using the risk quotient method, by considering the toxic equivalent concentration. The results indicated that the risk of estrogenic activity was the main ecological risk in the surface water of this rural area. Except for the reservoir site, the other sampling sites showed a moderate to high estrogenic activity risk, especially in the tributaries of River 2.

Keywords: multi-endpoint toxicity test; rural area; toxic equivalent concentration; ecological risk assessment

Citation: Li, F.; Tan, J.; Yang, Q.; He, M.; Yu, R.; Liu, C.; Zhou, X. Multi-Endpoint Toxicity Tests and Effect-Targeting Risk Assessment of Surface Water and Pollution Sources in a Typical Rural Area in the Yellow River Basin, China. *Chemosensors* **2022**, *10*, 502. https://doi.org/10.3390/chemosensors10120502

Academic Editor: Xiaolong Yang

Received: 11 October 2022
Accepted: 23 November 2022
Published: 28 November 2022

1. Introduction

As suppliers of cereals and other natural resources, rural areas occupy an important position in various countries worldwide [1]. Today, with the continuous acceleration of urbanization, rural communities face many problems due to economic structural changes, product transformations, technological innovations, and so on [2]. Rural areas are mostly dominated by agriculture and animal husbandry. While pursuing rapid economic development, more and more chemicals, such as pesticides, herbicides, and antibiotics, are used in agriculture and livestock and poultry breeding [3,4]. As a result, a large number of these substances are discharged into rural areas, which seriously threatens human health. In addition, the development of rural industry and the improvement of people's living standards also mean that emerging contaminants, such as perfluorinated compounds, microplastics, pharmaceuticals, and personal care products, enter the rural environment, causing huge environmental risks [5–8].

At present, all kinds of production activities require water, which also increases the risk of water pollution. Unlike in cities, water pollution in rural areas is more difficult to monitor and control, mainly due to the complex pollution sources. There are many types of pollution source, with an irregular distribution, which make it impossible for rural areas to detect certain pollutants purposefully and according to the potential pollution sources in the region, as is done in cities. In addition, due to gaps in some policies and a lack of awareness of environmental protection, the illegal discharge of pollutants in rural areas is serious, which makes it more difficult to correctly determine the pollution situation and source of pollution in water bodies [9]. Therefore, in order to effectively deal with the problem of water pollution in rural areas, it is necessary to conduct a comprehensive analysis of water bodies.

With the increase of diverse classes of pollutants in waterbodies, the environmental risks are also diverse [10], making water pollution assessment challenging but crucial [11]. With the increasing attention to health and the in-depth study of pollutants in the environment, toxicity indicators directly related to the health of organisms, such as acute toxicity, neurotoxicity, genotoxicity, and endocrine disrupting effects have become widely used [12–14]. Mixtures of pollutants may cause one or more toxic effects. Through the multi-endpoint toxicity analysis of water bodies, the comprehensive impact of complex mixtures of pollutants in water bodies can be evaluated. At present, multi-endpoint toxicity analysis has been widely used in urban rivers, wastewater treatment plants, industrial parks, and other sites, while there have been relatively few toxicity assessments and pollution analyses of waterbodies in rural areas, especially of surface water [15–17]. For the complex situation of water pollution in rural areas, a multi-endpoint toxicity test can reflect the overall problems of the water body and give insights about the existing pollution. Qualitative and quantitative analysis of the water body can be carried out by chemical analysis, to determine specific groups or pollutants. Moreover, using these multi-endpoint toxicity test results, the qualitative analysis of pollutants will be simpler and faster and more sensitive to potential toxicity.

We selected a typical rural area in the lower reaches of the Yellow River in China. This area is mainly dominated by agriculture and livestock and poultry breeding, with industry in the development stage. Acute toxicity, estrogenic activity, neurotoxicity, and genotoxicity tests of surface water were conducted in this area, to analyze the multi-endpoint toxicity of the surface water. At the same time, a multi-endpoint toxicity analysis was carried out for many types of possible pollution sources in the local area. Combined with the toxicity of the surface water, the ecological risks for water bodies in the rural area were analyzed. This study can provide a reference for future pollution assessments and the prevention of surface water pollution in rural areas.

2. Material and Methods

2.1. Study Sites and Sample Collection

The study area is situated in a rural area on the east side of the lower reaches of the Yellow River, China. There are two main rivers in the area, which are formed by a levee breach when the Yellow River flows through. The whole area is a plain area, with high terrain in the southwest and low terrain in the northeast. Therefore, the two main rivers flow from west to east, forming many small tributaries and ditches (Figure 1). At present, River 1 serves as a water source river, to help supplement domestic water and drinking water in this rural area; while River 2 flows through the whole rural area and receives all kinds of water from regional production and domestic activities.

The sampling campaigns were carried out in September, the autumn rainy season. A total of 27 water samples were collected in the study area, including 21 surface water samples and 6 possible pollution source samples. Table 1 displays the water sampling sites and sample types. Surface water samples were collected 0.5 m below the water surface with 1 L plexiglass water collector and then transferred to amber glass bottles. Each bottle was rinsed with the water sample in advance. Samples from pollution sources were collected directly in bottles. Then, 0.429 mL of concentrated hydrochloric acid (36%) was added to each 1 L water sample for acidification. Then the collected samples were transported to the laboratory in a low-temperature transport box and stored at 4 °C until analysis.

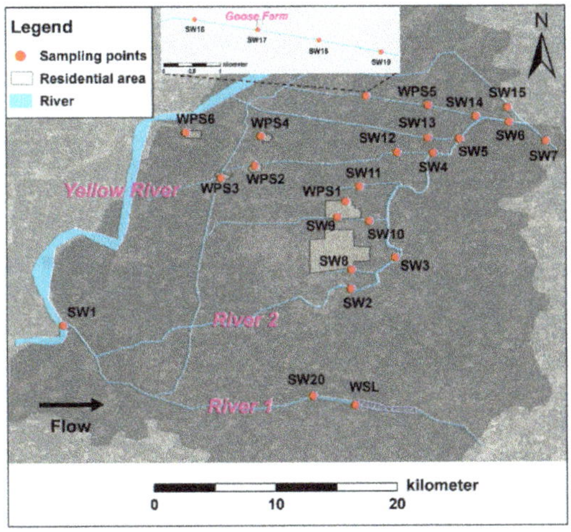

Figure 1. Map of the study area.

Table 1. Sampling sites in the selected study area.

Code	Sites	Types	Date
SW1	Yellow River diversion gate, headstream of River 1/2	Surface water	26 September
SW2-7	Main stream of River 2		26 September
SW8-15	Tributary of River 2		26 September
SW16-19	Tributary of River 2 adjacent to a goose farm		26 September
SW20	River 1		26 September
WSL	Reservoir on River 1	Surface water	26 September
WPS1	Chemical industrial park	Rainwater	27 September
WPS2	Planting farmland	Ditch water	27 September
WPS3	Duck farm	Fecal sewage	27 September
WPS4	Cattle farm		27 September
WPS5	Swine farm	Urine sewage	27 September
WPS6	sewage treatment plant	Effluent	27 September

2.2. Sample Pretreatment for Toxicity Analysis

The samples were pretreated as described below, to pre-concentrate them for the toxicity assays, except for the acute toxicity test. Then, 500 mL of each sample was filtered through a 0.7 μm glass fiber membrane (Whatman) under vacuum. The filtered samples were processed by solid phase extraction (SPE) using Oasis® HLB 6cc extraction cartridges (Waters). The HLB cartridges were conditioned with 5 mL methanol and equilibrated with 5 mL 5 mM HCl before use. The pre-filtered samples were flowed through a cartridge connected to a Visiprep DL (Supelco) under vacuum at a flow rate of 5–10 mL min^{-1}. Afterward, the cartridges were kept in vacuum aspiration until the residual liquid had drained. The cartridges were wrapped in aluminum foil and stored at −20 °C until elution.

The cartridges were successively eluted with 5 mL methanol and 5 mL mixed solution of n-hexane and acetone (1:1, *v*/*v*). The solution was added and allowed to stand for 5 min before elution. The flow rate of the whole elution process was maintained at 1–2 mL min^{-1} or slower. Finally, the eluent was blown dry under gentle nitrogen and reconstituted with 1 mL methanol. The sample solutions were stored at −20 °C until analysis.

2.3. Acute Toxicity Analysis

An acute toxicity analysis was conducted on Centro LB 960 luminescence microplate readers using *Vibrio fischeri* and according to the test protocol ISO 11348-3 [18]. The samples were filtered through a 0.45 μm filter to remove particles and then adjusted to a pH value of 6.0–8.5 before analysis. After mixing 100 μL sample solutions with 100 μL bacterial suspensions, the decrease of bacterial luminescence at 30 min compared with the initial value was recorded. In addition, 2% NaCl gradient diluted 3,5-dichlorophenol (DCP) was used as positive control and 2% NaCl was used as negative control for each test. The relative luminescence inhibition rate (%) of each sample was calculated, to determine toxicity using the following formula:

$$\text{Inhibition rate} = \frac{I_0 \times C_f - I_t}{I_0 \times C_f} \times 100 \quad (1)$$

where I_0 is the initial luminescence intensity of *Vibrio fischeri*, I_t is the luminescence intensity of *Vibrio fischeri* after 30 min of being mixed with the water samples, and C_f is the correction factor for the contact time of 30 min;

$$C_f = \frac{C_t}{C_0} \quad (2)$$

where C_0 is the initial luminescence intensity of negative control sample, and C_t is the luminescence intensity of *Vibrio fischeri* after 30 min of being mixed with the negative control sample. The acute toxicity test is valid when the C_f is between 0.6 and 1.8.

2.4. Estrogenic Activity Analysis

The estrogenic activity of the water sample after SPE could be determined using the yeast estrogen screening (YES) method [19]. Yeast strains stored at −80 °C were added to SD/-Trp-Leu medium and cultured with shaking at 30 °C for 24–30 h. Then, 180 μL of the bacterial suspensions was added to a 96 well plate until an absorbance level of 0.7–1.0 at 600 nm (OD600) was achieved. Concentrated samples were made into a series of twofold dilutions with 0.2 M phosphate-buffered saline (PBS), and 20 μL of diluted samples were added to a 96 well plate. 17β-estradiol (E2) was prepared as a positive control in a series of concentration gradients in methanol for each plate. The OD600 was measured after the mixing of the bacterial suspensions, and samples was incubated at 30 °C, 800 rpm for 4 h. Next, 120 μL of test-buffer (100 mL PBS, 270 μL β-Mercaptoethanol and 3.33 mL SDS 0.1%) and 20 μL of chloroform were added sequentially and pre-incubated for 10 min. The chromogenic reaction was started by adding 40 μL of 8 mg mL^{-1} chlorophenolred-β-D-galactopyranoside (ONPG) and terminated by adding 100 μL of 1 M Na_2CO_3 after 1 h. The supernatant was separated by centrifugation at 2500 rpm for 2 min, and the absorbance was measured at 420 nm. Combined with the OD600 value, the β-galactosidase activity (U) was expressed using the following formula [20]:

$$U = \frac{OD_{420s} - OD_{420b}}{t \times V \times OD_{600}} \times D \quad (3)$$

where t is the time of chromogenic reaction (min); V is the volume of the final test (mL); D is the dilution ratio between the final test and adding test-buffer; OD_{420s} and OD_{420b} are the absorbance of samples and blank at 420 nm, respectively; and OD_{600} is the absorbance of samples at 600 nm.

Dose–response curves for E2 and samples were plotted as the concentration and relative enrichment factor (enrichment factor multiplied by dilution factor) versus β-galactosidase activity, respectively. The comprehensive estrogenic activity of the water samples was measured by dividing the EC50 of the two curves and expressed as the 17β-estradiol equivalent (E2-EQ).

2.5. Neurotoxicity Analysis

Acetylcholinesterase (AChE) activity was used to indicate neurotoxicity and measured according to a modified Ellman method [21,22]. The enriched samples were added to 96 well plates and diluted step by step with 0.05 M PBS. Then, 10 μL of N-bromosuccinimide and 20 μL of the mixture of PBS and 4 g L^{-1} ascorbic acid (1:1, *v*/*v*) were added. Afterward, 10 μL of 800 U L^{-1} AChE solution was added and mixed on a plate shaker for 10 min;

140 μL of the test solution (2 mL S-Acetylthiocholine iodide 78 mM, 2 mL 5,5′-dithion-bis-2-nitrbenzoic acid 7.8 mM, and 10 mL PBS) were added; and the absorbance at 420 nm was immediately measured for 4 min (30 s intervals). The inhibition rate of enzyme activity was calculated from the enzyme velocity and fitted to the dose–response curve. Serial concentrations of parathion were used as a positive control, and the toxicity was expressed as parathion equivalent (PT-EQ) through the IC50 value.

2.6. Genotoxicity Analysis

CHK1 acts as a checkpoint in response to any DNA general damage [23,24]. We selected *Saccharomyces cerevisiae* (ATCC 201388) containing GFP-tagged CHK1 to detect neurotoxicity. Yeast strains were grown in clear-bottom black 384 well plates (Costa) with SD medium at 30 °C for 4–6 h, until reaching early exponential growth (OD600 about 0.2–0.4). Then, 10 μL of concentrated samples (diluted from 500× to 100× and 10× with medium) and controls (medium only) were added to each well. Plates were placed in a micro-plate reader to measure the absorbance (OD600 for cell growth) and GFP signal (excitation 485 nm, emission 525 nm for protein expression) every 10 min for 24 h.

The protein expression profiling data of yeast were processed as described in previous studies [25–28]. OD and GFP raw data were corrected using the background OD and GFP signal of the medium control. The protein expression P for each measurement was normalized by the cell number, as $P = (GFP_{corrected}/OD_{corrected})$. The alteration in protein expression of CHK1 at each time point due to chemical exposure was represented as $I = P_{experiment}/P_{control}$. The accumulative protein expression change over the exposure period was represented by the protein effect level index (PELI).

$$\mathrm{PELI} = \frac{\int_{t=0}^{t} I\ \mathrm{d}t}{\text{exposure time}} \tag{4}$$

where t is the exposure time. Genotoxicity was represented by PELI and genotoxicity positive was defined as having a PELI value greater than 1.5. [25,26].

2.7. Ecological Risk Assessment

The potential risks of toxic substances in the surface water to aquatic organisms were assessed using a risk quotient (RQ) method. The RQ values were calculated using the following formula:

$$\mathrm{RQ} = \frac{\mathrm{MEC}}{\mathrm{PNEC}} \tag{5}$$

where MEC is the measured environmental concentration, and PNEC is the predicted no effect concentration in water. In this study, the equivalent concentration of quality control substance corresponding to the different toxicities was used as the MEC value; that is, E2-EQ was used for estrogenic activity and PT-EQ was used for neurotoxicity. According to European Commission Technical Guidance Document, the PNEC value is usually obtained by dividing the toxicity data of the most sensitive species by the assessment factor (AF), as shown in the following equation:

$$\mathrm{PNEC} = \frac{\mathrm{E(L)C50}}{\mathrm{AF}}\ \text{for acute toxicity} \tag{6}$$

$$\mathrm{PNEC} = \frac{\mathrm{NOEC(LOEC)}}{\mathrm{AF}}\ \text{for chronic toxicity} \tag{7}$$

where EC50 or LC50 was the half maximum effect concentration or the median lethal concentration in acute toxicity, and the AF value was set as 1000; NOEC or LOEC was the no observed effect concentration or lowest observed effect concentration in chronic toxicity, with AF values of 10, 50, or 100 [29].

The PNEC values for aquatic organisms of different compounds are shown in Table 2. To avoid underestimating the potential ecological risks, the lowest PNEC was selected for calculation.

Table 2. Aquatic toxicity data and PNEC values of the target compounds.

Compound	Taxonomic Group	Toxicity	NOEC/LOEC or EC50/LC50 (ng L^{-1})	AF	PNEC (ng L^{-1})	Reference
17β-Estradiol	Algae	Acute	2,480,000	1000	2480	[30]
	Invertebrate	Chronic	1000	100	10	[31]
	Fish	Acute	2000	1000	2	[32]
Parathion	Invertebrate	Acute	2500	1000	2.5	[33]
	Fish	Acute	1,500,000	1000	1500	[33]

3. Results and Discussion

3.1. Acute Toxicity Assessment with Vibrio fischeri

Water samples from the study sites were evaluated for acute toxicity using the *Vibrio fischeri* luminescence inhibition rate, and samples with a luminescence inhibition rate of more than 10% were identified as having acute toxicity [18]. Acute toxicity was determined in some surface water samples and pollution source samples. As shown in Figure 2, no acute toxicity was detected in River 1, where the water supply reservoir was located, nor was acute toxicity detected in the main stream of River 2. In the surface water samples collected from the tributary of River 2, except for the goose farm, acute toxicity was only detected in SW10 and SW13, and the inhibition rate was 10% and 38%, respectively. SW10 was collected downstream of the chemical industrial park on the tributary of River 2, and acute toxicity was detected in the rainwater from the surface runoff of the chemical industrial park, with an inhibition rate of 26%. Due to a deficient rainwater collection system, the downstream rivers of the chemical industrial park were likely polluted by rainwater runoff [34], which was a possible reason for the acute toxicity of SW10. Among the surface water samples, SW13 had the strongest acute toxicity, with an inhibition rate of 38%. The tributary where the sampling site was located often received domestic sewage and garbage from the surrounding large villages, as well as rainwater carrying livestock and poultry fecal matter. These were the possible reasons for the acute toxicity of the site.

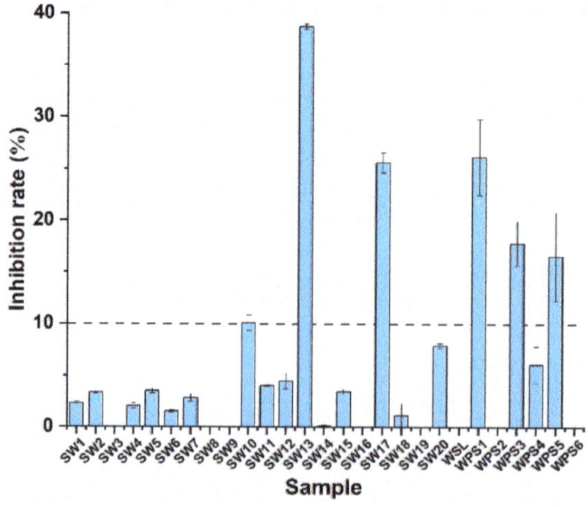

Figure 2. Acute toxicity of the samples determined by the inhibition of *Vibrio fischeri* with an exposure time of 30 min. SW1–SW20 show the surface water samples, among these, SW16–SW19 are adjacent to a goose farm. WSL is the sample collected from a water source reservoir, and WPS1–WPS6 are the samples collected from possible pollution sources.

Weak acute toxicity was also observed in the samples WPS3 and WPS5 collected from livestock and poultry farms, with inhibition rates of 17% and 16%, respectively. As livestock and poultry breeding sewage, feed additives, antibiotics, and their metabolites were the main factors causing acute toxicity [35]. As shown in Figure 2, the acute toxicity was detected in sample SW17 collected from the river adjacent to the goose farm, with an inhibition rate of 25%. The geese excrement during the period of geese stocking along the river would directly enter the river, and the wastes produced during the period of greenhouse breeding might also leak into the river. This suggests that large-scale livestock and poultry stocking or unregulated breeding may cause toxic substances to leak into the water body, thus generating water toxicity. However, this acute toxicity might be diluted by the river water or eliminated by the water body self-purification, because samples SW18 and SW19 collected downstream of the goose farm showed no obvious inhibition of luminescent bacteria.

3.2. *Estrogenic Toxicity Assessment by YES*

The collected water samples were tested for estrogenic effects using a YES assay, and the test results were expressed as the E2-EQ value. As shown in Figure 3, the E2-EQ value of sample SW20 collected from River 1 was 18.8 ng L^{-1}, which might have been related to the farmland and livestock and poultry breeding enterprises distributed on both sides of the river. No estrogenic activity was detected in sample WSL collected from the water supply reservoir on River 1. Referring to the latest Japanese drinking water quality standard, which stipulated that the concentration limit of 17β-estradiol was 80 ng L^{-1}, the water supply source in this area met the requirements on estrogen indicators of drinking water. The E2-EQ value of the seven samples from the main stream of River 2 was 6.6~25.2 ng L^{-1}, and the E2-EQ value of the eight samples from the tributaries was 6.6~39.6 ng L^{-1}. The estrogenic activity of the tributaries of River 2 was higher than that of the main stream. To analyze the estrogenic activity correlation between the main stream and the tributaries, further material analysis was required. Compared with the estrogenic activity in the lower Yellow River Basin reported by other studies, the activity in this study was relatively high; the highest estrogen activity was 0.21 ng L^{-1} in the lower reaches of Tongguan of the Yellow River in rainy season [36]; while the estrogen activity in Zhengzhou basin of the Yellow River was 0.72~1.19 ng L^{-1} [37]. However, it should be noted that the sampling sites reported above were located in the main stream of the Yellow River and not in rural areas, and there have been few reports on environmental surface water in the rural basin of the Yellow River.

In addition to farmland samples, different degrees of estrogenic activity were observed in samples collected from possible pollution sources. Strong estrogenic activities were observed in the livestock and poultry fecal sewage samples WPS3 and WPS4, with E2-EQ values of 74.1 and 32.9 ng L^{-1}, respectively. As the samples were filtered and pretreated, most of the estrogenic substances adsorbed by solids were removed, and untreated fecal sewage discharged into rivers might cause more serious pollution [38]. The strongest estrogenic activity was observed in swine urine sewage WPS5, with an E2-EQ value of 7265.2 ng L^{-1}. Previous studies have shown that human or animal urine exhibits a higher estrogenic activity than feces and that swine and poultry only excrete estrogens through urine, so compared to fecal sewage, urine sewage might contribute more environmental estrogens to rivers [39,40]. Estrogenic activity was also observed in the chemical park sample WPS1, with an E2-EQ value of 44.4 ng L^{-1}. However, compared with urban areas, the industrial enterprises in the rural area were relatively undeveloped, and were very concentrated in the study area, so the contribution of industrial production to the estrogenic activity in the surface water was low. The number of livestock and poultry breeding enterprises in the study area was extremely high, the locations were scattered, and their waste or sewage could enter surface water in many ways, such as runoff or seepage, accompanied by high concentrations of environmental estrogens [41–43]. The samples collected adjacent to the goose farm proved this to a certain extent. The E2-EQ value of sample SW16 collected upstream of the goose farm was only 4.7 ng L^{-1}, while the E2-EQ value of samples SW17 and SW18 collected from the farm and downstream were 131.0 ng L^{-1} and 77.5 ng L^{-1}, respectively. The geese in the farm were free-ranging by the river, during which their excreta would enter the riverside land or directly into the river, and it was also found that the sewage in the breeding shed was irregularly discharged into the river through pipelines during sampling; all these factors led to an increase in the estrogenic activity of the river water, which also affected the downstream waters. The E2-EQ value of sample SW19 collected from further downstream was 4.9 ng L^{-1}, indicating that the self-purification and dilution of the water effectively reduced the content of environmental estrogens in the water body.

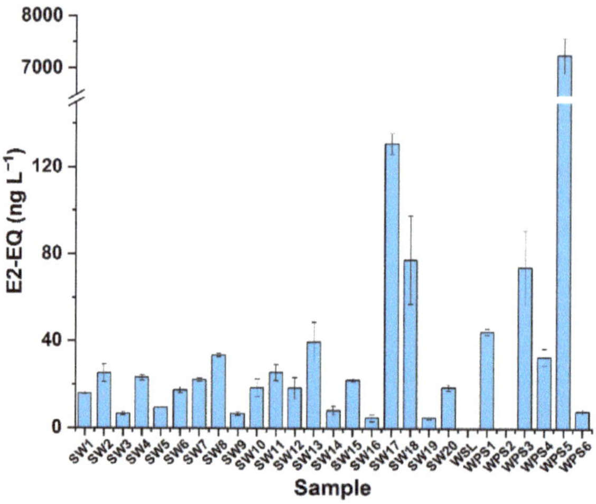

Figure 3. E2-EQ (ng L^{-1}) values of the samples determined using the YES assay. SW1–SW20 show the surface water samples; among these, SW16–SW19 were adjacent to a goose farm. WSL is the sample collected from a water source reservoir, and WPS1–WPS6 are the samples collected from possible pollution sources.

In addition, the E2-EQ value of 7.9 ng L^{-1} was detected in the effluent sample WPS6 of the sewage treatment plant, indicating that the sewage still contained some environmental estrogens after treatment, which was also reported in previous studies [44]. Although the situation of other sewage treatment plants in the study area might be different, this was still one of the possible factors that might have caused more environmental estrogens to enter the surface water, especially the sewage with high levels of environmental estrogens.

3.3. Neurotoxicity Assessment

Neurotoxicity was assessed using AChE activity, and the results were expressed as PT-EQ value. Among all the samples collected in River 1 and River 2, only sample WPS1 collected from the chemical industrial park was observed with neurotoxicity, with a PT-EQ value of 4.0 μg L^{-1}, as shown in Figure 4. AChE can be inhibited by many substances, such as organophosphorus pesticides, organochlorine pesticides, heavy metals, bisphenol-based compounds, etc. [45–47]. For rural areas, the pesticides and herbicides used in agricultural production activities are the main sources of neurotoxic substances. However, according to the test results, no neurotoxicity was detected in sample WPS2 collected from a farmland ditch and the surface water sample SW16 collected near farmland. The pesticides and herbicides used in local agricultural activities were mainly emamectin benzoate, bifenthrin, acetamiprid, and propisochlor, which would not produce neurotoxicity within the specified range [48–50]. There were many factories in the chemical industrial park, such as a fertilizer factory and wig factory, whose wastes such as heavy metals or dyes would cause inhibition of AChE, which might explain the neurotoxicity detected in WPS1; however, no neurotoxicity was detected in samples SW9 and SW10 collected downstream of the chemical industrial park, indicating that the sewage or rainwater of the chemical industrial park did not cause neurotoxicity in the surface water.

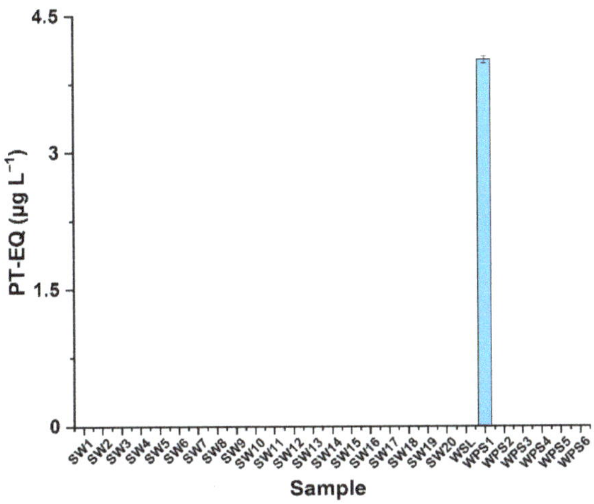

Figure 4. PT-EQ (μg L^{-1}) mean values of the samples determined using the AChE inhibition assay. SW1–SW20 show the surface water samples; among these, SW16–SW19 are adjacent to a goose farm. WSL is the sample collected from a water source reservoir, and WPS1–WPS6 are the samples collected from possible pollution sources.

3.4. Genotoxicity Assessment

The genotoxicity of the samples was determined using the cell cycle checkpoint kinase CHK1 as a biomarker. The occurrence of genotoxicity was reflected by the up-regulation of CHK1, and the determined result was expressed as the PELI. As shown in Figure 5, no genotoxicity was detected in the samples collected from River 1 and the main stream of River 2. Genotoxicity was detected in five of the eight samples collected from the tributaries of River 2, SW11–SW15, with PELI ranging from 1.56 to 6.06, all of which were downstream of River 2. Compared with the middle reaches flowing through relatively developed counties, the downstream basin was composed of several large villages. More extensive pollution sources and the relatively basic sewage collection and treatment facilities made the surface water more prone to toxicity.

For samples collected from possible pollution sources, only the rainwater sample WPS1 collected from the chemical industrial park had genotoxicity, with a PELI of 5.40; the surface water samples SW9 and SW10 collected downstream of the park had no detected genotoxicity, indicating that the sewage and rainwater in the park did not cause genotoxicity risks to the relevant rivers. Some herbicides and insecticides used in agricultural activities can cause genotoxicity, so the impact of farmland planting requires a follow-up investigation. In addition, livestock and poultry breeding was an important source of hormone substances, and some also showed genotoxicity. Both endogenous and exogenous hormones can show genotoxicity, manifested in causing DNA damage, which has been confirmed in previous studies [51–54]. Although the livestock and poultry sewage contained high concentrations of hormone substances, no genotoxicity was detected in the collected sewage samples.

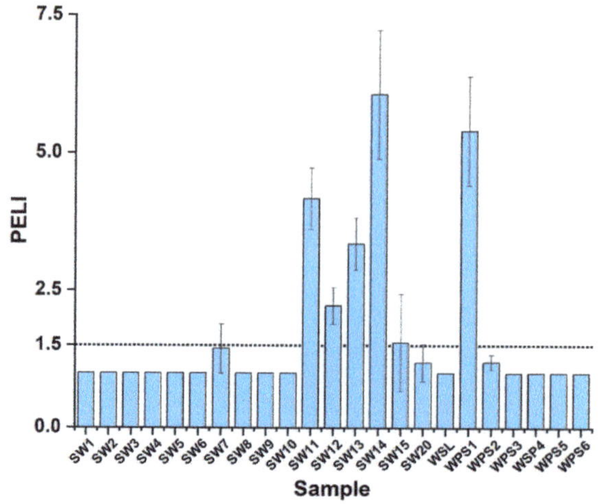

Figure 5. PELI values of the samples with CHK1 as a biomarker. SW1–SW15 and SW20 are the surface water samples. WSL is the sample collected from a water source reservoir, and WPS1–WPS6 are the samples collected from possible pollution sources.

The discharge of domestic sewage was also a possible source of genotoxicity. The sewage collection and treatment facilities in the area are not perfect, resulting in untreated domestic sewage being directly discharged into the river, and there was also an accumulation of garbage. In previous reports, domestic sewage showed the same genotoxicity as mixed sewage containing industrial sewage [55], and the liquid produced by domestic wastes can also cause serious genotoxicity [56]. In addition, only one marker, CHK1, was selected in this study. CHK1 is activated for DNA repair after DNA damage, which could reflect general DNA damage [24]. Its upregulation indicated that genotoxicity existed, but there are many DNA damage and repair pathways, and judgment by CHK1 alone might lead to false negatives [57]. Therefore, the genotoxicity test method and sample test results used in this paper were not sufficient to make a complete assessment and judgment of the surface water and pollution sources in the study area, and follow-up research is needed.

3.5. Risk Assessment

The ecological risks for estrogen activity and neurotoxicity, as well as the degree of toxicity and acute toxicity were assessed for all 27 samples. According to the calculated RQ value, the environmental risks were divided into four levels: no risk ($0 \leq RQ < 0.1$), low risk ($0.1 \leq RQ < 1$), medium risk ($1 \leq RQ < 10$), and high risk ($RQ \geq 10$) [58]. According to the inhibition of luminescent bacteria, acute toxicity was divided into four levels: no toxicity ($0 \leq$ inhibition rate $< 10\%$), low toxicity ($10\% \leq$ inhibition rate $< 20\%$), medium toxicity ($20\% \leq$ inhibition rate $< 50\%$), and high toxicity ($50\% \leq$ inhibition rate $\leq 100\%$). The RQ values and inhibition rate of water samples in the rural area are shown in Figure 6.

Among the surface water samples, the sample from River 1 had a medium risk for aquatic organisms from estrogen activity, but the sample collected from the water source reservoir had no risk. The ecological risk to aquatic organisms of River 2 was mainly from estrogenic activity. Three of the seven samples collected from main stream were high risk, with RQ values ranging from 11.1 to 12.6, and the rest were medium risk. Six of the twelve samples collected from the tributary were high risk, with RQ values ranging from 11.1 to 65.5, and the rest were medium risk. Compared with the main stream, the tributary had more high risk points and higher risks, which might be related to the small water volume and slow flow rate, making the concentration of toxic substances relatively higher and more easily enriched. According to previous reports, the reproduction and development of aquatic organisms were affected by low concentrations of estrogens. For example, a decrease in egg production was observed in zebrafish exposed to 10 ng L^{-1} ethinylestradiol [59], and female grayling exposed to 1 ng L^{-1} 17β-estradiol were shown to ovulate earlier [60]. In this paper, the main risk to surface water in the study area was found to be estrogen risk, through the assessment of ecological risk for the most sensitive aquatic organisms. In addition, both industrial rainwater and livestock sewage showed a high risk of estrogenic activity, indicating that both industrial production and livestock and poultry breeding were important sources of environmental estrogen in the surface water. There were many livestock and poultry breeding enterprises in the study area, being widely spaced in the River 1 and River 2 basins, especially on the tributaries. Surface water samples adjacent to livestock and poultry breeding sources, such as SW13, SW17, and SW18, all showed a high risk, indicating that the contribution of livestock and poultry breeding to environmental estrogen in the surface water might be significant. In view of the high risk of estrogenic activity in surface water in rural areas, appropriate measures should be taken to reduce environmental risks, such as strengthening the supervision of wastewater treatment and sewage discharge in production activities, reducing the occurrence of illegal discharge, and reducing environmental estrogen substances used in production activities, especially livestock and poultry breeding.

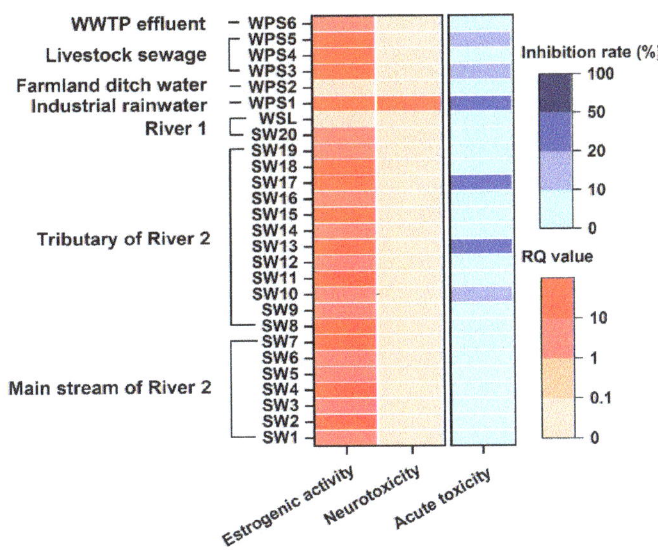

Figure 6. Acute toxicity, estrogenic activity, and neurotoxicity risk assessment of water samples in the study area.

4. Conclusions

In this study, a multi-endpoint toxicity analysis was carried out for environmental water bodies and sewage from possible pollution sources in rural areas of the lower Yellow River basin. Of the two rivers studied, River 1, which was the water source river, had no obvious toxicity risk; while River 2, which flowed through most of the region, presented different degrees of acute toxicity, estrogenic activity, and genotoxicity risks. The estrogenic activity of River 2 was generally high, with E2-EQ values ranging from 4.8 to 131.0 ng·L^{-1}, and both acute toxicity and genotoxicity occurred in the tributary of River 2. Untreated industrial sewage was considered to be one of the sources of acute toxic substances, domestic sewage was inferred to have contribution to the acute toxicity and genotoxicity, and livestock and poultry breeding sewage was found to be the source of acute toxicity and estrogenic activity, especially for the environmental estrogens in surface water. Through a risk assessment, most samples showed high risk of estrogenic activity, and the risk in tributaries was higher than that of the main stream. This study provides a reference for subsequent research on water bodies in rural areas.

Author Contributions: Investigation, Validation, Formal analysis, Writing—Original Draft, F.L.; Methodology, J.T.; Investigation, Formal analysis, Q.Y.; Resources, M.H.; Writing—Review and Editing, R.Y. and C.L.; Project administration, Writing—Review and Editing, X.Z. All authors have read and agreed to the published version of the manuscript.

Funding: This project was supported by National Key Research and Development Project of China (2019YFD1100505 and 2022YFF0609100).

Institutional Review Board Statement: Not applicable.

Informed Consent Statement: Not applicable.

Data Availability Statement: The datasets generated during the current study are available from the corresponding author on reasonable request.

Acknowledgments: We thank Yihan Yang and Boyuan Xue (School of Environment, Tsinghua University) for their support and scientific resources.

Conflicts of Interest: The authors declare no conflict of interest.

References

1. Ellis, F.; Biggs, S. Evolving Themes in Rural Development 1950s–2000s. *Dev. Policy Rev.* **2001**, *19*, 437–448. [CrossRef]
2. Tian, T.; Speelman, S. Pursuing Development behind Heterogeneous Ideologies: Review of Six Evolving Themes and Narratives of Rural Planning in China. *Sustainability* **2021**, *13*, 9846. [CrossRef]
3. Withers, P.J.A.; Neal, C.; Jarvie, H.P.; Doody, D.G. Agriculture and Eutrophication: Where Do We Go from Here? *Sustainability* **2014**, *6*, 5853–5875. [CrossRef]
4. Ongley, E.D.; Xiaolan, Z.; Tao, Y. Current status of agricultural and rural non-point source Pollution assessment in China. *Environ. Pollut.* **2010**, *158*, 1159–1168. [CrossRef]
5. Bhagirath, B.; Reddy, V.R. Environment and Accountability: Impact of Industrial Pollution on Rural Communities. *Econ. Political Wkly.* **2002**, *37*, 257–265.
6. Li, X.; Zheng, W.; Kelly, W.R. Occurrence and removal of pharmaceutical and hormone contaminants in rural wastewater treatment lagoons. *Sci. Total Environ.* **2013**, *445–446*, 22–28. [CrossRef]
7. Chen, S.; Jiao, X.-C.; Gai, N.; Li, X.-J.; Wang, X.-C.; Lu, G.-H.; Piao, H.-T.; Rao, Z.; Yang, Y.-L. Perfluorinated compounds in soil, surface water, and groundwater from rural areas in eastern China. *Environ. Pollut.* **2016**, *211*, 124–131. [CrossRef] [PubMed]
8. Yin, L.; Wen, X.; Du, C.; Jiang, J.; Wu, L.; Zhang, Y.; Hu, Z.; Hu, S.; Feng, Z.; Zhou, Z.; et al. Comparison of the abundance of microplastics between rural and urban areas: A case study from East Dongting Lake. *Chemosphere* **2020**, *244*, 125486. [CrossRef]
9. Geping, Q.; Jinchang, L. *Population and the Environment in China*; Rienner: Boulder, CO, USA, 1994.
10. Walker, D.H. Decision support, learning and rural resource management. *Agric. Syst.* **2002**, *73*, 113–127. [CrossRef]
11. Su, Y.; Qi, Y.; Xiao, Z.; Wei, Y. Contributing institutional factors of rural environmental pollution in the process of modernization in China —In the perspective of the efficiency of environmental management system. *Front. Environ. Sci. Eng. China* **2009**, *3*, 75–90. [CrossRef]
12. Tan, J.; Liu, L.; Li, F.; Chen, Z.; Chen, G.Y.; Fang, F.; Guo, J.; He, M.; Zhou, X. Screening of Endocrine Disrupting Potential of Surface Waters via an Affinity-Based Biosensor in a Rural Community in the Yellow River Basin, China. *Environ. Sci. Technol.* **2022**, *56*, 14350–14360. [CrossRef] [PubMed]
13. Do Nascimento, M.T.L.; de Oliveira Santos, A.D.; Felix, L.C.; Gomes, G.; de Oliveira e Sá, M.; da Cunha, D.L.; Vieira, N.; Hauser-Davis, R.A.; Baptista Neto, J.A.; Bila, D.M. Determination of water quality, toxicity and estrogenic activity in a nearshore marine environment in Rio de Janeiro, Southeastern Brazil. *Ecotoxicol. Environ. Saf.* **2018**, *149*, 197–202. [CrossRef] [PubMed]
14. Hong, Y.; Huang, Y.; Yan, G.; Yin, H.; Huang, Z. DNA damage, immunotoxicity, and neurotoxicity induced by deltamethrin on the freshwater crayfish, *Procambarus clarkii. Environ. Toxicol.* **2021**, *36*, 16–23. [CrossRef] [PubMed]
15. Esteban, S.; Gorga, M.; Petrovic, M.; González-Alonso, S.; Barceló, D.; Valcárcel, Y. Analysis and occurrence of endocrine-disrupting compounds and estrogenic activity in the surface waters of Central Spain. *Sci. Total Environ.* **2014**, *466–467*, 939–951. [CrossRef] [PubMed]
16. Surujlal-Naicker, S.; Gupta, S.K.; Bux, F. Evaluating the Acute Toxicity of Estrogen Hormones and Wastewater Effluents Using Vibrio fischeri. *Hum. Ecol. Risk Assess. Int. J.* **2015**, *21*, 1094–1108. [CrossRef]
17. Suthar, S.; Sharma, J.; Chabukdhara, M.; Nema, A.K. Water quality assessment of river Hindon at Ghaziabad, India: Impact of industrial and urban wastewater. *Environ. Monit. Assess.* **2010**, *165*, 103–112. [CrossRef]
18. *ISO I. 11348-3*; Water Quality-Determination of the Inhibitory Effect of Water Samples on the Light Emission of Vibrio Fischeri (Luminescent Bacteria Test)-Part 3: Method Using Freeze-Dried Bacteria. International Organization for Standardization: London, UK, 2007.

19. Routledge, E.J.; Sumpter, J.P. Estrogenic activity of surfactants and some of their degradation products assessed using a recombinant yeast screen. *Environ. Toxicol. Chem.* **1996**, *15*, 241–248. [CrossRef]
20. Rehmann, K.; Schramm, K.-W.; Kettrup, A.A. Applicability of a yeast oestrogen screen for the detection of oestrogen-like activities in environmental samples. *Chemosphere* **1999**, *38*, 3303–3312. [CrossRef] [PubMed]
21. Ellman, G.L.; Courtney, K.D.; Andres, V.; Featherstone, R.M. A new and rapid colorimetric determination of acetylcholinesterase activity. *Biochem. Pharmacol.* **1961**, *7*, 88–95. [CrossRef] [PubMed]
22. Guilhermino, L.; Lopes, M.C.; Carvalho, A.P.; Soared, A.M.V.M. Inhibition of acetylcholinesterase activity as effect criterion in acute tests with juvenile Daphnia Magna. *Chemosphere* **1996**, *32*, 727–738. [CrossRef]
23. Li, G.-M. Mechanisms and functions of DNA mismatch repair. *Cell Res.* **2008**, *18*, 85–98. [CrossRef] [PubMed]
24. Sancar, A.; Lindsey-Boltz, L.A.; Ünsal-Kaçmaz, K.; Linn, S. Molecular Mechanisms of Mammalian DNA Repair and the DNA Damage Checkpoints. *Annu. Rev. Biochem.* **2004**, *73*, 39–85. [CrossRef] [PubMed]
25. Lan, J.; Rahman, S.M.; Gou, N.; Jiang, T.; Plewa, M.J.; Alshawabkeh, A.; Gu, A.Z. Genotoxicity Assessment of Drinking Water Disinfection Byproducts by DNA Damage and Repair Pathway Profiling Analysis. *Environ. Sci. Technol.* **2018**, *52*, 6565–6575. [CrossRef] [PubMed]
26. Lan, J.; Gou, N.; Rahman, S.M.; Gao, C.; He, M.; Gu, A.Z. A Quantitative Toxicogenomics Assay for High-throughput and Mechanistic Genotoxicity Assessment and Screening of Environmental Pollutants. *Environ. Sci. Technol.* **2016**, *50*, 3202–3214. [CrossRef] [PubMed]
27. Lan, J.; Gou, N.; Gao, C.; He, M.; Gu, A.Z. Comparative and Mechanistic Genotoxicity Assessment of Nanomaterials via a Quantitative Toxicogenomics Approach across Multiple Species. *Environ. Sci. Technol.* **2014**, *48*, 12937–12945. [CrossRef] [PubMed]
28. O'Connor, S.; Lan, J.; North, M.; Loguinov, A.; Zhang, L.; Smith, M.; Gu, A.; Vulpe, C. Genome-Wide Functional and Stress Response Profiling Reveals Toxic Mechanism and Genes Required for Tolerance to Benzo[a]pyrene in *S. cerevisiae*. *Front. Genet.* **2013**, *3*, 316. [CrossRef] [PubMed]
29. European Commissions. *Technical Guidance Document on Risk Assessmen*; In Support of Commission Directive 93/67/EEC on Risk Assessment for New Notified Substances, Commission Regulation (EC) No 1488/94 on Risk Assessment for Existing Substances, and Directive 98/8/EC of the European Parliament and of the Council Concerning the Placing of Biocidal Products on the Market; European Commission: Brussels, Belgium, 2003.
30. Lu, S.; Lin, C.; Lei, K.; Xin, M.; Wang, B.; Ouyang, W.; Liu, X.; He, M. Endocrine-disrupting chemicals in a typical urbanized bay of Yellow Sea, China: Distribution, risk assessment, and identification of priority pollutants. *Environ. Pollut.* **2021**, *287*, 117588. [CrossRef]
31. Huang, L.; Xi, Y.; Zha, C.; Zhao, L.; Wen, X. Effects of dieldrin and 17β-estradiol on life history characteristics of freshwater rotifer Brachionus calyciflorus Pallas. *J. Freshw. Ecol.* **2012**, *27*, 381–392. [CrossRef]
32. Caldwell, D.J.; Mastrocco, F.; Anderson, P.D.; Länge, R.; Sumpter, J.P. Predicted-no-effect concentrations for the steroid estrogens estrone, 17β-estradiol, estriol, and 17α-ethinylestradiol. *Environ. Toxicol. Chem.* **2012**, *31*, 1396–1406. [CrossRef] [PubMed]
33. Zhou, Y.; Wu, J.; Wang, B.; Duan, L.; Zhang, Y.; Zhao, W.; Wang, F.; Sui, Q.; Chen, Z.; Xu, D.; et al. Occurrence, source and ecotoxicological risk assessment of pesticides in surface water of Wujin District (northwest of Taihu Lake), China. *Environ. Pollut.* **2020**, *265*, 114953. [CrossRef] [PubMed]
34. Vlastos, D.; Antonopoulou, M.; Lavranou, A.; Efthimiou, I.; Dailianis, S.; Hela, D.; Lambropoulou, D.; Paschalidou, A.K.; Kassomenos, P. Assessment of the toxic potential of rainwater precipitation: First evidence from a case study in three Greek cities. *Sci. Total Environ.* **2019**, *648*, 1323–1332. [CrossRef]
35. Wang, H.; Xu, J.; Liu, X.; Sheng, L.; Zhang, D.; Li, L.; Wang, A. Study on the pollution status and control measures for the livestock and poultry breeding industry in northeastern China. *Environ. Sci. Pollut. Res.* **2018**, *25*, 4435–4445. [CrossRef] [PubMed]
36. Wang, L.; Ying, G.-G.; Chen, F.; Zhang, L.-J.; Zhao, J.-L.; Lai, H.-J.; Chen, Z.-F.; Tao, R. Monitoring of selected estrogenic compounds and estrogenic activity in surface water and sediment of the Yellow River in China using combined chemical and biological tools. *Environ. Pollut.* **2012**, *165*, 241–249. [CrossRef]
37. Song, W.T.; Wang, Z.J. Occurrence and Biological Effects of Endocrine Disrupting Chemicals in the Yellow River (Zhengzhou Section). *Bull. Environ. Contam. Toxicol.* **2016**, *97*, 763–769. [CrossRef]
38. Khanal, S.K.; Xie, B.; Thompson, M.L.; Sung, S.; Ong, S.-K.; van Leeuwen, J. Fate, Transport, and Biodegradation of Natural Estrogens in the Environment and Engineered Systems. *Environ. Sci. Technol.* **2006**, *40*, 6537–6546. [CrossRef]
39. Hanselman, T.A.; Graetz, D.A.; Wilkie, A.C. Manure-Borne Estrogens as Potential Environmental Contaminants: A Review. *Environ. Sci. Technol.* **2003**, *37*, 5471–5478. [CrossRef] [PubMed]
40. Palme, R.; Fischer, P.; Schildorfer, H.; Ismail, M.N. Excretion of infused 14C-steroid hormones via faeces and urine in domestic livestock. *Anim. Reprod. Sci.* **1996**, *43*, 43–63. [CrossRef]
41. Yao, B.; Li, R.; Yan, S.; Chan, S.-A.; Song, W. Occurrence and estrogenic activity of steroid hormones in Chinese streams: A nationwide study based on a combination of chemical and biological tools. *Environ. Int.* **2018**, *118*, 1–8. [CrossRef] [PubMed]
42. Cavallin, J.E.; Durhan, E.J.; Evans, N.; Jensen, K.M.; Kahl, M.D.; Kolpin, D.W.; Kolodziej, E.P.; Foreman, W.T.; LaLone, C.A.; Makynen, E.A.; et al. Integrated assessment of runoff from livestock farming operations: Analytical chemistry, in vitro bioassays, and in vivo fish exposures. *Environ. Toxicol. Chem.* **2014**, *33*, 1849–1857. [CrossRef]

43. Alvarez, D.A.; Shappell, N.W.; Billey, L.O.; Bermudez, D.S.; Wilson, V.S.; Kolpin, D.W.; Perkins, S.D.; Evans, N.; Foreman, W.T.; Gray, J.L.; et al. Bioassay of estrogenicity and chemical analyses of estrogens in streams across the United States associated with livestock operations. *Water Res.* **2013**, *47*, 3347–3363. [CrossRef] [PubMed]
44. Qiang, Z.; Dong, H.; Zhu, B.; Qu, J.; Nie, Y. A comparison of various rural wastewater treatment processes for the removal of endocrine-disrupting chemicals (EDCs). *Chemosphere* **2013**, *92*, 986–992. [CrossRef] [PubMed]
45. Olson, D.L.; Christensen, G.M. Effects of water pollutants and other chemicals on fish acetylcholinesterase (in vitro). *Environ. Res.* **1980**, *21*, 327–335. [CrossRef] [PubMed]
46. Neale, P.A.; Escher, B.I. Coextracted dissolved organic carbon has a suppressive effect on the acetylcholinesterase inhibition assay. *Environ. Toxicol. Chem.* **2013**, *32*, 1526–1534. [CrossRef]
47. Yuan, L.; Qian, L.; Qian, Y.; Liu, J.; Yang, K.; Huang, Y.; Wang, C.; Li, Y.; Mu, X. Bisphenol F-Induced Neurotoxicity toward Zebrafish Embryos. *Environ. Sci. Technol.* **2019**, *53*, 14638–14648. [CrossRef] [PubMed]
48. Wang, S.-l.; Liu, F.-m.; Jin, S.-h.; Jiang, S.-r. Dissipation of propisochlor and residue analysis in rice, soil and water under field conditions. *Food Control* **2007**, *18*, 731–735. [CrossRef]
49. Xiao, J.-J.; Wang, F.; Ma, J.-J.; Xu, X.; Liao, M.; Fang, Q.-K.; Cao, H.-Q. Acceptable risk of fenpropathrin and emamectin benzoate in the minor crop Mugua (*Chaenomeles speciosa*) after postharvest processing. *Environ. Pollut.* **2021**, *276*, 116716. [CrossRef]
50. Matioli, T.F.; da Silva, M.R.; de Bastos Pazini, J.; Barroso, G.; Vieira, J.G.; Yamamoto, P.T. Risk Assessment of Insecticides Used in Tomato to Control Whitefly on the Predator *Macrolophus basicornis* (Hemiptera: Miridae). *Insects* **2021**, *12*, 1092. [CrossRef]
51. Russo, J.; Russo, I.H. Genotoxicity of steroidal estrogens. *Trends Endocrinol. Metab.* **2004**, *15*, 211–214. [CrossRef]
52. Balabanič, D.; Filipič, M.; Krivograd Klemenčič, A.; Žegura, B. Genotoxic activity of endocrine disrupting compounds commonly present in paper mill effluents. *Sci. Total Environ.* **2021**, *794*, 148489. [CrossRef]
53. Martínez-Paz, P.; Morales, M.; Martínez-Guitarte, J.L.; Morcillo, G. Genotoxic effects of environmental endocrine disruptors on the aquatic insect Chironomus riparius evaluated using the comet assay. *Mutat. Res. Genet. Toxicol. Environ. Mutagenes.* **2013**, *758*, 41–47. [CrossRef]
54. Cavalieri, E.L.; Rogan, E.G. Unbalanced metabolism of endogenous estrogens in the etiology and prevention of human cancer. *J. Steroid Biochem. Mol. Biol.* **2011**, *125*, 169–180. [CrossRef] [PubMed]
55. Jolibois, B.; Guerbet, M. Evaluation of industrial, hospital and domestic wastewater genotoxicity with the Salmonella fluctuation test and the SOS chromotest. *Mutat. Res. Genet. Toxicol. Environ. Mutagenes.* **2005**, *565*, 151–162. [CrossRef] [PubMed]
56. White, P.A.; Rasmussen, J.B. The genotoxic hazards of domestic wastes in surface waters1Summary of material presented at the workshop Sources, Effects and Potential Hazards of Genotoxic Complex Mixtures in the Environment held at the annual meeting of the Environmental Mutagen Society, April 20, 1997, Minneapolis, MN.1. *Mutat. Res. Rev. Mutat. Res.* **1998**, *410*, 223–236. [CrossRef]
57. Hercog, K.; Maisanaba, S.; Filipič, M.; Sollner-Dolenc, M.; Kač, L.; Žegura, B. Genotoxic activity of bisphenol A and its analogues bisphenol S, bisphenol F and bisphenol AF and their mixtures in human hepatocellular carcinoma (HepG2) cells. *Sci. Total Environ.* **2019**, *687*, 267–276. [CrossRef] [PubMed]
58. Ågerstrand, M.; Rudén, C. Evaluation of the accuracy and consistency of the Swedish Environmental Classification and Information System for pharmaceuticals. *Sci. Total Environ.* **2010**, *408*, 2327–2339. [CrossRef] [PubMed]
59. Coe, T.S.; Hamilton, P.B.; Hodgson, D.; Paull, G.C.; Stevens, J.R.; Sumner, K.; Tyler, C.R. An Environmental Estrogen Alters Reproductive Hierarchies, Disrupting Sexual Selection in Group-Spawning Fish. *Environ. Sci. Technol.* **2008**, *42*, 5020–5025. [CrossRef] [PubMed]
60. Lahnsteiner, F.; Berger, B.; Kletzl, M.; Weismann, T. Effect of 17β-estradiol on gamete quality and maturation in two salmonid species. *Aquat. Toxicol.* **2006**, *79*, 124–131. [CrossRef] [PubMed]

MDPI AG
Grosspeteranlage 5
4052 Basel
Switzerland
Tel.: +41 61 683 77 34

Chemosensors Editorial Office
E-mail: chemosensors@mdpi.com
www.mdpi.com/journal/chemosensors

www.ingramcontent.com/pod-product-compliance
Lightning Source LLC
Chambersburg PA
CBHW061614210926
51850CB00009B/2321
* 9 7 8 3 7 2 5 8 1 9 3 5 5 *